Cognitive Engineering and Safety Organization in Air Traffic Management

Cognitive Engineering and Safety Organization in Air Traffic Management

By

Tom Kontogiannis and Stathis Malakis

CRC Press
Taylor & Francis Group
Boca Raton London New York

CRC Press is an imprint of the
Taylor & Francis Group, an **informa** business

CRC Press
Taylor & Francis Group
6000 Broken Sound Parkway NW, Suite 300
Boca Raton, FL 33487-2742

International Standard Book Number-13: 978-1-138-04972-7 (Hardback)

Library of Congress Cataloging-in-Publication Data

Names: Kontogiannis, Tom, author. | Malakis, Stathis, author.
Title: Cognitive engineering and safety organization in air traffic management / Tom Kontogiannis, Stathis Malakis.
Description: Boca Raton : CRC Press, Taylor & FrancisGroup, 2017.
Identifiers: LCCN 2017018537| ISBN 9781138049727 (hardbook : acid-free paper) | ISBN 9781315168814 (ebook)
Subjects: LCSH: Air traffic control–Decision making–Case studies. | Aeronautics–Safety measures. | Air traffic controllers–Psychology.
Classification: LCC TL725.3.T7 K66 2017 | DDC 387.7/404260684–dc23
LC record available at https://lccn.loc.gov/2017018537

Visit the Taylor & Francis Web site at
http://www.taylorandfrancis.com

and the CRC Press Web site at
http://www.crcpress.com

MIX
Paper from responsible sources
FSC® C013056

Printed and bound in Great Britain by
TJ International Ltd, Padstow, Cornwall

To all air traffic controllers, who keep our skies safe
although working hard behind the scenes
and away from the media spotlights

Contents

List of Abbreviations

A

ACC	Area control center
ACS	Area control surveillance
ADS	Automatic dependent surveillance
AFTN	Aeronautical Fixed Telecommunications Network
AIP	Aeronautical information publication
AMAN	Arrival manager
ANSP	Air navigation service provider
AoR	Area of responsibility
APP	Approach control unit
APW	Area proximity warning
ASM	Airspace management
ATC	Air traffic control
ATCO	Air traffic controller
ATFCM	Air traffic flow and capacity management
ATFM	Air traffic flow management
ATM	Air traffic management
ATS	Air traffic service
ATZ	Airport traffic zone

C

CAA	Civil Aviation Authority
CAM	Conflict alert message

CC	Coordinating controller
CDO	Continuous descend operations
CFIT	Controlled flight into terrain
CLAM	Cleared level adherence monitoring
CNS	Communication, navigation, and surveillance
CSE	Cognitive systems engineering
CTA	Cognitive tasks analysis
CWP	Controller's working position

D

DMAN	Departure manager
DME	Distance measuring equipment
DSS	Decision support systems

E

EASA	European safety aviation agency
EAT	Expected approach time
EATMP	European air traffic management program
EC	Executive controller
EEC	Experimental center
ELT	Emergency locator transmitter
ESARR	EUROCONTROL Safety Regulatory Requirement
ETTO	Efficiency–thoroughness trade-off
EUROCONTROL	European Organization for the Safety of Air Navigation

F

FAA	Federal Aviation Administration
FAB	Functional airspace block
FCOM	Flight crew operating manual
FDPS	Flight data processing system
FIS	Flight information service
FL	Flight level
FMS	Flight management system
FOM	Flight operations manual
FPS	Flight progress strip
FRAM	Functional resonance accident model
Ft	Feet
FTA	Fault tree analysis
FUA	Flexible use of airspace

G

GA	General aviation
GAT	General air traffic
GPS	Global positioning system
GRD	Ground controller
GS	Glide slope

H

HERA	Human error in ATM
HF	High frequency
HFCAS	Human factors analysis and classification system
HFE	Human factors engineering
HMI	Human machine interaction
HRO	High reliability organization
HTA	Hierarchical task analysis

I

I/O	Input/Output
IANS	Institute of Air Navigation Services
IAP	Instrument approach procedure
ICAO	International Civil Aviation Organization
IFR	Instrument flight rules
ILS	Instrument landing system
IM	Inner marker
IRP	Integrated risk picture

J

JCS	Joint cognitive systems

L

LLWS	Low level wind shear
LoA	Letter of agreement
LOC	Local controller

M

MDI	Minimum departure interval
MEL	Minimum equipment list
METAR	Meteorological terminal air report
MM	Middle marker

MONA Monitoring aids
MSAW Minimum safe altitude warning
MSL Mean sea level
MTCD Medium term conflict detection

N
NATS National Aviation Traffic Services
NDM Naturalistic decision making
NextGen Next generation air transportation system
Nm Nautical mile

O
OAT Operational air traffic
ODS Operator input and display system
OJT On-the-job training
OJTI On-the-job training instructor
OM Outer marker
ORM Operational risk management

P
P/A Public announcement
PANS Procedures of air navigation services
PPL Private pilot license
PSR Primary surveillance radar

Q
QNH Code for altimeter setting indicating altitude above Mean Sea Level
QRH Quick reference handbook

R
RA Resolution advisory
RADAR Radio detection and ranging
RAM Route adherence monitoring
RAT Risk analysis tool
RCF Radio communication failure
RDP Radar data processor

RIMCAS	Runway incursion monitoring and conflict alert system
RTF	Radio telephony
RVA	Radar vectoring area
RVSM	Reduced vertical separation minima
RX	Reception

S

SA	Situation awareness
SADT	Structured analysis and design technique
SAR	Search and rescue
SARPs	Standards and recommended practices
SES	Single European skies
SESAR	Single European sky ATM research program
SID	Standard instrument departure
SME	Subject matter expert
SMR	Surface movement radar
SNET	Safety net
SOPs	Standard operating procedures
SSR	Secondary surveillance radar
STAMP	Systems-theoretic accident model and processes
STAM	Short-term ATFCM measures
STAR	Standard arrival route
STCA	Short term conflict alert
Std	Standard

T

T²EAM	Taskwork/teamwork for effective and adaptive management
TA	Traffic advisory
TCAS	Traffic alert and collision avoidance system
TEM	Threat and error management
TMA	Terminal maneuvering area
TRACEr	Technique for the retrospective and predictive analysis of cognitive error
TRM	Team resource management
TWR	Tower
TX	Transmission

U

UAC	Upper area control
UACC	Upper area control centre
UCS	Unit competency scheme
UHF	Ultra high frequency
UNL	Unlimited
UTP	Unit training plan

V

VFR	Visual flight rules
VHF	Very high frequency
VSM	Viable system model
VMC	Visual meteorological conditions
VOR	Very high frequency omni directional radio range

List of Figures

List of Tables

Preface

Air traffic management has been a highly reliable system for some decades now. However, modern systems are continuously seeking new challenges and are rarely content with the current state of affairs. The initiatives of NextGen and SESAR may indicate that the new aviation environment will become more complex and tightly coupled in order to cope with increased traffic, minimize delays, operate in adverse weather, smooth out aircraft trajectories, and so on. As the air traffic system is being stretched to its capacity limits, safety challenges may increase in the near future.

As the complexity of the operating environment becomes higher, so should the capabilities of organizations expand. This would require the introduction of new technologies, safety nets, and supporting tools, which may affect the operating practices of controllers, their team coordination, and the organizational procedures and regulation rules. In this sense, people, organizations, and artifacts should learn how to adapt to the challenges and demands posed by new situations and new technologies. This view of a joint cognitive system has been the focus of cognitive engineering, a discipline of human factors where practitioners' activities are understandable in relation to the properties of the environment in which they work; for instance, conflict resolution strategies are affected by traffic constraints, availability of job aids, team composition, collaboration with flight crews, and regulations. In turn, a work environment can be understood in terms of what it demands and affords to the people at the sharp end.

There are several reasons why we have chosen to focus on cognitive engineering rather than on human factors in general. Cognitive engineering looks beyond the performance of individuals in laboratory experiments and tests theories in the context of real work environments or task simulations that reproduce the environment in which work is performed, the tools that are used to support people, and the organizational policies and rules that guide human control. In this sense, the first part of this book presents an overview of the work environment of ATM to provide a basis for understanding how practitioners interact with others, how they use their tools, and how the organizational processes shape human performance.

Until recently, many human factors studies focused on human performance problems while fewer studies looked into how practitioners succeed despite the weaknesses in the system they control. Cognitive engineering has accepted new paradigms of human performance in which people are seen as assets in their organizations as their variability enables them to compensate for inadequate organizational practices and rules. It is now increasingly recognized that expertise and errors are two sides of the same coin and that human variability may lead to success in most situations but it could lead to failures in certain circumstances. For this reason, the second part of this book focuses on decision-making and sensemaking strategies of air traffic controllers (ATCOs) as well as on their strategies in monitoring work, detecting errors, and developing new practices to cope with complex events.

Cognitive engineering has been driven largely by the requirements derived from the regulatory demands and from a need to operate effectively and safely. In this sense, theory is elaborated after the operational problem has been addressed in a real or simulated environment. As human factors and safety practitioners in the aviation domain are users of theory, we have become aware that the principles of cognitive engineering should be cast in practical terms (e.g., behavioral markers of poor and good performance) and should be illustrated in the context of ATM applications. In this respect, the third part of the book examines the relationship between the complexity and the workload of controllers in managing the modern ATM environment as well as possible ways to match people and technology through training and system design.

Recent developments in cognitive engineering have tried to look at the opportunities and constraints set by the wider organization on the performance of people at the sharp end. Particular emphasis has been on the safety requirements and constraints set by organizations and regulation authorities that affect the way people organize their cognitive resources and strategies. For instance, decision trade-offs regarding production and protection at higher levels may shape the work of practitioners who have to do their best to reconcile different goals simultaneously. Hence, in the fourth part of the book, an effort has been made to look at new ways of combining cognitive engineering with new systems thinking and resilience approaches that would enhance human and organizational performance.

There have been several excellent books on human factors in air traffic management but the majority are edited volumes. Also several studies of cognitive engineering have been published in scientific journals that have enhanced our understanding of how ATCOs organize their performance. The challenge we faced in this book was to try and see how different models fit together in order to provide a comprehensible and practical overview of human performance in modern ATM systems. In this sense, the results represent our personal knowledge of the subject and experience with the ATM domain. The first author has been in the academia for 25 years while the second author has been an active ATCO for 18 years. Both of us have spent a lot of time observing the work of ATCOs and talking to them to understand the way they develop and adapt their practices in the context of work pressures and complex scenarios.

To increase the credibility of the book chapters, we have also attended some refresher training courses where ATCOs had to cope with several familiar and novel scenarios under time pressure. Furthermore, we have used an ATM training simulator on many occasions in order to test theory and propose a model of human performance that would be relatively easy to put into practice. Our work with the training simulator has been very helpful in building a model of Taskwork/Teamwork for effective and adaptive management. In this regard, the proposed T^{2}EAM model has built on existing approaches of cognitive engineering and experimental work on a training simulator. The book reflects our continuous efforts, since 2005, to understand the work of ATCOs and develop practical models of human performance that could support the design and safe operation of the joint cognitive system. In the fourth part of the book, T^{2}EAM has been elaborated in order to provide a framework of performance for both humans and organizations.

Our target readership would include not only researchers and practitioners in human factors, but also people who manage or carry out everyday activities in air traffic management. We hope that the book will be of interest to human factors specialists, safety practitioners, incident investigators, ATM regulators, system designers, and, in general, the wider aviation community.

Our main aim has been to develop a model of human and organizational performance that would integrate several principles that can be adapted to the needs of individual organizations. We discuss a variety of applications in training, system design, and safety, but this book is not intended to provide a comprehensive list of off-the-shelf tools. Emphasis has been given to elaborating theoretical principles that could be tailored to particular organizations by paying attention to a range of workplace and organizational factors. Practitioners in the ATM domain are ingenious and adaptable people who can adapt principles to their own work and devise their own solutions.

In summary, the wealth of ideas and approaches presented throughout this book reflects the advances in human factors, cognitive engineering, and system safety that have been achieved over the last three decades. Our main challenge in this book has been to take a scientific approach to the architectures of cognition and organization but also bind theory to the application tasks of ATCOs. Part I of the book aims at describing the ATC system from the perspective of joint cognitive systems of people, organizations and artifacts that are adapted to the different demands posed by unfamiliar situations and new technologies. Part II focuses on the cognitive functions of decision-making, sense-making, problem detection, error recovery, and work adaptation derived from the literature in cognitive engineering and tested in the context of the ATC domain. Part III presents domain applications in the analysis of complexity and workload, the use of cognitive task analysis in training design, and the implications for new operational designs (e.g., SESAR and NextGen). In a joint cognitive system (JCS), cognitive functions and strategies operate within a larger organizational context that shapes both the cognitive strategies of controllers and the way that they use their technologies and artifacts. Although organizations work at a different level of abstraction and time frame from operating teams, an effort was made in Part IV to present the organization of safety structures in terms of similar functions to the ones utilized in the taskwork and teamwork activities.

Because the book addresses a wide audience of researchers and practitioners, different readers may dip into different chapters to find answers to specific questions they have. In this sense, Part I may be useful for people who are not very familiar with the complexities of the operating environment of ATCOs. Part II presents the backbone of cognitive engineering and it could be of interest to practitioners to see how their everyday strategies fit or differ with those of other colleagues. Researchers may be interested to see how the T^2EAM model has "twisted" existing models of human performance to fit the work of practitioners in the ATM domain. Part III is of general interest as it presents applications of the T^2EAM model in the areas of controller training, complexity and workload, and system design of the new operating environment. Finally, Part IV reflects our effort to adapt T^2EAM to organizational performance with emphasis on safety organization. The final chapter looks into the particular area of "safety risk management" to show how improvements can be made in system safety.

Our thanks in the production of the book go out to Cindy Carelli and Renee Nakash who supported this publication. Finally, we would like to acknowledge our gratitude to all the ATCOs who gave their valuable time and expertise to participate in the simulator sessions and explain the complexities of their work. Without them this book would have not been possible.

About the Authors

Tom Kontogiannis, PhD, is professor in Human Factors and Industrial Safety in the Department of Production Engineering & Management, Technical University of Crete. He holds degrees in industrial engineering, system safety, and human factors. While working as a human factors consultant in England (1988–1997), he participated in more than 30 industrial projects regarding human reliability, decision-support for process control skills, risk assessment, and business process management. Since 1997, he has headed the Cognitive Ergonomics & Industrial Safety (CEIS) Lab at the Technical University of Crete and currently works on human performance modeling, simulator training, error recovery processes, human reliability assessment, organizational safety, systems dynamics, and business process engineering. He has published many peer-reviewed journal papers and two books (*Guidelines For Preventing Human Error In Process Safety* published by CCPS, AIChE and *Ergonomic Approaches to Safety Management* published by Tziolas). Recent work with the European Commission includes virtual reality systems for industrial training, information tools for human performance modeling, quantification of human reliability, safety performance indicators, and information systems for safety management.

Stathis Malakis, PhD, is an active air traffic controller working for the Hellenic Civil Aviation Authority (HCAA) since 1999. He holds tower, approach procedural, and approach radar and instructor/assessor/examiner ratings. He led the development of HCAA's safety management system and he is the safety focal point of Rhodes/Diagoras International Airport ATC section. He holds a BSc in mathematics, an MSc in air transport management, and a PhD in cognitive engineering. A few years ago, he completed an experimental analysis of unusual occurrences in air traffic control in Eurocontrol and developed tools for training needs analysis. He has extensive experience in designing complex scenarios in ATC simulators and in refreshing training. He has published many peer-reviewed journal papers on cognitive strategies in air traffic control, error detection, simulator training, and accident investigation. Recent work with the European Commission includes modeling and design of safety management systems in aviation.

Cognitive Engineering in a Changing World

Air traffic control (ATC) is a work domain that relies on the cognitive functions of controllers and their collaboration with flight crews, airport operators, and other aviation stakeholders to control the airspace, manage safety and adapt to the changing demands of new technological initiatives (e.g., SESAR and NextGen). Cognitive engineering is a discipline of human factors that studies how practitioners organize their cognitive functions to build resilient strategies to a variety of situations that may be encountered in ATC. Cognitive engineering refers to the cognitive functions of controllers and their operating teams, such as problem detection, sensemaking, decision-making, replanning, and adaptation to changing situations. Particular emphasis has been placed on the interactions between cognitive functions, computer support artifacts (e.g., conflict detection aids, resolution aids, procedures, etc.), and teamwork processes (e.g., common understanding and coordination).

Cognitive Engineering (CE) views the ATC work domain as a joint cognitive system (JCS) where cognition is distributed to different agents (e.g., ATCOs, flight crews, flight dispatchers, flow control supervisors, aerodrome operators, ramp agents, and so on) who need to collaborate and join their efforts to achieve the overall system objective. Working in isolation, cognitive agents may achieve efficient local performance but this may be at the expense of converging toward common organizational goals. For example, direct routings and vector shortcuts are always welcomed by the flight crews. However, an approach controller who expedites arriving aircraft to land at a congested airport may exert unnecessary pressure on the tower controller, who may not be able to find any parking stands for this aircraft. In other situations, direct routings and vector shortcuts may end up with flight crews approaching too high and fast in a different runway to the one originally planned.

The ATC domain is a tightly-coupled system where cognitive work performed by controllers affects the work of others and hence, the overall performance of the system. An ATCO, for instance, may request an aircrew to make a small diversion from their route for separation purposes then resume navigation to a specific point in the airspace. However, this simple request may demand significant activity on the flight-deck because the aircrew needs to reprogram the flight management system (FMS). Route modifications require some interaction with the FMS because this system has been developed with concerns for accuracy and efficiency on routes in the airspace. As a result, the FMS is quite "brittle" in accepting new route changes that have some potential to produce unintended side effects (Sarter and Woods 1995). This example shows that controllers, flight crews, and artifacts (e.g., FMS) constitute a JCS and that the interaction between any two agents (e.g., controllers and flight crews) is affected by the third agent (e.g., how brittle the FMS can become).

To understand the theoretical models and methods in this book, it is important to present a short overview of how the discipline of CE has emerged and what sort of problems have been addressed. CE has been developed to deal with three particular problems that became apparent as computer-based systems came into widespread use in the 1980s:

1. The increasing complexity of sociotechnical systems that was largely due to a large-scale substitution of human functions with automation.
2. The inadequate deployment of new technologies that gave rise to unanticipated problems.

3. The limitations of existing theoretical models and tools that seemed to be inadequate in analyzing and designing complex systems.

CE set out to bring about a paradigm shift in the way that people thought about interactive systems. The traditional human factors engineering (HFE) approach has viewed human–machine systems as two-agent systems with certain interactions among them. This focus on the people and the technology undermined the constraints and pressures that were present in the actual field of practice (Woods and Sarter 2000). The traditional HFE approach was applied to system design in the 1970s and 1980s, producing machines and artifacts that seemed to work well in normal conditions. However, this design approach was out of the context of the work and soon became brittle when the human–machine interaction took place in unfamiliar or abnormal situations.

A number of ironies of automation have been described in the field of process control (Bainbridge 1983) and in the context of cockpit automation (Woods and Sarter 2000). One consequence of automation has been that humans were given a nearly impossible task as they had to understand not only the weaknesses of the technological system, but also the circumstances where automation was not working properly. Furthermore, removing humans from on-line control made it difficult to maintain a good understanding of the problem, particularly in cases where automation masked early symptoms of the problem (Norman 1990). In the ATC domain for instance, the air traffic flow and capacity management (ATFCM) system provides capacity enhancement by safeguarding that en-route sectors, approach control units, and airport towers are not saturated in heavy traffic conditions. The ATFCM system allocates departure slots to flights in order to delay them on the ground. In practice, however, the slot allocation algorithms rely mostly on physical factors (e.g., prevailing winds three hours earlier than the actual flight) but fail to take account of actual aircraft performance, airline flight profile preferences, and direct routings given by controllers. Hence, some ATC units may be stretched above their capacity, or others may be underutilized, as a result of under-specifications in the ATFCM algorithms.

There is growing evidence that machines and artifacts may constrain people's decisions and cognitive strategies in responding to unfamiliar situations. Designs that fail to address human–computer interaction in an unfamiliar context of work (e.g., new surprising events, increasing time pressure, and task constraints) may produce artifacts and decision aids that constrain practitioner's search activities and narrow the set of options that they consider. This may influence the likelihood that people miss critical information or fail to generate the correct solution, when the recommendations of the decision aids are inaccurate.

The term "artifact" refers to a range of job tools or aids that represent the work environment (e.g., controller's working position), alert them of imminent threats (e.g., short term conflict alert), provide overviews of tasks or planning sequences (e.g., flight strip display), or make intelligent recommendations (e.g., departure manager). In this sense, artifacts can represent knowledge about the world, or even can process this knowledge and make intelligent recommendations. For this reason, artifacts represent knowledge in the world that is distinct from knowledge in the head of practitioners. Therefore, an artifact is a cognitive agent, as it can perform several cognitive functions, such as structuring the work environment, perceiving threats, and communicating possible ways to recover problems (e.g., standard operating procedures and intelligent decision aids).

The traditional HFE approach has focused on specific actions of people performing well-defined tasks with well-specified goals; it has also treated computers as artifacts that can be designed independently of the needs and capabilities of the human controllers. In highly complex systems, however, human and organizational activities shape the ways that artifacts are used by people while the artifacts themselves affect the ways that people work and coordinate their efforts. The artifacts, cognitive strategies, coordinated activities, and organizational policies do not present themselves to the researchers, one at a time; they rather come in a wrapped package of interdependent, time-sensitive, and changing factors. While agent communication and human–machine interaction remain important topics, they may miss the more important goal of understanding how the system performs in a joint fashion and how it achieves its goals and functions.

A JCS approach that considers the joint role of multiple human and machine agents may change the question from how to compute better solutions to how to determine what assistance is useful, how to deliver it in the interface, and what explanations should be provided to humans. Roth et al. (1997) provided some insights into three general features in the design of intelligent artifacts that support human problem-solving:

1. Analysis of the work demands to identify aspects of human performance that require support (i.e., analysis of cognitive and decision requirements of the activities in context)
2. Analysis of the cognitive strategies used by practitioners to do their job
3. Translation of cognitive strategies into system design and user interface requirements.

While HFE considers the human–computer dyad as the basic unit of study, the CE also includes the work settings within which this interaction takes place. For instance, the interaction of controllers with their computers cannot be viewed independently of how controllers coordinate with flight crews, from how they are affected by traffic, and from how they are constrained by organizational policies. This third element refers to the "work conditions" or the "context of work" within which the controller–computer interaction takes place.

Woods and Roth (1988) coined the term "cognitive system triad" to emphasize that three interconnected elements determine the overall system performance:

1. The challenges to be met in an external world or domain of interest
2. The expertise and capabilities of human and machine agents who act on the world
3. The artifact or external representation through which the agents experience and learn about the world.

First, the characteristics of the work domain may contribute to the complexity of the situation and determine the cognitive demands and the range of situations that controllers are likely to face in performing their tasks. Examples of work demands may include: the number and complexity of tasks (e.g., how many aircraft must be controlled), the interactions and constraints of work (e.g., how many aircraft need separation instructions), the hazards in the world (e.g., areas of bad weather and military exercise), the coupling between systems, and so on.

Second, the information search and decision strategies of controllers are very important in meeting the challenges of the work environment. For instance, controllers may choose to delegate responsibilities to their colleagues as traffic increases, which implies that the computer interfaces should provide a good representation of airspace and air traffic to the whole team. In this sense, CE relies on human performance models to identify the cognitive strategies that controllers use in different circumstances. CE specialists should then have to translate the cognitive strategies into principles of human–computer interface design.

Third, the artifacts and their representation of the domain may also affect human performance by making certain aspects more accessible at the expense of others. It is well documented that the way a problem is represented may affect the cognitive work that is needed to solve the problem (Zhang and Norman 1994; Norman 1993). For instance, the representation of traffic on the radar screens and the availability of job aids may affect how a traffic sequencing problem can be solved.

Traditionally, we are used to thinking that cognition is an activity of individual minds, but from the CE perspective, it is a joint activity that is distributed across the members of the work domain and their artifacts. In ATC, cognition is distributed spatially between controllers, flight crews, and traffic planning or alerting artifacts. It is also distributed temporally so that controllers can build on earlier decisions and anticipate future actions. Most significantly, what may appear as an individual decision could emerge via the coordination among distributed agents.

In a JCS, practitioner activities are understandable only in relation to the properties of the environment within which they work; also, the work environment is understood in terms of what it demands and affords to people at work. The environment and the practitioners are mutually adapted in order to establish a dynamic equilibrium. A new change in the environment may trigger a new sort of adaptation until a new equilibrium is reached.

In a JCS, mutual adaptation is a three-way interplay of the strategies of agents (humans and machines), the affordances of artifacts, and the demands of the field of practice (Woods 1988; Woods et al. 2002). This implies that CE methods should be able to address this three-way interplay in a practical manner. For example, controllers may develop skills at overcoming work obstacles and balancing dilemmas posed by the wider organization. However, this adaptation may be short-lived since work obstacles may continue to exist if they are not removed by organizational policies; hence, organizational policies to address the systemic causes of obstacles should also take place in the long term. For example, controllers may develop a shortcut procedure in a lengthy instrument departure that works in daylight and good weather conditions. Flight crews may normalize this procedure and exert pressure to controllers to use this shortcut in marginal weather conditions, eventually leading to near CFIT events (controlled flight into terrain); instead, the organization should have designed and implemented a precision based navigation (PBN) procedure. In this respect, the cognitive engineering approach is situated within a larger organizational context that shapes both the strategies of controllers and the ways that they use their artifacts and technologies. The fourth part of the book has been devoted to research on how organizational practices and dilemmas interact with the work of practitioners.

Consequently, research in cognitive engineering relies on naturalistic field studies that look at agent interactions in the real world or in the context of simulations that are representative of the real world. The emphasis has been on understanding practitioner behavior in the presence of a rich informational environment and in the context of a collaborative environment rather than strict experimental or laboratory conditions. This allows an analysis of how practitioners actually perform tasks, how they use their tools to search for useful information, how they make sense of information for problem solving, and how they support each other as a team. In other words, the focus has been on examining what practitioners do, successfully or erroneously, in the context of work demands and available resources. By observing skilled practitioners and novices, one can develop human performance models that specify the knowledge and cognitive strategies of practitioners. In the following chapters, it would be possible to present several performance models that can be used to understand the work demands and the required cognitive strategies to control the system.

In the past, many human factors methods have been developed to analyze work demands and represent the work domain in terms of goals, constraints, and available means; examples include methods of task analysis (Kirwan and Ainsworth 1992; Shepherd 2001) and work domain analysis (Vicente 1999; Ahlstrom 2005). In general, goal-means analysis techniques are best suited to domains where the goals and methods are well specified and documented. This is often the case in engineered systems where the methods for achieving goals are strongly constrained by the characteristics of the physical system (e.g., process control industries, manufacturing, etc.). ATM is a domain where goals and constraints are more fluid and cognitive strategies of practitioners are not well understood. Hence, performance models of ATCOs should be developed on the basis of field studies and cognitive task analysis to understand how practitioners interact with their artifacts in a work environment that is supported by collaboration but constrained by work demands and restricted capacity.

For this reason, field studies and cognitive task analysis are needed to examine how practitioners change their priorities, what practices they use to produce approximate solutions, and how they manage conflicts between goals. This adaptation of people and organizations may produce nonlinear phenomena that are difficult to understand without a good grasp of empirical models of performance. Therefore, a mixture of analytic and empirical techniques becomes necessary to examine human performance within the constraints of a real-world environment.

For many years, people believed that the higher the traffic complexity, the higher the workload, leading to severe decrements in performance. Designers expected that human performance might collapse beyond a cutoff level of complexity or workload and this led them to propose new systems of computerized support for practitioners. However, a review of recent studies has shown that practitioners resolve trade-offs of efficiency and safety differently as the situation changes (Brooker 2003; Loft et al. 2007; Kontogiannis and Malakis 2013b). As complexity increases, for instance, controllers may tend to reduce the quality of service offered to air crews by adapting planning and performance criteria to the situation. Alternatively, controllers may provide air crews with broader instructions to allow themselves more space for replanning later on. By better understanding the way that practitioners adapt their cognitive strategies to complex systems, designers and organizations can develop better expectations when and how performance may fall as well as what type of decision aids might be appropriate for supporting humans in complex situations.

Finally, time becomes an important consideration in cognitive engineering because the performance of people and organizations is affected by temporal changes of the situation. For instance, flight crews may withhold information that can be revealed to controllers at later times, a piece of information may not be passed to another airspace sector until it is too late, or the attention of controllers may be captured by other urgent events. The risk here is that practitioners may be stuck in outdated behaviors, failing to reframe their understanding. To avoid this risk, practitioners must be in a state of alertness and mindfulness to detect problems and recover errors at an early stage.

Time makes the most important difference between the work of practitioners and organizations. Performance loops at the sharp end are a lot faster than organizational loops at the blunt end that process risk information and reinforce new rules. The different time dynamics across organizational levels increase the coordination problems between higher and lower system levels. For instance, a safety manager may need a considerable amount of time in collecting safety critical information that may be promptly available to practitioners. Also, the travelling of orders and feedback between organizational scales is much slower than the actions and feedback at the sharp end. For this reason, aspects of cognitive engineering should be closely related to the wider organizational environment.

Organization of the Book

Cognitive engineering considers the architecture of cognitive agents and artifacts that perform cognitive functions (e.g., perception, sensemaking and choices) in order to achieve goals and tasks within the opportunities and the constraints set by the work organization. The basic elements of this architecture are the cognitive functions undertaken by practitioners, teams, and artifacts as well as the organizational context of work. In this sense, this book attempts to take a practical view of the human–task–artifact interaction in terms of a number of essential cognitive functions (e.g., decision-making, sense-making, problem detection, error recovery, and human adaptation). Equally important, the book considers the larger organizational context within which human work takes place and examines the organizational functions that shape the work of controllers and the way that they use their artifacts.

Part I of the book aims at describing the ATC system from a CE perspective as a JCS of people, teams, and artifacts that are adapted to the different demands posed by unfamiliar situations and new technologies (see Chapter 1). This human–task–artifact interaction takes place within a changing context of work that is characterized by situational demands and organizational constraints. Chapter 2 presents an overview of system factors that affect how controllers organize their work, collaborate, and use their computer artifacts. Finally, Chapter 3 looks at the organizational context in terms of four safety management processes that provide orientation, assess risks and safety barriers, manage human performance, and promote safety. Safety policy communicates organizational goals to the sharp-end practitioners and establishes a degree of autonomy. Safety risk management identifies work hazards, defines acceptable risk levels, and provides resources for managing risks (e.g., procedures, tools, and alerting systems). Safety assurance establishes channels of feedback to monitor performance, defines safety indicators to measure safety, and assesses any new programs of change. Finally, safety promotion looks at controller competences to overcome problems and achieve a satisfactory level of performance. The safety management processes provide safety constraints for lower levels in the organization so that local human–machine interactions comply with system and organizational requirements.

Part II looks into the cognitive functions, their interaction and their adaptation in normal and abnormal situations. In this sense, four chapters have been written for the following cognitive functions: (1) decision-making, (2) sensemaking, (3) problem detection and replanning, and (4) adaptation to new situations. These functions are addressed both at the individual level and the team level that shape a common understanding and coordination. To make the book a practical field guide, theoretical models are illustrated with behavioral markers that provide observable indicators of poor and good practices. The behavioral markers for the cognitive functions have been derived from several empirical studies from ATC simulator training, empirical observations from the authors, and reviews of the literature.

Part III presents CE applications of cognitive task analysis of controller competences for abnormal situations (see Chapter 8) and the analysis of work complexity encountered in the ATC environment (see Chapter 9). Metrics of air traffic complexity provide a useful basis for adjusting the planning and the regulation of traffic as well as for gauging new technologies that change the work environment. Chapter 8 examines how cognitive functions adapt to abnormal and emergency situations and makes useful suggestions for controller refreshing training. Chapter 9 provides useful insights on how controllers adapt their cognitive functions to manage workload and complexity. The cognitive functions discussed in Chapters 8 and 9 provide a framework for applying a sort of cognitive task analysis to identify cognitive requirements and training needs. Finally, the relationship between complexity and competences is considered in the new operating context advocated in SESAR and NextGen approaches (see Chapter 10).

Part IV looks at the organizational context, that is, the organizational policies, plans, and safety management systems that shape the cognitive work of controllers. Chapter 11 provides an overview of three approaches to the way that safety-critical organizations structure their functions to remain within a safety envelope. Chapter 12 attempts to elaborate the control dynamics of the systems-theoretic accident model and process (STAMP) technique on the basis of a theoretical model of organizational viability (i.e., the viable systems model). The joint framework can help analysts to rethink the safety organization, model new information loops and constraints, look at the adaptation and steering functions of the organization, and finally, develop high leverage interventions. Chapters 13 and 14 look at the organizing aspects that make organizations resilient institutions and provide a framework for integrating human and organizational performance. Finally, Chapter 15 presents several cases that illustrate how to apply the theoretical models and methods of previous chapters to provide practical guidance in safety risk management.

This book provides practical ways to understand how complex systems behave and how people adapt to changes, as meaning-seeking, context-sensitive, and coordinating agents (Woods and Hollnagel

2006). In the study of the ATC domain, we embark on a process of discovery, wherein errors are considered interesting openings for further inquiry. Errors can often be traced into misleading information technology, contradictory processes, or competing goals. The models of cognition and organization presented in this book can guide this process of discovery so that the specific challenges of a work domain can be captured and addressed throughout the design process.

Cognition and organization enable modern organizations to anticipate, cope with, and recover from unexpected demands. In order to succeed at that, we need to pose, and answer, substantive questions.

- How sources of brittleness can be recognized in a system before, rather than after, a misadventure occurs?
- What features comprise brittleness, what are their sources, and how can they be represented?
- How do controllers make sense of a changing situation and make decisions under time pressure and uncertainty?
- Is it possible for controllers to detect errors and recover them before critical consequences are ensued?
- How do controllers and their teams adapt to their work situations by modifying their practices and what factors can limit this adaptability?
- Can CE assess the increasing complexity of ATC situations and the work adaptation of controllers?
- What competences are needed for managing traffic complexity and how can these be trained?
- How can we model the ways that organizational constraints affect the work of sharp-end practitioners?
- And finally, how can cognition and organizational work be integrated into a practical framework?

Models of cognitive engineering and safety organization can bring about an improved understanding of human cognition and expertise that evolve within a work system as well as the constraints that shape the work system. This book highlights a design process where practitioner needs, expertise, cognitive demands, constraints, and goals are considered throughout the design. This support of attending, perceiving, remembering, and reasoning can be contrasted to projects in which features to support these cognitive functions are added as fixes near the end of the design process or after technologies have been fielded. Hence, system designs that use the proposed methods and principles are more likely to feature the flexibility required to accommodate a changing world.

In this book, an effort has been made to show how to operationalize or utilize theory in cognitive engineering and safety organization in order to provide practical guidance in the areas of safety management, training, personnel assessment of cognitive and technical skills, evaluation of traffic complexity, and system design. Particular emphasis has been placed on theoretical models and techniques for managing safety in organizations. The aim was not, however, to present a field guide to risk management but rather to show how the book material can be used to expand existing approaches and techniques that constitute current practice in ATM. Part I, for instance, addresses how the higher organizational levels can develop their safety functions to achieve their safety requirements. At the operational level, the cognitive engineering perspective (Parts II and III) provided a good basis for modeling, the interactions of controllers, pilots, and airport staff in an organizational context of opportunities and constraints. Finally, it is hoped that the integrated models of human and organizational models in the last chapters of the book will provide safety analysts with a condensed form of knowledge to apply in the management of safety.

PART I

An Overview of Management of Operations and Safety

1

The Air Traffic Management System

1.1 Introduction

The air traffic management (ATM) system is a "joint cognitive system" of people, teams, and artifacts (e.g., radar consoles, procedures, automated support functions, and alerting services) that adapts to the challenges and demands posed by unfamiliar situations and new technologies. In a joint cognitive system, practitioner activities are understandable in relation to the properties of the environment in which they work; for instance, conflict resolution strategies are affected by traffic constraints, availability of job aids, team composition, collaboration with flight crews, and regulations. In turn, a work environment can be understood in terms of what it demands and affords to the people at the sharp end. This chapter presents an overview of the work environment of air traffic controllers (ATCOs) in order to provide a basis for understanding human performance challenges.

The demand for increased throughput has often stretched ATM resources and capabilities to a point where accident rates may be difficult to control. For instance, the introduction of radar, a few decades ago, resulted in lower separation distances between aircraft, which increased the degree of coupling in the system and thus the likelihood of errors and conflicts. In general, the ATM system has witnessed the introduction of new technologies and procedures that gradually resulted in higher throughputs and standards of performance, hence, stretching the system closer to the safety boundary. The ATM system is a prime example of the "law of stretched systems," as it is stretched to operate close to its capacity; as soon as there is some improvement, it will be exploited to achieve a new intensity and tempo of activity (Woods and Hollnagel 2006). Past, present and future changes in the ATM system have been dominated by two competing goals:

1. *Higher throughput*: Demand requirements for transporting people and freight usually implicate a greater capacity for air transportation.
2. *Lower accident rates*: Requirements for lower accident rates are more pressing than ever. Even if the accident rate remained the same, an increase in air traffic would give rise to an unacceptable number of accidents.

In fact, the demand for increased air transportation has never ceased although, in certain periods, some temporal decreases in demand were observed as compensating responses to major events and crises (e.g., the sharp decline of air traffic after the 9/11 terrorist attacks).

Before embarking on an exploration of cognitive and organizational functions, it is necessary to provide a concise description of the multiplicity of elements of the ATM system. Due to its functional and operational complexity, it is neither possible nor desirable to cover all aspects of the ATM system. However, a selective presentation of some functional elements of the ATM system and their interactions can provide a practical framework for understanding the cognitive strategies of controllers and the organizational functions of the ATM system. The following questions can be seen as drivers in our exploration of the ATM domain in this chapter:

- What are the basic elements of the ATM system?
- What are the operating characteristics of the ATM operational units?
- How do controllers manage challenging and unusual situations?
- How do ATM units interact in the management of traffic?
- What are the major interactions between controllers and pilots in different phases of flight?
- What safety assessment and incident investigation systems are in place in the ATM domain?

1.2 The ATM System

The ATM system is a complex, highly interactive engineering system that involves many organizations and a large number of subsystems and components onboard and in the ground. According to the International Civil Aviation Organization (ICAO), the purpose of the ATM system is the provision of a set of airborne and ground functions to ensure the *safe, orderly and expeditious* movement of aircraft in all operational phases.

The three main functions of the ATM system include (ICAO 2001, 2007a):

1. *Air traffic service*: ATS is the primary functional component of the ATM system that relies on Flight Information Services (FIS), alerting services, air traffic advisory services and air traffic control (ATC) services (i.e., area control, approach control, and airport tower services).
2. *Airspace management*: ASM refers to airspace utilization strategies and policies including, management activities for achieving the most efficient use of airspace while avoiding airspace segregation.
3. *Air Traffic Flow Management*: ATFM enables the safe, orderly and expeditious flow of air traffic by ensuring that ATC capacity is effectively utilized and traffic volume is compatible with the capacities declared by the appropriate authorities. ATFM flow controllers are employing efficient airspace management tactics and policies by directly interacting with the ATS units.

Figure 1.1 shows a high-level functional representation of the ATM system, where the airborne-based and ground-based parts interact to attain the main system objectives. The airborne-based ATM system includes many onboard systems that provide communication, navigation, and surveillance (CNS) capabilities as well as alerting services (e.g., traffic alert and collision avoidance system [TCAS]). In case that the separation distances between aircraft fall below certain critical values, TCAS generates traffic advisories (TAs) and resolution advisories (RAs) – i.e., collision avoidance advisories in the vertical plane. In complying with a TCAS RA, for example, a pilot may deviate from ATC instructions and assume responsibility for traffic separations since the controllers are no longer responsible for this event.

A vertical profile of flight phases and associated ATC services is depicted in Figure 1.2. In the following sections, the main duties and challenges faced by controllers and pilots will be described with reference to the different ATC units and flight phases.

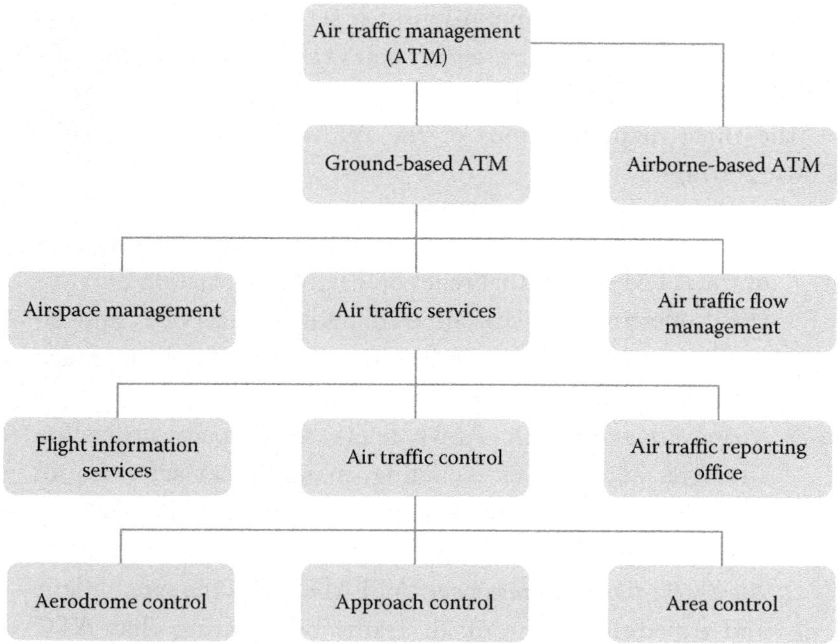

Figure 1.1 A high-level functional representation of the air traffic management (ATM) system.

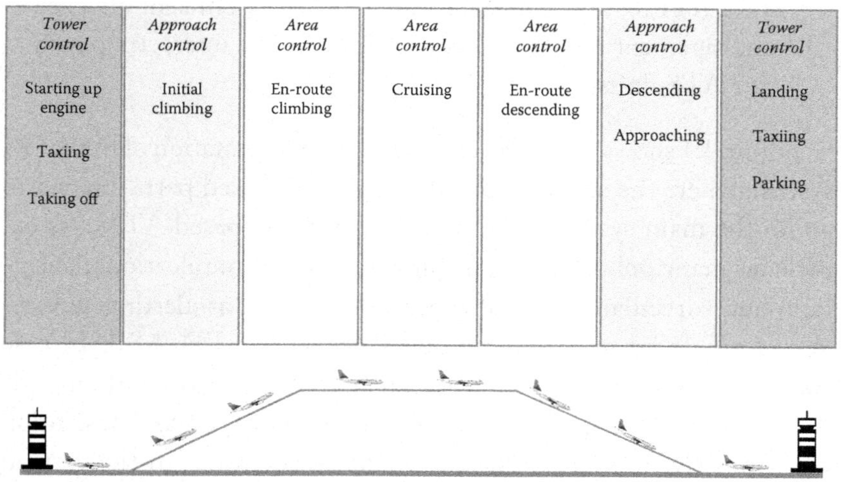

Figure 1.2 Vertical profile of flight phases and Air Traffic Control Services.

To meet the goals of safe, orderly, and expeditious traffic, the ATM system relies on the smooth interaction between adequately trained and licensed practitioners, highly automated systems, and international regulations and procedures. The practitioners at the sharp end, the technology, and the regulations that compose the ATM system can be assigned into six discrete control elements:

1. *Procedures and regulations*: They refer to the national and international legislation (ICAO, European Union, EASA, and State legislation) according to which the ATM system operates.
2. *Air traffic controllers (ATCOs)*: The properly trained and licensed practitioners responsible for the provision of ATM services.
3. *Automation systems*: The computers, displays, Controllers' Working Positions (CWPs), and the special-purpose software that provides information related to the status, position, and separation of aircraft.
4. *Communication systems*: Air–ground, ground–ground, and air–air voice communications as well as data exchange systems.
5. *Navigation systems*: They provide real-time 3D positional information to aircraft in order to support navigation through the airspace and movement on the airport.
6. *Surveillance systems*: They provide near real-time positional and other information to controllers for tracking aircraft and monitoring hazardous weather conditions.

The last three elements are collectively referred to as communication navigation and surveillance (CNS) systems. All control elements of the ATM system are briefly presented in the following sections.

1.2.1 Procedures and Regulations

The operation of the ATM system is described in detail through the annexes, documents, and other guidance material published by ICAO. The overriding goal of ICAO legislation is the harmonization of international procedures and regulations in order to provide a smooth integration of national and international ATM systems. Hence, many efforts have been made for national differences to be kept to a minimum and to be communicated adequately to airspace users.

The ATC system comprises three interconnected levels supported by the following units:

1. Airport control tower (TWR) that provides control services to departing and arriving aircraft.
2. Approach control unit (APP) that provides services to aircraft approaching a terminal maneuvering area (TMA).
3. Area control center (ACC) that provides services to overflying aircraft.

All ATC units incorporate standardized control positions, areas of responsibility (AoRs), CNS systems, standard operating procedures (SOPs), operations manuals, contingency plans, unit training plans (UTP), unit competency schemes (UCS), and letters of agreement (LoAs); they also apply specific separation minima that are determined by the class of the airspace that they control. These elements are elaborated in the aeronautical information publication (AIP) issued by or with the authority of a State which contains aeronautical information essential to air navigation (ICAO 2007a).

The fundamental operating characteristics of the three ATC units are shown in Table 1.1 (ICAO 2001, 2005a, 2007a).

The ATC units provide air traffic services for instrument flight rules (IFR) flights conducted in accordance with instrument flight rules in instrument meteorological conditions (IMC) and

Table 1.1 Fundamental Characteristics of the ATC Units

UNIT CHARACTERISTICS	AIRPORT CONTROL (TWR)	APPROACH CONTROL (APP)	AREA CONTROL (ACC)
Control positions	Airport controller	Executive/tactical controller	Executive/tactical controller
	Ground controller	Coordinating/ planner controller	Coordinating/planner controller
	Delivery controller		
Area of responsibility	Airport traffic zone (ATZ)	Terminal maneuvering area (TMA)	Control sector
Airspace classification	Class D,E	Class C,D	Class A,B,C
Applicable legislation and procedures	ICAO, European commission, EASA, national legislation, local SOPs, LoAs, operation manual, contingency plan.		
	Visual	Radar	Radar
Applicable separation minima	Wake turbulence	2.5–5 Nm horizontally	5–10 Nm horizontally
	Time based	1000 ft vertical	1000–2000 ft vertical

visual flight rules (VFR) flights in accordance with visual flight rules in visual meteorological conditions (VMC). In IFR flights, controllers are responsible for safe separation from obstacles and other aircraft by providing appropriate services in accordance with the ATS type, available CNS systems, and airspace classification. In VFR flights, flight crews are responsible for visually separating their aircraft from obstacles by remaining outside clouds.

1.2.2 Air Traffic Controllers (ATCOs)

Controllers remain the cornerstone of the ATM system as they manage to adapt the operation of the system to many irregularities that have not been foreseen in the initial design. Controllers work at the sharp end of the system to ensure safe, efficient, and expeditious traffic. The training of controllers is extensive and meticulously structured in accordance with international standards. All European Union ATC units usually base their training regimes on two documents:

1. The unit competency scheme (UCS), which indicates the methods by which the units maintain the competence of all licensed controllers
2. The unit training plan (UTP), which details the processes and time frames that allow the unit procedures to be applied under the supervision of an on-the-job training (OJT) instructor.

In general, controller training is divided into four phases that correspond to the progression of student controllers to licensed controllers and to special roles (e.g., on-the-job instructor (OJTI), unit assessor, and supervisor). By successfully completing the first two phases, the student becomes a licensed controller. European commission regulation 2015/340 provides a detailed framework of technical requirements and administrative procedures relating to air traffic controller licenses and certificates. Table 1.2 briefly explains the four phases of controller training.

1.2.3 Automation Systems

The improved reliability and computational power of modern digital computer systems and their networking capabilities allowed a

Table 1.2 The Four Phases of Controller Training

TRAINING PHASE	DESCRIPTION
Initial training	Training on technical subjects, ATC theory and simulator. The objective is to prepare Ab-Initio students for training at ATC units.
Unit training	Transitional training between pre-on-the-job (OJT) and OJT training, leading students to obtain an ATCO license, with appropriate rating and unit endorsements.
Continuation training	Training for licensed controllers in order to augment their knowledge and skills. It includes refresher training in abnormal situations and conversion training that provides knowledge and skills appropriate to changes in the operational environment.
Development training	Training to provide additional knowledge and skills for specific job profiles (e.g., OJTI, Unit Assessor, Unit Supervisor, Team Resource Management, Safety Occurrence Investigation, Safety Assessment and Safety Surveys).

large scale introduction of automated features in the ATM system. The uniqueness of digital computers over other machines stems from the fact that practitioners end up having a powerful special purpose machine (Leveson 2012). For instance, a TCAS is a special purpose machine built on a set of algorithmic instructions to accomplish an advisory service. The same applies to a vast array of automation systems built under the simple principle of writing appropriate software for digital computers.

The automated systems in the ATM domain can be divided into two broad systems:

1. *The controller's working positions (CWPs)*, which constitute the working environment and the tools through which controllers practice their profession. A CWP consists of a range of standard voice and data input/output (I/O) devices (e.g., keyboards, displays, mouse, VHF headsets, and telephones) and special-purpose software that enable controllers to perform the following tasks:
   a. Communicate with aircraft
   b. Communicate with other units
   c. Monitor the functionality status of CNS systems
   d. Monitor meteorological data
   e. Manage flight progress and other type of information
   f. Manage CWP displays and data presentation

g. Operate aeronautical ground lighting systems (e.g., precision approach path indicators, taxiways, stop bars, runway lighting)

2. *The decision support systems (DSS),* which support decision-making in managing air traffic. DSS can be classified into three main categories:
   a. *Sequencing managers*: Automation systems designed to provide controllers with suggestions about the optimal management of departure and arrival traffic flows under normal conditions. For instance, the departure manager (DMAN) provides information on a calculated departure sequence of aircraft to the runway while the arrival manager (AMAN) provides an arrival plan that is monitored and updated regularly by the system.
   b. *Monitoring aids (MONA)*: Automation systems that assist controllers in track monitoring and routine clearance tasks. Examples are: the route adherence monitoring (RAM) that verifies whether aircraft are adhering to their routes and the cleared level adherence monitoring (CLAM) system that verifies whether aircraft are adhering to their cleared flight levels.
   c. *Air traffic flow and capacity management (ATFCM) aids*: These decision support systems are available to flow controllers and include: enhanced tactical flow management system (ETFMS), integrated initial flight plan processing system (IFPS), and central airspace and capacity database (CACD).

Safety nets are important affordances in the provision of ATM services and work closely together in a control loop that is shown in Figure 1.3. The loop starts when a controller issues an instruction to the flight crew (e.g., a flight level change) using the communication systems. The crew acknowledges the instruction and makes an appropriate input into the autopilot system. The aircraft initiates the commanded change (e.g., a level change) which is captured by ATC surveillance sensors (e.g., the radar). The data on the radar screen are processed in combination with other relevant data (i.e., flight level changes of aircraft in the vicinity) through special computer algorithms. The resulted information is depicted on the CWP screens

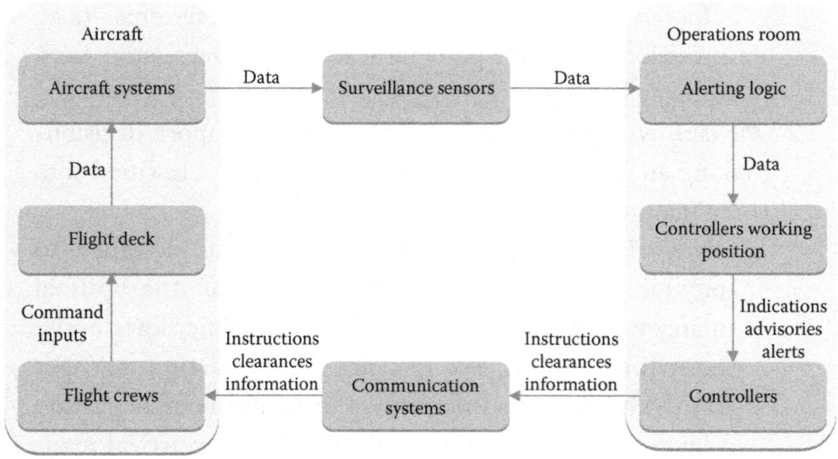

Figure 1.3 Safety nets loop in the ATM system.

(e.g., visual and/or aural warnings) when the prescribed horizontal and/or vertical distance may be infringed in a certain period. Finally, the controller detects and acknowledges the warnings and intervenes in order to resolve aircraft conflicts.

Although the technology for automatic interventions already exists, controllers are kept in the loop because automation is not allowed to intervene in an autonomous fashion. This is not the case however in aviation, where airlines may allow automation to intervene autonomously in order to keep aircraft within a safe flight envelope (e.g., avoiding to exceed a certain bank angle or speed, preventing a stall, etc.). In many cases, following flight automation advisories may be regulatory mandatory. For instance, the TCAS system generates resolution advisories that flight crews are obliged to follow by regulation, irrespective of ATC clearances (ICAO 2007a).

Safety nets are subject to false alarms and technical problems that reduce their reliability. Although reliability engineers may recognize these failures, other subtle problems can escape their attention, especially when the interactions between the safety nets are hidden or implemented in unexpected ways. A case in point concerns the interaction between the short term conflict alert (STCA) and TCAS systems. In the ATC domain, the STCA is designed to warn controllers of imminent separation minima infringements. Normally, controllers are warned by STCAs before the activation of TCAS in the aircraft cockpit. However, in some rare cases, the flight crew may get a TCAS TA or RA in the cockpit before the identification of the conflict by the STCA in the radar screen of controllers. In certain conflict geometries, the information update rate of the TCAS system can be faster than the STCA update rate. This implies that controllers may unexpectedly have to manage a TCAS RA while they were certain that their traffic planning was appropriate. The subsequent vertical movement of the aircraft that responds to the TCAS RA may cause a significant disruption in traffic management as well as secondary activations of TCAS on other aircrafts.

1.2.4 Communication Systems

Controllers communicate with flight crews, directly using voice communications, or indirectly using data links. Air–ground communications include very high frequency (VHF) systems as well as data links for information exchanges. Every ATC unit is assigned a set of frequencies that enable controllers to communicate verbally with aircraft using standard radio telephony (RTF) procedures. Communications are vital to the safe and expeditious operation of aircraft while many incidents occurred due to the use of nonstandard procedures and phraseology (ICAO 2007b). The crew–controller communication loop constitutes a confirmation-correction process and includes some degree of redundancy, as illustrated in Figure 1.4. Controllers also communicate with other ATC units or services via land lines. For this purpose, ground voice and data communication networks are installed that enable them to communicate virtually with any other ATM facility in the world using the aeronautical fixed telecommunications network (AFTN).

In VHF RTF communications, crews and controllers cannot use the same frequency simultaneously because when one is transmitting, the other is receiving and vice versa; hence, controllers and crews cannot transmit and receive simultaneously. Even though this technical shortcoming is well known and properly documented, it remains a causal factor in a large number of incidents.

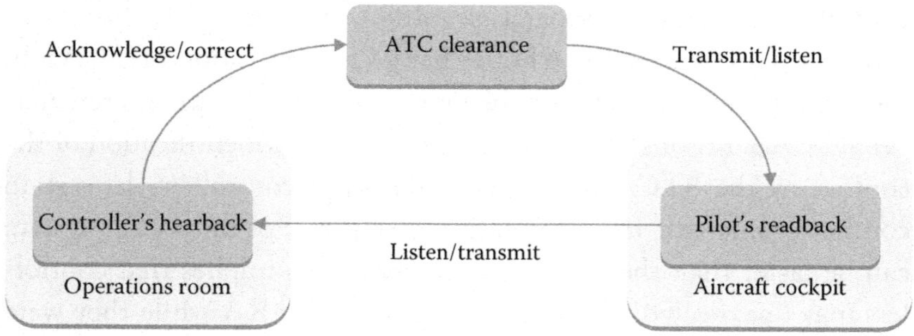

Figure 1.4 Flight crew–controller communications loop.

1.2.5 Navigation Systems

Navigation systems (i.e., commonly termed navaids) refer to a group of land and space based systems that enable pilots to know their exact position in the airspace or, in the vicinity of an airport. The en-route navigation depends on airways that essentially form a network of "highways" in the sky. An airway is a control area that forms a sort of corridor in the airspace (ICAO 2007a). In the vicinity of airports, navigation depends on creating funnels for approach and landing with the routes that connect the airport with the surrounding airways. For instance, a Standard Instrument Departure (SID) is created when an IFR departure route links a runway with a specified significant point normally in an airway. With the aid of appropriate flight instrumentation systems, the flight crew can make use of an Instrument Approach Procedure (IAP) in order to maneuver from an initial approach fix, or the beginning of an arrival route, to a point for landing in the runway.

Navaids can be used either in the vicinity of an airport for the purpose of approach and landing or for en-route navigation. The main characteristics of mainstream ground navaids are described in the Table 1.3 (ICAO 2006a, b, c).

The quality of the required navigation information may differ in each phase of the flight. For the approach and landing phases, the requirements for signal accuracy are the most stringent due to the close proximity of the ground and the limited maneuvering potential of the aircraft. Moreover, the requirements for availability, reliability, and integrity are higher in the approach and landing phases than the en-route phase.

Table 1.3 Characteristics of Mainstream Ground-Based Navaids

NAME	ILS (INSTRUMENT LANDING SYSTEM)	VOR/DME(VERY HIGH FREQUENCY OMNI DIRECTIONAL RANGE/ DISTANCE MEASURING EQUIPMENT)
Operational description	Enables vertical and lateral guidance during approach and landing	Enables short range en-route navigation and approach
Guidance	Vertical and lateral	Lateral (range and bearing)
	Three levels of precision approaches	En-route navigation
	(ILS category I – II – IIIa, IIIb, IIIc)	(VOR-VOR)
Operational use	Approach and landing	En-route & approach
Instrument procedures	Instrument approach procedures (IAPs)	IAPs, SIDs, STARs, holdings

1.2.6 Surveillance Systems

Surveillance is the function that provides controllers with aircraft information about range, bearing, and altitude. ICAO (2007a) provides detailed guidance for the surveillance function which can be accomplished as follows:

- *Pilot reporting*: Using voice communications, flight crews can report aircraft position in reference to certain navaids.
- *Responses of primary surveillance radar*: A PSR is a sort of radar system that utilizes a rotating antenna in a ground station that emits electromagnetic pulses that are reflected by the metallic exterior of the aircraft and returned back to the antenna. This is a noncooperative form of surveillance because it does not require the cooperation of the aircraft (carriage of a transponder device). PSR can also provide significant weather data such as storm cells positions and areas of precipitation. The PSR is useful in cases of detecting noncooperative aircraft that affect traffic planning (e.g., military traffic, aircraft with nonfunctioning transponders).
- *Returns from secondary surveillance radar*: The SSR uses a rotating antenna in a ground station that emits interrogation messages in the form of electromagnetic train pulses that trigger automatic responses from the transponder of the aircraft and are subsequently received by the antenna.

- *Automatic dependent surveillance*: ADS is a data-link that periodically broadcasts the state vector of the aircraft and other flight information (e.g., estimated time over the next waypoints, weather data, and navigation data). The ADS-B system improves the use of airspace, reduces ceiling/visibility restrictions, improves surface surveillance, and enhances conflict management.

Most of the previous systems represent the legacy surveillance function currently used in ATM systems while the ADS represents the near future.

Powerful radar data processors (RDPs) transform raw data from PSR returns and SSR responses received via radar antennas into digitized aircrafts tracks on radar displays. The complex progression of signal reception and processing is referred to as surveillance processing chain.

Most radar systems provide controllers with the following information for all aircraft carrying a transponder:

- Identification derived from SSR Mode A
- Callsign of the aircraft derived from SSR Mode A and flight plan correlation
- Altitude derived from SSR Mode C
- Velocity Leader derived from RDP Processing
- Ground Speed derived from RDP Processing
- Attitude Indicator derived from SSR Mode C and RDP processing

Figure 1.5 shows an example of a correlated track (e.g., a SSR return coupled with flight plan details) depicted on the radar screen of controllers. The surveillance system displays an integrated picture related to aircraft position and other information known as correlated track. In essence, the track is a digital representation of the aircraft state vector information.

1.3 ATC Units

Every ATC unit is assigned a controlled airspace that defines its area of responsibility (i.e., an airspace of certain vertical and lateral dimensions that is classified accordingly). A controlled airspace can cover

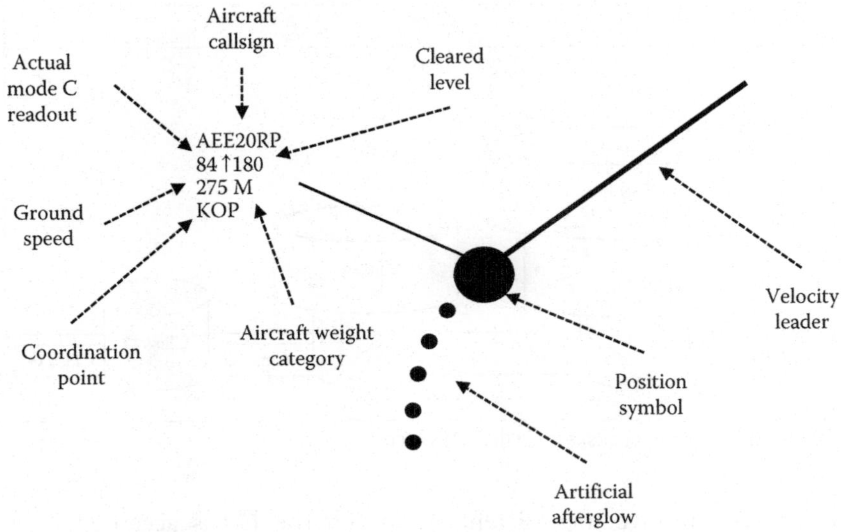

Figure 1.5 A correlated track of an aircraft shown on the radar screen of controllers.

the classification areas A, B, C, D and E, including airways, control areas, TMAs, and control zones. A simplified schematic depiction of the basic geometric shapes that correspond to different airspace classes is provided in Figure 1.6.

The controlled airspace in the first two levels (i.e., airport and terminal approach areas) has the shape of a cylinder with the center being the terminal navigation aid of the airport. The area control is divided into upper and lower areas that are above the terminal approach areas and normally have a rectangular shape. Uncontrolled airspace is outside the controlled airspaces and corresponds to categories F and G.

The three control levels of the ATC system interact through lateral and vertical coordination procedures during the regulation of flights. The central coordination element is the flight plan (FP), which comprises specific information provided to ATS in relation to the intended flight course of the aircraft (ICAO 2007a).

In the FP, the flight captain fills in a form with all necessary information relevant to the flight. The FP is submitted to the ATC unit in the departure airport. Controllers check the FP for compliance with the format and data conventions, information completeness, and the degree of accuracy. In the event of errors or discrepancies, they take necessary actions by coordinating with the originator to correct

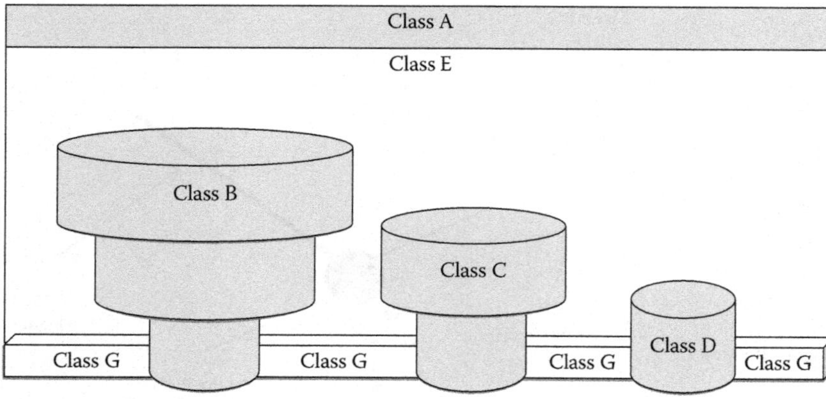

Figure 1.6 Classes of airspaces in the ATM system.

errors and resolve inconsistencies. When the FP is accepted, it is transmitted to all relevant units. The FP is activated in the departure airport, when the aircraft request start-up, and it is closed with the landing phase in the destination airport. During the flight, all ATC units are authorized to change parts of the FP but efforts are made to adhere to the original version. Flight progress through the ATC system is monitored by consulting information on the FP. More information about essential flight characteristics is provided in electronic flight progress strips (e.g., callsign, type of aircraft, SSR code, route, time estimates over significant points of the route).

1.4 Airport Control Tower (TWR) Operations

The airport control tower (TWR) is where all flights begin and eventually terminate. The AoR of a TWR is an ATZ in the shape of cylinder, with 5–10 Nm radius and 2,000–10,000 ft height (ICAO 2007a). The functions of an TWR are normally performed by two control positions:

1. *The airport or local controller (LOC)*, who is responsible for operations on the runway and all aircraft flying within the AoR of the control tower.
2. *The ground controller (GRD)*, who is responsible for all traffic on the maneuvering area with the exception of runways.

Airport or tower (TWR) controllers issue information and clearance to aircraft in order to achieve a safe, orderly, and expeditious flow of traffic in the vicinity of an airport with the purpose of preventing collision(s) between:

1. aircraft flying within the designated AoR of the control tower
2. aircraft operating in the maneuvering area
3. aircraft landing and taking off
4. aircraft and vehicles operating in the maneuvering area
5. aircraft in the maneuvering area and obstructions on that area

ICAO (2007a) defines the maneuvering area as the part of the airport that is used for takeoff, landing, and taxiing of aircraft, excluding aprons. ICAO clearly states that only the maneuvering area is within the jurisdiction of the airport controllers.

A simplified version of airport operations comprises six phases, as shown in Figure 1.7. In Position 1, a departing aircraft in the apron initiates a call, requesting an engine start-up. The ground controller normally approves the start-up except in cases where the aircraft is subject to ATFM restrictions or locally imposed restrictions (e.g., another unit may restrict the departure flow due to heavy workload). From the same position, the aircraft is provided with taxi clearance information regarding the runway and weather information. In Position 2, the departing aircraft is held at the runway holding position where the pilots accomplish the engine run-up procedures. At this position, the ground controller transfers control to the local

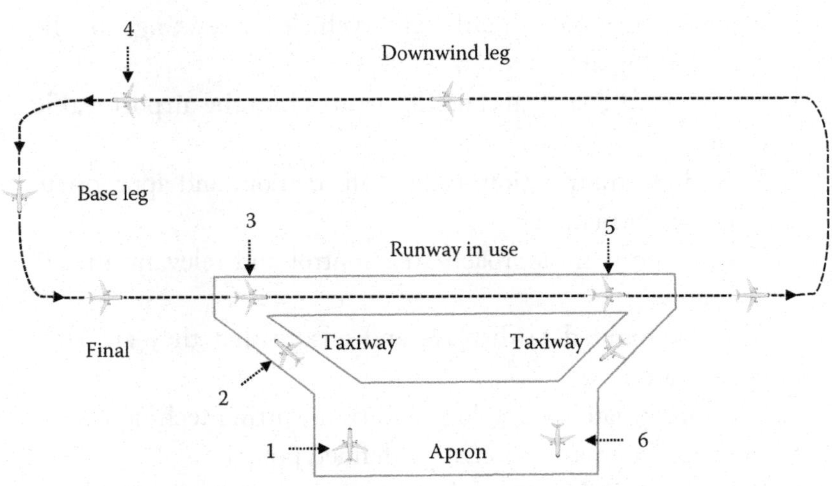

Figure 1.7 Typical operations in airports.

controller who issues a clearance to the aircraft to enter the runway. Subsequently, in Position 3, the local controller issues a takeoff clearance if not practicable in Position 2 (e.g., if there are conflicts in the arriving traffic on final). After the aircraft is airborne, the local controllers transfer control to the approach controllers.

Control of airport traffic is based on visual observations of the maneuvering area and the vicinity of the airport. In low visibility conditions, a surface movement radar (SMR) can be used to augment visual observation of traffic in the maneuvering area and surveillance of traffic in those areas that cannot be observed visually. The duties and challenges faced by controllers and pilots in airport operations are briefly presented below.

1.4.1 Airport Controller Duties

The description of the duties of airport or tower controllers is based on the training and operational manuals of TWR units. The general duties of the local, ground, and delivery positions are:

- Ensure the integrity of the working position and use automation tools as appropriate
- Issue inbound and outbound instructions to aircraft
- Obtain and issue IFR clearances to departing aircraft and ensure correct readbacks
- Ensure that flow management procedures are met
- Select runway in use
- Correctly handle aircraft and vehicles operating in the maneuvering area
- Integrate VFR arrivals and departures into the airport traffic circuit
- Issue flight information, traffic information, and appropriate airport information
- Coordinate with approach/area control and relevant airport operators
- Monitor flight data displays and ensure that they are kept up-to-date
- Use appropriate phraseology and transmitting techniques
- Transfer communication at predefined points
- Prioritize and delegate tasks when appropriate

- Communicate with aircraft and colleagues in a clear and precise manner
- Ensure that all coordination is in accordance with prescribed procedures
- Manage air-ground and ground-ground communication failures
- Assist and give priority to aircraft in unusual situations and take all actions necessary to ensure safety

1.4.2 Pilot Duties during Taxiing, Start up, or Landing

Upon arriving at the aircraft, the flight crew must complete a large number of tasks in a short period of time. One pilot normally sets up the flight deck while the other goes outside to check the aircraft and look for defects, bird-strike damages, and obstructions of the sensors and probes. The flight deck is set up for the particular departure including arrangements for standard instrument departure, levels, and frequencies. The flight management system (FMS) is programmed to receive feedback about fuel requirements, flight time, maximum flight level, take-off speed, and so on. Once the clearance to start up is received, the flight crew begins the departure briefing, which also covers responses to several unusual situations.

During the preflight stage, the flight crew is supplied with many data about the course of the flight (e.g., navigation waypoints, flight levels and associated speeds, and meteorological data). In addition, NOTAMs are distributed by means of telecommunication that contains information concerning the establishment, condition, or change of any aeronautical facility, service, procedure, or hazard that is essential to aviation practitioners (ICAO 2007a). Company and ATFM delays (e.g., slots and late arrivals) are taken into account at this phase. The flight crews perform basic weight calculations and decide how much fuel, passengers, and cargo they can afford. Fuel calculations must take into account several external factors, such as weather conditions at destination and the location of alternate airports. The crews need to verify that the aircraft's final weight and balance remain within predefined limits before entering them into the FMS. At this stage, the ATC units are notified about the flight through the flight data processing system and about any ATFM measures imposed on the particular flight.

Once the engines are running and all after-start checklists are completed, a taxi clearance is obtained from the ground controller. In large airports, taxiing may take several minutes while proceeding though various taxiways and intersections may be a complex task that requires constant monitoring and coordination with ground controller. Both crew members are monitoring the taxi route and try to avoid the "heads down" syndrome in the cockpit.

1.5 Approach Control (APP) Operations

The approach control (APP) unit provides control services to arriving and departing aircraft from one or more airports (ICAO 2007a). The area of responsibility of the APP unit is normally a TMA in the shape of cylinder with 60 Nm radius and 24,000 ft height. The functions of the APP unit are normally performed by two control positions:

1. *The tactical or executive controller (EC)* responsible for the direct control of aircraft and for carrying out the overall plan established by the coordinating controller.
2. *The planner or coordinating controller (CC)* responsible for establishing the overall plan for the entry and exit of aircraft and for assisting the executive controller.

A simplified version of arrival operations is shown in Figure 1.8 where two flows of arriving aircraft coming from the ALPHA and BRAVO entry points are merged and put in sequence to establish the instrument landing system (ILS). At the two entry points, the aircraft are transferred from the ACC unit to the APP unit at the position, level, speed and heading agreed between the coordinating controllers of the two units. The executive controller (APP) provides information about the runway in use and the relevant instrument approach procedure (IAP) and vectors the aircraft with a series of heading and altitude instructions to establish on the ILS final approach course.

Subsequently, the executive controller decides on the approach sequence and merges the traffic in the final leg of the traffic circuit.

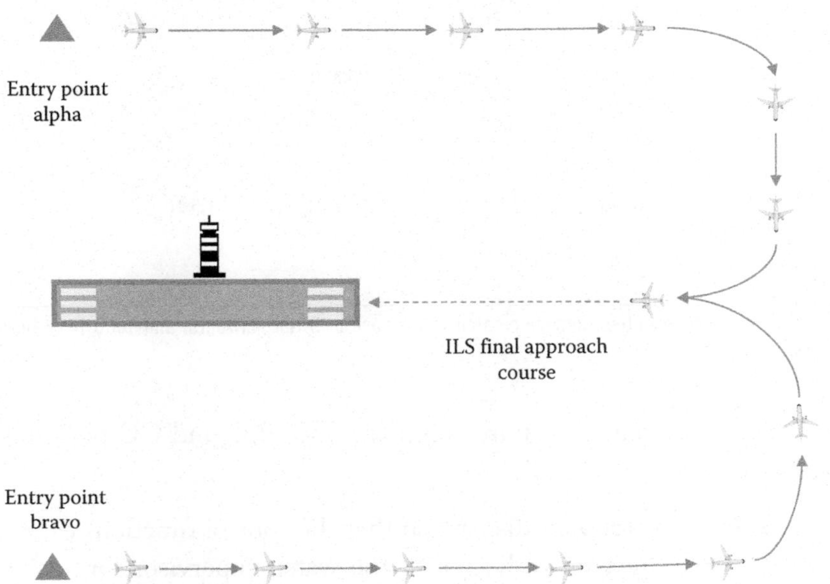

Figure 1.8 A schematic of two arriving flows merged for landing in a runway.

The selection of the approach sequence is a difficult task since the selection is made from a large number of alternative sequences (e.g., 5 arrivals generate 5! or 120 sequences). The flight crews are guided to establish the ILS at the appropriate altitude, speed, and angle with the advice of the APP controllers who later transfer control to the TWR unit.

The role of the APP unit is to provide separation instructions and maintain a safe and expeditious traffic flow. The normal separation minima are 1000 ft vertically and between 2.5 to 5 Nm horizontally. In this way, a protected airspace is created around each aircraft and traffic should be regulated to prevent overlaps between the protected spaces (Figure 1.9).

1.5.1 Approach Controller Duties

The description of the duties of APP controllers is provided in the training and operational manuals of the APP units and include common and specialized duties for the EC and CC positions.

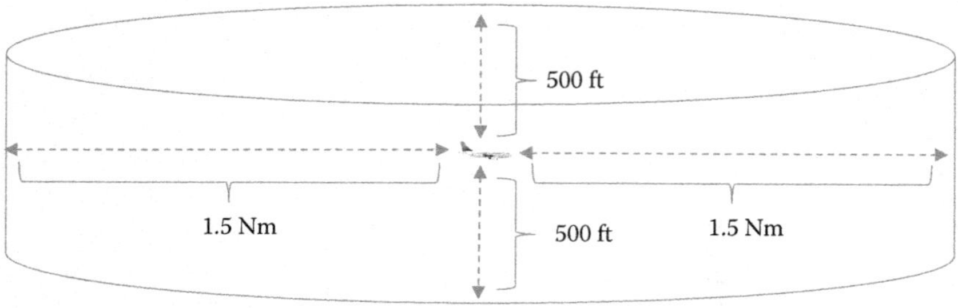

Figure 1.9 A protected airspace created in the case of 3 Nm horizontal and 1000 ft vertical separation.

The generic duties that are common to the EC and CC positions are as follows:

- Adjust the relevant displays so that the control functions can be performed properly; notify the watch supervisor for any technical failures
- Analyze, plan, and control the flow of traffic by using information from the radar and other systems
- Detect potential conflicts between aircraft
- Provide and maintain separation between aircraft between aircraft and airspace boundaries and between aircraft and terrain
- Manage concurrent tasks
- Monitor and ensure that flight data displays and flight strips are maintained up-to-date
- Prioritize and delegate tasks when appropriate
- Communicate with aircraft and colleagues in a clear and precise manner
- Ensure that coordination is in accordance with prescribed procedures
- Manage air-ground and ground-ground communication failures
- Assist and give priorities to aircraft in emergency.

The specific duties for the EC position are:

- Maintain a continuous listening watch on the unit frequencies and carry out RTF communications with the aircraft
- Take all necessary actions within the area of responsibility to comply with the coordination plan established by the CC

- Liaise with the CC position when planned exit levels or other arrangements cannot be achieved
- Ensure that the CC position is warned about traffic situations that overload the area of responsibility of the unit
- Ensure that the CC position is informed of any unusual situations accruing in the unit

The specific duties for the CC position are as follows:

- Plan and accept aircraft into the AoR in accordance with prescribed procedures
- Plan exit conditions in accordance to the planning standards or as agreed with the accepting unit
- Ensure that the coordination with the adjacent units is effected prior to the transfer of aircraft
- Coordinate with EC position to accept aircraft that does not comply with navigation or communication requirements
- Transfer radar identity of aircraft to the EC position
- Ensure that the EC position is aware of any coordinated aircraft climb or descent made with adjacent units
- Inform the watch supervisor of any unusual situations in their area of responsibilit

1.5.2 Pilot Duties during Take-off, Climb, Descent, and Approach for Landing

Most flight procedures for takeoff, climb, descent, and approach for landing are carried out by the flight crews while traffic is regulated by the approach control unit. It makes sense then to provide a short description of the duties of flight crews and the challenges they face.

The takeoff procedure begins when the aircraft enters the runway, once the crew completes the before takeoff checklist. When lined up on the runway and cleared for takeoff, the pilot flying (PF) starts to advance the thrust levers once the engines have spooled up. During the takeoff roll, the pilots cannot reject the takeoff unless specific conditions prevail (e.g., runway incursions, low level windshear warning, engine failure, crew incapacitation).

During climb, the flight crew positions the aircraft to a safe height away from terrain and obstacles. For this reason, the engine thrust is set to a high "takeoff power" setting and the aircraft attitude is pitched up to maintain a specific speed. Once the safe height is reached, the engine thrust can be reduced to a more efficient setting. In order to provide minimum noise disturbance to the area surrounding the airport, regulatory procedures require the aircraft to fly specific profiles (i.e., noise abatement procedures).

The descending process starts with the determination of the top of descent (TOD) point in order to achieve a continuous idle power descent from the cruising level until a certain point on the final approach. In many cases, the calculation of the TOD point is performed by the FMS on the basis of several factors (e.g., tail/head winds and the descent speed given by the cost index) and accounts for the distance required to effectively manage a final approach speed.

During final approach, the workload of the flight crew peaks as multiple tasks may be required at the same time. At this high workload phase, the crew should also arrange the landing configuration of the aircraft. This is a safety-critical task since the aircraft flies at a low level, with minimal speed, close to terrain, and with little margin for maneuvering.

If a nonprecision approach is flown, the workload of the flight crew gets higher because the pilot not flying (PN) becomes very busy since, at each mile, s/he must state the aircraft's position relative to the required vertical profile and predict the altitude of the flight at the next mile checkpoint. In addition, the automatic aircraft systems are not capable of "locking onto" a nonprecision path in final approach. This is different from the case of an ILS approach where the autopilot can "lock onto" the localizer and glide-slope signals allowing the flight crew to monitor the flight path. The flight crew is obliged to fly a "stabilized approach" avoiding any problems due to deviations from the correct flight parameters. In the stabilized approach, the SOPs ensure that the aircraft is on the correct flight path, the flight parameters are within limits and other controls are put on specific positions (e.g., the engines are set at an appropriate power setting, the gear is down, and the flaps are set for landing).

To accomplish a successful landing, the aircraft must be "flared" prior to its touchdown where the engines are commanded to idle power. Sometimes, upon touchdown, the aircraft may not be in a "roll-out" mode which requires the autopilot to be disconnected so that the pilot flying can control the aircraft and maintain the runway center line during the landing roll. After landing, the flight crew must accomplish many actions, such as change radio frequency, receive taxi clearance, and check the taxi route on the airport charts. After vacating the runway, the aircraft may continue the taxi to an assigned stand.

In this sensitive flight phase, emphasis is given to the removal of nonessential verbal communication between crew members that may distract from essential tasks. In general, most operators employ "sterile cockpit" procedures that call for a "verbal silence" below a certain level during climb or descent.

1.6 Area Control Center (ACC) Operations

The area control center (ACC) unit handles the regulation of the cruising phase of the flight that is the longest phase of all. The area of responsibility of the ACC units is extensive with dozens of controllers working in every watch and hence it is divided into a number of sectors and subsectors. One or more ACC units are usually responsible for the airspace of a country. An ACC unit may be responsible for a flight information region (FIR) within which flight information and alerting services are provided (ICAO 2007a). The control functions of the ACC units are performed by the executive and coordinating controllers who have similar roles to those of the APP units.

1.6.1 Area Controller Duties

A description of the duties of ACC controllers is provided in the training and operational manuals of ACC units, which include common and specialized duties for the EC and CC positions. It is evident that the duties of the two positions are similar to those of the TWR and APP units. It is quite common in large ACCs, area controllers with terminal control endorsements to provide approach service as in major airports. Given that the CNS systems and the regulations are also similar, it can be safely assumed that there is a high degree of transfer of cognitive skills from APP to ACC control.

However, there are certain differences between ACC and APP units, such as the following:

- APP controllers provide more vertical, lateral, and speed instructions than their ACC colleagues.
- An area control sector is much wider in dimensions than an APP sector.

- VFR flights are minimal or even nonexistent in ACC higher sectors, compared to a large number of VFR flights in APP sectors.
- Terrain is a major factor in APP operations while it has a minimal effect on ACC operations.
- Weather is a major concern for APP controllers while it has a minor effect on ACC operations.
- APP controllers participate, to a certain degree, in the design of traffic flows into their TMAs while the network of flows in ACC sectors is more or less static.
- APP controllers make more use of surveillance information while ACC controllers rely also on flight plan data to resolve aircraft conflicts.

In general, the number of aircraft that can be simultaneously controlled by APP controllers is significantly smaller than the number that can be controlled by ACC controllers. This is mainly due to the fact that, as aircraft enter into a complex phase of their flight, the number of crossings of flight paths is greater, the effect of weather or terrain restrictions becomes a concern, and VFR flights are a factor, while the mixture of runways in use tends to complicate traffic flows.

1.6.2 Pilot Duties at the Cruising Phase

Once the aircraft reaches its preferred final flight level, the cruise phase begins and the flight crew provides a monitoring function while managing ATC instructions and any paperwork. During the cruise phase, the aircraft flies on autopilot, which operates either in a strategic mode (i.e., following the programmed route entered into the FMS) or in a tactical mode (i.e., allowing aircraft changes in response to direct inputs from the flight crew).

For efficiency reasons, the aircraft often cruises at levels that are close to the maximum level with regard to the aircraft's performance as this would require close monitoring of flight parameters since acceptable margins become very narrow. From an ATM perspective, the flight is controlled leveled at the cruise level by ACC controllers and transferred between adjacent sectors. Vertical changes are minimal while controllers try to keep lateral changes to a minimum within the limits specified by the flight route requested by the flight crews.

1.7 Air Traffic Flow and Capacity Management (ATFCM) Operations

In the 1980s, it became apparent that the three level representation of the ATM system was inadequate to maintain control of heavy traffic in congested sectors and major airports. A need emerged to develop a strategic flow management system to reduce delays and the probability of saturating controllers with high workload. As a result, aviation organizations initiated an effort to design and operate advanced air traffic flow management (ATFM) systems. Despite their differences in their degree of centralization and their proactive or reactive character, all ATFM systems serve two goals:

1. Prevent an overload or over-delivery of aircraft to all ATC units and airports to maintain safety.
2. Minimize economic penalties and other business-related deficiencies for the operators of the ATM system.

The European Commission has nominated Eurocontrol as the network manager that essentially runs the ATFM system (EU 2011a). The EU legislation covers in detail the role of the network manager of the ATM system (EU 2010, 2011b). Through successive updates, a complex system has been evolved that was extended into an air traffic flow and capacity management (ATFCM) system that maintains a balance between demand and capacity by optimizing available resources and by coordinating adequate responses (Eurocontrol 2016). The ATFCM system comprises four functions as follows (Eurocontrol 2016):

1. *The strategic flow management* function that includes research, planning, and coordination activities at least seven days ahead of traffic arrangements. Eurocontrol works together with all ATM stakeholders (i.e., Air Navigation Service Providers, airport operators, airspace users and the military) to produce a single document that incorporates information on traffic demands and capacity plans as well as possible bottlenecks and countermeasures suggested by the ATFCM system.
2. *The pretactical flow management* function of advance planning and coordination activities. This function aims at selecting the best way of managing available capacity resources and putting in action a wide range of appropriate ATFCM measures.

3. *The tactical flow management* function that updates the daily plan according to the actual traffic, capacity and monitoring requirements (i.e., the number of flights entering a sector that triggers the initial traffic assessment in rolling one-hour periods from which coordinated actions may be considered).
4. *The postoperational analysis* function that is performed right after the day of operation. It constitutes the closing part of the loop, as the day of operation is analyzed in detail and feedback is provided into the earlier functions.

At the heart of the ATFM system lies the collaborative decision-making (CDM) system that allows decisions to be taken on the basis of the most comprehensive, up-to-date and accurate information (Eurocontrol 2016). The CDM process is a key enabler of ATFCM that allows sharing of all relevant information between stakeholders involved in the decision making process.

The ATFCM functions cover a range of solutions for managing demands and available capacities as follows (Eurocontrol 2016):

1. Optimize the use of available capacity.
   a. Sector management (e.g., sector configuration and number of sectors)
   b. Balancing arrival/departure capacity
   c. Flight list assessment (flights of minor workload)
   d. Negotiate extra capacity (e.g., monitoring values and occupancy counts)
   e. ATFCM/ASM (civil/military coordination)
   f. Reduce traffic complexity
   g. Holding pattern
2. Use other available capacity in order to shift traffic demands into areas where capacity is available.
   a. Rerouting of flows and flights
   b. Flight level management
   c. Advancing traffic
   d. FMP tactical ATFCM measures
3. Regulate the demand.
   a. Regulation
   b. Network cherry-pick regulation
   c. FMP tactical ATFCM measures
   d. Constrain airborne capacity

Restrictions in the form of regulations are typically applied when a mismatch between demand and capacity is anticipated. The ATFM measures encompass a wide range of techniques aimed at resolving a mismatch that may originate from temporary excess demand or reduced capacity (Figure 1.10). In this example, the ACC sector has a declared capacity of 20 aircraft per hour. However, the controllers may end up handling 32 aircraft in a given hour in the event that 8 aircraft from the previous hour are delayed and 4 aircraft from the next hour entered the sector earlier. The ATFM anticipates such problems in order to balance demand with capacity. Typical reasons for capacity reductions are strong crosswinds, severe weather conditions, staffing issues, CNS or airport equipment failure, and industrial actions.

Although ATFCM becomes the strategic agent of ATC operations, inherent design limitations may actually result in an increase of the workload of controllers that was supposed to safeguard. For example, flights receiving a slot (i.e., a delay in their departure) may result in time-consuming coordination with tower controllers, handling agencies, and the network manager in order to get an earlier slot. This may increase the workload of tower controllers who have to monitor the situation for this flight, coordinate with flow controllers and, eventually, plan the departure of this flight in a narrower time frame than the original one.

1.8 Safety Regulatory Framework

Safety is the overriding consideration in all aviation activities which is reflected in Article 44 of the Convention on International Civil Aviation (ICAO 2006d). From a regulatory perspective, the role of ICAO is to provide aviation stakeholders with procedures and guidance for the safe conduct of aircraft operations and to foster planning and development of the air transportation system. To this end, standards and recommended practices (SARPs) are developed and contained in the Annexes of the Chicago Convention. The procedures for air navigation services (PANS) contain practices beyond the scope of SARPs, where a measure of international uniformity is highly desirable for safety and efficiency. In other words, these documents define an international framework for promoting safety and efficiency in the aviation system.

The provision of safe air traffic services remains the main objective of member states and air navigation service providers (ANSPs). Every

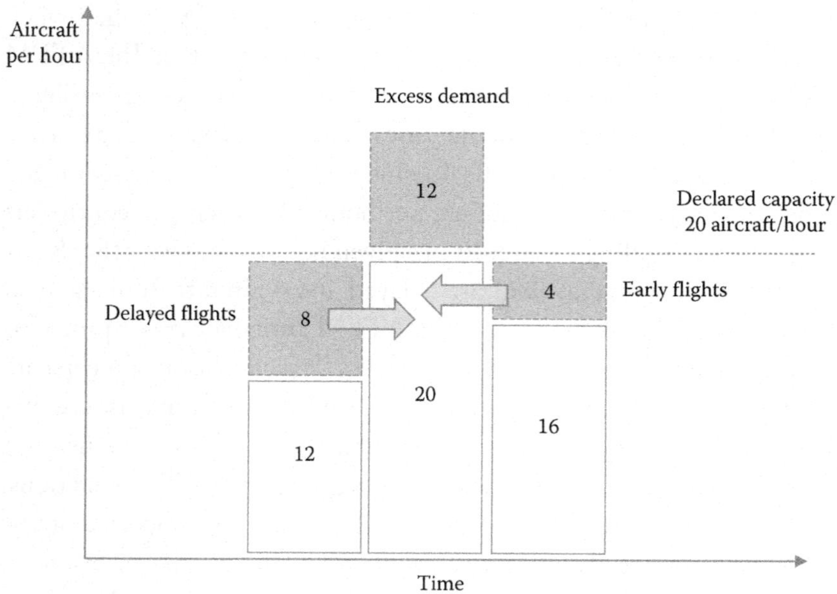

Figure 1.10 An example of imbalance between demand and capacity in flow management.

member state is expected to adopt the ICAO legislation and to notify others of any differences. In recent years, the European Aviation Safety Agency (EASA) has been gradually charged with the development of appropriate legislation (EU 2008). The main role of EASA is the harmonization of international procedures and regulations in order to provide a 'seamless' transition between national and international ATM systems.

As mandated by the European Union, ANSPs shall implement safety management systems (SMS) for the air traffic services under their jurisdiction. The provisions included in the SMS manual for all ANSPs take into consideration:

1. The national legislation
2. The EU Single European Sky (SES) regulations and Common Requirements as well as any other relevant EU legislation
3. The provisions of the ICAO Convention and the relevant annexes

In the European context, the primacy of safety is recognized by all stakeholders while safety performance schemes have been coordinated with EASA.

In general, an ATM related incident means that it is relevant to ATM; however, it may not necessarily have an ATM contribution. In some cases, an incident may be classified in more than one category (e.g., a runway incursion or a deviation from an ATC clearance). An important aspect of incident analysis is the classification of incidents into distinct categories for which statistical data can be collected and analyzed on an international basis. Although there may be overlaps between categories, it is useful for all stakeholders to rely on a common classification scheme.

According to EASA, only a fraction of ATM related incidents may have an ATM contribution in the causation of events. For each ATM related incident, the risk analysis tool (RAT) can be used to assess and classify the associated risk; this involves an assessment of its severity and its likelihood to repeat itself in the near future. The RAT tool provides a method for a consistent and coherent identification of risk elements and aims to support end-users in prioritizing actions to reduce safety repercussions (Eurocontrol 2009).

The most severe incidents are classified as serious incidents (severity A) and major incidents (severity B). Lower severity classes include significant (severity C), no safety effect (E), and not determined (D). As expected, the incident category with the largest proportion of risk bearing incidents (severity A and B) regards separation minima infringements (i.e., occurrences where certain minimum separations were not maintained).

1.9 Incidents and Accidents

The general public may use terms such as "incidents" and "accidents" to refer to adverse events with severe consequences for the system and the people involved. A more subtle distinction between incidents and accidents has been provided by the ICAO Annex 13 (2010a). *Accidents* are defined as occurrences in the operation of aircraft that entail serious injuries or fatalities, substantial damages to equipment requiring major repairs, and missing or inaccessible aircraft. In contrast, *incidents* refer to occurrences with safety repercussions that have not evolved into more severe situations. Hence, an incident or a safety occurrence may be seen as a situation that could have resulted in an accident in slightly different circumstances.

This classification scheme provides a standard set of codes and information on incidents and accidents worldwide. It includes all main categories of safety occurrences that ANSP organizations may encounter. Many elements are further classified into subcategories that are described in detail in the SMS manuals.

In the European Union, a detailed classification of occurrences, quoted in mandatory reporting systems, is set out in regulation (EU) 1018/2015. In particular, Annex III provides the following occurrence classification for air navigation services providers:

1. A ground or air collision between two aircraft or between aircraft and terrain or obstacle (including near-controlled flight into terrain, CFIT)
2. Separation minima infringement
3. Inadequate separation
4. ACAS RAs
5. Wildlife strike including bird strike
6. Taxiway or runway excursion
7. Actual or potential taxiway or runway incursion
8. Final Approach and Take-off Area (FATO) incursion
9. Aircraft deviation from ATC clearance
10. Aircraft deviation from applicable ATM regulations
11. Callsign confusion related occurrences
12. Inability to provide ATM services or to execute ATM functions
13. Missing or incorrect, corrupted, inadequate or misleading information from any support service
14. Failure of communication, navigation and surveillance services
15. Failure of data processing and distribution function or service
16. Failure of ATM system security which had or could have a direct negative impact on the safe provision of services
17. Significant ATS sector/position overload leading to a potential deterioration in service
18. Incorrect receipt or interpretation of communications that could have a negative impact on safety
19. Prolonged loss of communication with an aircraft or with other ATS unit
20. Declaration of an emergency ("Mayday" or "PAN" call)

21. Significant external interference with Air Navigation Services
22. Fuel dumping
23. Bomb threat or hijack
24. Fatigue impacting the ability to perform safely the air navigation or air traffic duties
25. Any occurrence where human performance has contributed to an accident

This taxonomy is by no means exhaustive and further elaborations can be provided by ANSP organizations. The two most significant incidents in the field of ATM regard separation minima infringements and runway incursions. Separation is the generic term used to describe action on the part of ATC in order to keep aircraft separated at distances where that the risk of collision is reduced (ICAO 1984). Separation minima are specified horizontally (in nautical miles or degrees of angular displacement) and vertically (in meters or feet). A separation minima infringement is any occurrence where the distances between two aircraft are reduced below certain minima. For example, Figure 1.11 shows an ACC sector with a horizontal minima of 5 Nm and two aircraft flying on the same level, converging in space and time in a conflicting course. When two circles surrounding the aircraft overlap, a separation minima infringement may be observed.

A runway incursion can be defined as any occurrence at an airport involving the incorrect presence of an aircraft, a vehicle or a person on the protected area of a surface designated for the landing and takeoff of aircraft (ICAO 2007c). The incorrect presence may be a consequence of a failure of a pilot to comply with a valid ATC clearance or a compliance with an inappropriate ATC clearance. Typical runway incursion scenarios include: a departing aircraft that enters the runway contrary to an ATC clearance while another aircraft is rolling for takeoff or landing, an aircraft that commences a crossing of a runway contrary to ATC clearance while another aircraft is rolling for takeoff or landing. A well-known example of runway incursion was the Milan Linate accident (ANSV 2004).

1.10 Concluding Remarks

This chapter has presented an overview of the work environment in which controllers regulate air traffic. In a sense, the work environment

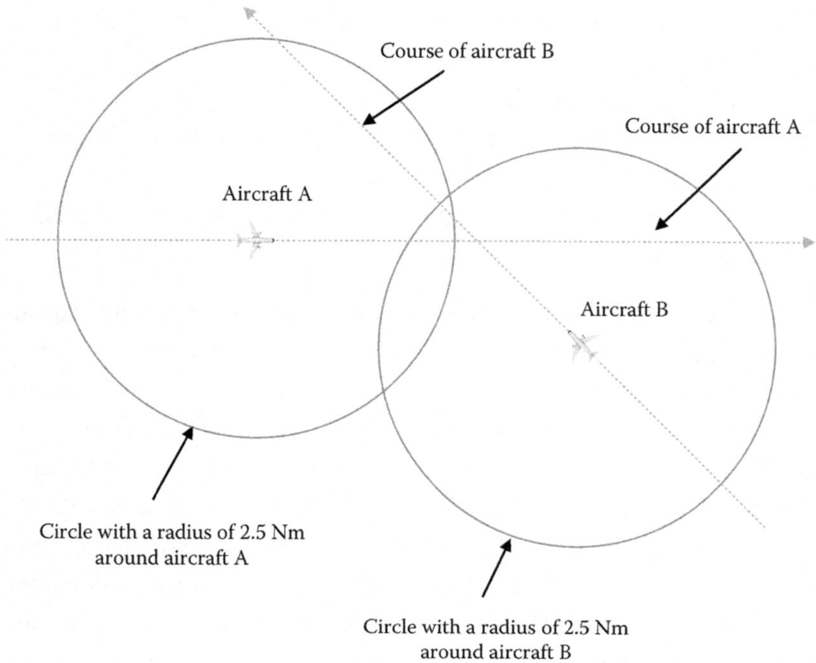

Figure 1.11 An example of horizontal separation minima infringement.

specifies the opportunities, constraints, assistance, and facilities available to controllers. The actions of other practitioners (i.e., pilots, ground personnel, and other colleagues), the safety rules and the automated artifacts may be seen as additional knowledge agents that should work closely with air traffic controllers. The ATM system is a joint cognitive system where knowledge is distributed to many human and technical agents that need to coordinate their work. For this reason, this chapter has presented all elements of the work environment of controllers to provide a good basis for understanding the factors that affect human performance.

The duties of controllers have been presented in relation to the duties of flight crews at different operational phases because their interaction is as important as their individual competences. Hence, crew-controller interactions could provide a framework for addressing the multiple perspectives, the goal trade-offs, the coordination costs, and the allocation of task roles that influence the overall performance of the joint cognitive system. This is particularly important in the new initiatives in aviation (e.g., single European sky ATM research program [SESAR] and next generation air transportation system [NextGen]) where existing task roles may change between crews and controllers. For instance, flight crews may feel more uncertain in managing their new separation tasks or may feel uncertain whether they can still rely on controllers as a last resort when control of separation breaks down in the cockpit. On the other side, controllers may be uncertain about the new roles of the flight crews since their decision criteria may differ from their own criteria and culture. Hence, some familiarization with the roles of controllers, pilots, job tools or aids, and organizational procedures is essential in putting in context the main material of this book that focuses on the cognitive functions and safety organization of the ATC domain.

2

Factors Affecting ATM Performance

2.1 Introduction

Air traffic management (ATM) is a complex, dynamic and highly automated system that provides a variety of air navigation services. In many cases, controllers have to process a large amount of data from a complicated communication system and strive to balance many trade-offs that stem from diverse requirements of the broad community of ATM users (e.g., commercial airlines, general aviation, military, unmanned aerial vehicles, etc.). Controllers manage many challenging situations that require a sequence of critical decisions regarding the safety of passengers and flight crews. During their professional careers, controllers are expected to handle successfully a wide range of events from separation minima infringements to complex emergencies, or even cases of total communication navigation surveillance (CNS) system failure. In the terrorist attack of 9/11, for instance, the ATM system in the U.S. airspace was ordered to an abrupt halt while thousands of flights were in the air, hence requiring controllers to manage numerous flight emergencies and other abnormal situations.

This chapter looks at the work factors or demands that affect the performance of controllers and their operating teams in normal and abnormal situations. An introduction is made first of the challenging nature of emergencies and abnormal situations as they receive extensive coverage in the public media. The main focus of this chapter is on the demands and the context of work in which ATM performance takes place. Work demands are presented at the individual, team, and organizational levels of the ATM system. Certain popular models of human performance are presented as an introduction to the cognitive strategies of controllers that are elaborated in later chapters. Two other paradigms of work factors and demands are also presented in this chapter as they bring up the importance of traffic complexity and information uncertainty in assessing and handling complex situations.

2.2 Challenges in Coping with Abnormal Situations

Controllers have to make vital decisions pertaining to aviation safety in situations of time pressure, uncertainty and minimal error tolerance. Although flight crews take credit for their skills in managing abnormal events—as they are the actors who face grave and imminent danger—it is also widely established that controllers play a significant role in the prevention of the incident trajectory and the recovery from disasters. There are ample cases where controllers timely and accurately mitigated or prevented disasters, although they did not receive the same publication from mass media as other cases of mishandling and poor judgment.

Emergencies are critical situations close to the margins of safety that present many challenges to controllers, requiring competence in problem-detection and replanning. As soon as the relevant cues are detected, a problem is formulated and the need to replan for the situation becomes critical. In the evolution of an occurrence, new threats may appear while current ones may change their demands. This amplifies the need for gathering new information to fill in gaps, clarify assumptions and correct explanations. All this calls for cognitive strategies for how to assess new demands, how to manage uncertainty, and when to engage other team members in the situation.

On another level, team performance is also challenged because emergencies require teamwork strategies such as synchronization of activities, exchange of critical information, and reallocation of roles as new tasks are added and priorities are altered. Furthermore, emergencies are not tolerant of errors and require competence in managing error detection and correction. Taskwork and teamwork strategies are important elements of effective human performance in air traffic control (ATC) that are elaborated in Chapter 4.

Even in everyday situations, controllers encounter various normal threats that have the potential to increase work complexity, such as:

- Adverse weather conditions
- Degradation of CNS systems
- Distractions in the operations room

- High-complexity traffic
- Military exercise areas affecting the area of responsibility
- Special handling traffic
- Diversion of flights
- Air traffic flow and capacity management (ATFCM) failures that result in over deliveries of flights and high workload

Everyday threats can combine in different ways, escalate in steep patterns, or become difficult to anticipate, hence making it difficult to manage them.

In the beginning of the twenty-first century, two fatal accidents that were attributed to human error (the Linate runway collision and the Ueberlingen midair collision) triggered considerable interest in ATM safety. Plane crashes are by their nature high-profile events that attract extensive media coverage that eventually has an impact on air transportation policies. The active involvement of controllers in the causation of accidents is usually pointed out by an ever-present blame culture. In hindsight, it is easy to attribute blame to controllers since all the information about a mishap is revealed after the fact, although it was possibly unavailable in the actual course of events. Furthermore, media intervention and blame attribution may lead directly to extreme actions as manifested in the assassination of the controller involved in the Ueberlingen accident. In contrast, limited media coverage is devoted to everyday operations where controllers successfully manage to avert numerous hazards and adverse events.

The ATC units are responsible for the provision of alerting services in their area of responsibility (AoR). An alerting service aims at notifying appropriate clusters of aircraft in need of search and rescue. In general, an alerting service is provided through the declaration of three emergency phases as follows (ICAO 2007a):

1. *The uncertainty phase* where uncertainty exists as to the safety of the aircraft and the passengers.
2. *The alert phase* where apprehension exists as to the nature of the emergency, which allows more units to be called in the situation.
3. *The distress phase* where there is reasonable certainty that an aircraft is threatened by grave and imminent danger or requires immediate assistance.

In the most straightforward case, flight crews may declare an emergency to the ATC units by directly stating the exact nature of the problem (e.g., explosive decompression) and their intention to act (e.g., immediate landing). In this case, it is obvious that a distress phase is declared directly and the aircraft in emergency is compelled to land at the nearest airport. Normally there is no precise guidance to the transition from one phase to the other apart from generic rules in the operational manuals. ICAO (2007a) has explicitly acknowledged that the wide range of circumstances surrounding an emergency precludes the establishment of detailed instructions for how to respond in the operating procedures.

The constant demand for greater traffic capacity has increased the range of possible emergencies and created more possibilities for new classes of complicated emergencies. Although complex occurrences may be less frequent than typical or textbook emergencies, the consequences of human mishandlings could be extremely important. Controllers are expected not only to handle everyday traffic efficiently, but also to manage complex emergencies that may occur unexpectedly and follow a steep escalation pattern.

Official investigation reports and field observations of practitioners have shown that controllers employ a range of cognitive strategies to meet their work demands successfully by relying on their professional knowledge, acquired skills, and trained competencies. Cognitive strategies—such as decision-making, sensemaking, replanning, and adaptation—are the building blocks of cognition that are elaborated in Parts II and III. Mainstream ATM research has focused mainly on methods, tools, and taxonomies of errors that can assist the investigation of mishaps and the design of automated support tools. With respect to training, most aviation organizations have devoted their resources to technical skill training so that practitioners acquire and perfect their skills in the full range of tasks. Although operational teams are the basic functional blocks of the ATC system, a relatively small appreciation has been made for training controllers as effective team members. Only recently, we have witnessed courses on team resource management (TRM) introduced in the development phase of training. There is, however, an increasing recognition that team decision-making should become an integral part of ATC training, together with other conventional knowledge and technical skills in the management of abnormal situations.

2.3 Work Demands and Stress in the Operating Environment

The operating environment of flight crews and controllers exposes them to numerous work demands, ranging from physical stress (e.g., noise, heat, cold, vibration, and altitude) to time pressure, workload, and negative feedback on performance. The perception of practitioners of the imbalance between work demands and coping abilities is an important factor in the incidence of stress (Cox 1987). Figure 2.1 shows a transactional model where stress is viewed as a process by which certain work demands evoke an appraisal process in which perceived demands exceed coping resources and result in undesirable physiological, emotional, cognitive and social changes.

This section looks at the work demands of the ATM environment and the sort of stressors that are likely to degrade the performance of aviation practitioners. It is interesting to note that the reaction to stress is not always dysfunctional since experienced people can manage to adapt their priorities as they start to perceive an imbalance between demands and resources. This matter is treated more thoroughly in the examination of models of human performance (Chapter 4) and in the management of work complexity (Chapter 9).

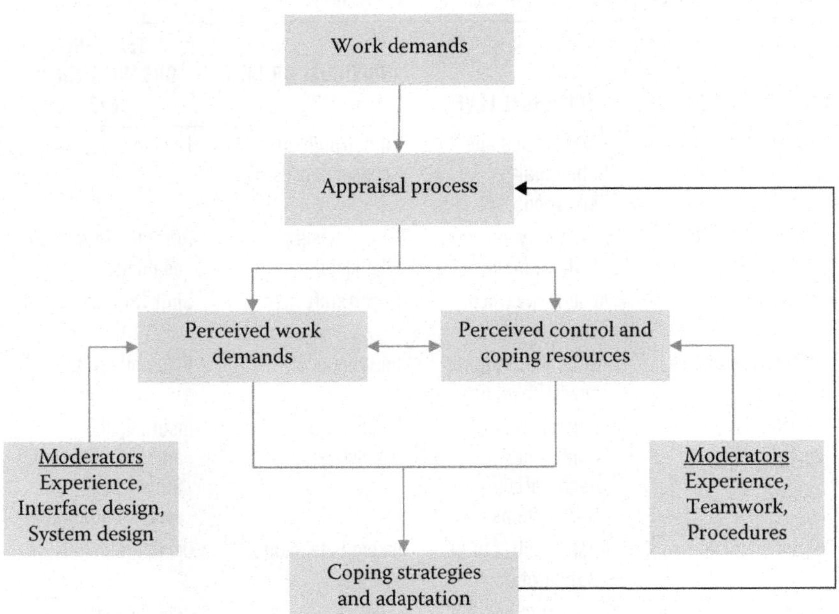

Figure 2.1 A transactional model of stress.

Early approaches to stress at work have focused on the characteristics of the technical environment that could have a high potential for causing stress (e.g., noise, heat, cold, vibration, and altitude). The introduction of new technology has shifted the focus to work demands at the task level such as time pressure, workload, information uncertainty and negative feedback on performance. With the increasing recognition of the importance of the balance between resources and demands has come an awareness of a range of team-level factors that could act either as stressors or as stress-moderators. Team characteristics—such as cohesion, communication, supervision, and allocation of roles—could affect our perception of resources and may either exacerbate or alleviate the effects of task-level stressors. A similar argument can be made for what Ivancevich and Matteson (1980) called "organizational-level factors" that derive from the general climate and the organizational processes. Table 2.1 shows examples of a range of stressors that seem to have general applicability to the ATM domain (Kontogiannis 1999a).

Table 2.1 also provides a useful context for examining the role of technology either as a source of stress or as a moderator of stress. On

Table 2.1 Work Stressors at Different Levels in the ATM System

WORK ENVIRONMENT	TECHNICAL LEVEL	INDIVIDUAL OR TASK LEVEL	TEAM AND ORGANIZATIONAL LEVEL
Data overload, noise	Emergencies and abnormal situations	Threat/high error consequences	Lack of team cohesion
Humidity conditions	Complexity and traffic volume	Time pressure/ Workload	Intra and inter team conflicts
Low visibility	Adverse weather conditions	Uncertainty	Shift work/night shifts
Lighting conditions	Traffic that requires special handling	Inexperience	Role ambiguity
Interruptions in the operations room	Malfunctions/ limitations/ degradations of CNS systems	Incomplete knowledge	Insufficient/ ambiguous/missing SOPs/LoAs, contingency plans
Design inefficiencies	Airspace structural complexity	Incomplete mental modes	Occupational stress
Tower cabin windows (dirty, spots, etc.)	Automated handoff failures	Illness	Communication difficulties

the one hand, technology can give rise to many stressors ranging from environmental to team-level stressors. Poor organization of alarms, for instance, can increase noise levels while unfriendly interfaces can increase information uncertainty and workload in navigating through the computer screens. The design of technology may also reduce the "horizon of observation" (Hopkin 1995) by restricting access to the work of other colleagues. On the other hand, the design and use of technology can function as moderators of stress. Prioritization of alarms and logging devices, for instance, can minimize noise levels while user-friendly interfaces can minimize diversion of attention to secondary tasks. In addition, error-tolerant technologies may increase the response time and provide opportunities for error recovery. At the team level, appropriate design of technologies can facilitate distributed cognition and enable timely hand-over of tasks to other work shifts.

The work demands or stressors in Table 2.1 can be used to define a set of work characteristics in the ATM system as follows:

- *Rapidly escalating situations*: The transition between normal and high tempo operations can be rapid. In an explosive decompression, for example, the crew may initiate a rapid descent (e.g., 5000–6000 feet per min) from its cruising level that could affect many other aircraft without any prior notice.
- *Multiple information resources*: ATC operations rooms are information rich environments but can also be noisy at other times. There are multiple sources of information including radar screens, CNS systems, pilot communications, and other adjacent traffic units.
- *Uncertainty*: Weaknesses in the presentation of information can increase uncertainty (i.e., missing data, unreliable data, and inconsistent data).
- *Severe time pressure*: Available time for decision-making and coordination may be severely constrained; in cases of loss of separation, for instance, an avoiding action must be issued within seconds.
- *Errors with high consequences*: Errors may have disastrous effects especially in cases where safety nets may be malfunctioning. For example, a loss of separation may lead to a

mid-air collision simply because the traffic alert and collision avoidance system (TCAS) may be malfunctioning while weather conditions may prevent visual maneuvering.

- *Multifaceted decisions and conflicting goals*: The goals of safe, orderly, and expeditious traffic may be in conflict and their consequences could cascade into the tactical level. Expediting traffic implies working close to separation minima which, in turn, could erode safety margins.
- *Multiple stakeholders*: Airlines, airport operators, government agencies, and other organizations can interact in complex ways with the ATC system. For example, an airline may reroute a flight to avoid an overcrowded sector but may saturate a previously unaffected sector that was not in the initial flight plan.
- *Physiological stressors*: ATC operations require 24 hours of service, seven days a week, resulting in long and unsocial hours of work. Working in shifts may desynchronize human biological and circadian rhythms, hence causing some physiological stress.

The transactional approach to stress has gained wider recognition with its emphasis on the appraisal process of work demands and stress coping strategies (Salas et al. 1996). Confidence in one's own coping resources, experience with handling other emergencies, team support and availability of procedures may influence one's perception of the ability to cope and, hence, the appraisal of the situation. This implies that training for emergency responses should both specify the conditions that optimize coping resources and increase confidence in managing emergencies.

The experimental literature has documented many disorganizing effects of stress on human performance (Kontogiannis 1999a). It can be generally said that stress narrows the perceptive field, decreases vigilance, reduces the capacity of working memory, causes premature closure of options, and may result in task shedding. However, there is little evidence that these dysfunctional reactions observed in laboratory tasks could transfer to real-life emergencies that involve highly experienced teams. In fact, field research within the paradigm of naturalistic decision-making (Klein et al. 1993; Zsambok and Klein 1997; Flin et al. 1997) claims that the reactions of experienced practitioners are adaptive rather than dysfunctional. Perceptual narrowing, for instance, may support a more selective use of cues when there is insufficient time to examine all information; especially for experienced practitioners with the skill to prioritize cues, narrowing of attention would seem to make sense. The same holds true for task shedding since some tasks may have to be deferred under time pressure; recognizing high priority tasks to start with could make task shedding a quick and efficient response. In the same sense, premature closure of options may not be so dysfunctional since delays in making decisions could be proved a graver problem. Moreover, taking precautions for some predictable side effects could alleviate any problems due to premature closure of options.

Field studies have shown that experienced people can maintain performance under stress by establishing priorities, adapting their decision strategies, and changing their communication patterns (Serfaty et al. 1993; Cohen et al. 1996; Lipshitz 1997; Xiao et al. 1997). For example, with the Federal Aviation Administration (FAA) mandate for grounding all traffic due to U.S. airspace closure following the 9/11 terrorist attacks, controllers managed to redirect and safely land an enormous volume of en-route traffic. They were faced with the unthinkable scenario of a complete closure of the U.S. airspace and the pressing goal of redirecting and landing of en-route traffic at a short notice. Dozens of aircraft over-flying the Atlantic Ocean inbound to the United States were redirected to other airports while others were in critical fuel conditions. Controllers responded to this stressful scenario by effectively managing traffic without any accidents.

Further research in coping with stress and complexity is reviewed in Chapter 9, showing that performance can decline gracefully when controllers work under stress. In fact, controllers adapt their criteria of performance, their plans, and their coordination patterns in ways that they manage a complex situation satisfactorily. Under stress, for instance, controllers may give priority to safety over efficiency, may develop plans that are looser and increase chances of error recovery, or may become more sensitive and supportive to the needs of their teams. Of course, these skills in stress management require extensive experience and specialized training. Chapter 9 describes a study that recorded several complexity-mitigation strategies.

2.4 Classical Performance Models in Aviation

Aviation is a highly scripted environment where appropriate responses to normal and abnormal situations are specified in operating procedures, manuals, and checklists. Sometimes, organizations devise proper mnemonics that summarize a series of responses in terms of a few keywords (e.g., ASSIST: Acknowledge—Separate—Silence—Inform—Support—Time). During training, practitioners are provided with mnemonic aids and procedures while sufficient practice is allowed in a series of simulated scenarios. Although a thorough consideration of models of human performance is provided in Chapter 4, this section presents a few simple models of human strategies and responses to abnormal situations that have provided the basis of emergency training in many organizations.

The broader context of abnormal situations can be best described using a classic aviation axiom: "when aviators are confronted with abnormal situations, they will normally prioritize their tasks according to the following simple axiom:"

- Aviate
- Navigate
- Communicate

2.4.1 Aviate

From single-engine aircraft to four-engine airliners, flight crews have an immediate priority to fly the aircraft which includes not only maintaining a stable flight path, but also performing some emergency checklists. In a typical two-person flight deck, one person is responsible for flying duties and radio communications (i.e., the pilot flying) while the other one is responsible for completing relevant checklists (i.e., the pilot nonflying). Workload during emergencies is inevitably high and flight crews may choose not to communicate with ATC units, or delay the provision of any additional information. Upon completion of the emergency checklists, flight crews can reassess the situation. This activity generally follows the simple mnemonic format (CAA 2005) that is used by many major airlines as in the case of DODAR (Walters 2002).

- **D** Diagnose the problem :*What is the problem?*
- **O** Organize options :*What are the options available?*
- **D** Decide what to do :*What are we going to do?*
- **A** Allocate tasks :*Who does what?*
- **R** Review the situation :*What has happened and how will we continue?*

Other airlines use similar mnemonics to support decision-making in abnormal situations. For example, some airlines have adopted the acronym FOR-DEC (Hormann 1995):

- **F** Facts :*What are the facts?*
- **O** Options :*What are the options?*
- **R** Risk and Benefits :*What are the risks/benefits for each option?*
- **D** Decision :*What will be the decision?*
- **E** Execution :*How can it be executed?*
- **C** Check :*What has happened and how will we continue?*

DODAR and FOR-DEC are two simple decision-making models that follow the same principle of providing flight crews with a simple linear process of steps for making decisions in normal and abnormal situations.

2.4.2 Navigate

In an emergency, the second element of the aviation axiom corresponds to answering the question "where to go next?" In this stage, flight crews decide whether to continue the flight to its original destination or initiate a diversion to a suitable airport. This decision depends on several factors, such as the following:

- *Nature of emergency*: This may become the most critical factor in the selection of a diversion airport. In a cabin fire, for instance, the crew may be compelled to land as soon as possible at the nearest airport, regardless of any other considerations.
- *Functionality status of the aircraft systems*: In general, the more critical the systems that fail, the more critical the emergency is. Sometimes, a seemingly innocent failure may result in a

serious emergency due to high coupling, mode proliferation, lack of redundancy, and high complexity.

- *Length of runway*: All flight crews would prefer a long runway and, occasionally, this may be the overriding factor. For instance, in the case of a failure that prevents the flaps to be fully deployed during landing, the length of the runway is the main consideration.
- *Airport distance and bearing from present position*: Normally, flight crews would prefer airports that are close to their position and in their front quarter in order to avoid long turns.
- *Airport navaids availability*: In IMC conditions, an airport equipped with a precision approach navaid with vertical and lateral guidance is preferable to an ill-equipped airport.
- *Familiarity with airport*: Flight crews prefer airports that they have visited before and know their instrument approach procedures, runway orientation, and surrounding terrain.

The time frame for making critical decisions may be restricted and sensitive to earlier assessments of the emergency. In many cases, decisions may involve extensive coordination with many other agents such as airline dispatchers and maintenance personnel.

2.4.3 Communicate

Finally, flight crews are expected to communicate the nature of an abnormal situation to all relevant agents and inform them of their intentions. In general, flight crews are expected to use the following mnemonic in their communications:

- **N** :*Nature of the emergency*
- **I** :*Intentions of the crew*
- **T** :*Time available*
- **S** :*Supplementary information*

Furthermore, flight crews can use one of the two classes of emergency messages that are defined in ANNEX 2: Rules of the Air (ICAO 2005a):

- *Distress*: When the crew and the passengers are threatened by grave and imminent danger that requires immediate assistance.

- *Urgency*: When the safety of an aircraft or a person on board is at stake but it does not require immediate assistance.

It is not certain that flight crews will provide adequate information to controllers regarding the situation they encounter. A crew may elect not to declare an emergency but instead request a certain amount of time to react (e.g., enter a holding pattern to carry out diagnosis, or assess and stabilize the situation). The controller then faces a dilemma whether to carry out appropriate actions or not. In most cases, the controller can perform a minimal set of actions in anticipation of a possible escalation of the problem without informing the crew. For instance, the controller may inform the watch supervisor, estimate distances and bearings to the nearest airport, and coordinate with adjacent units with regard to the possibility of a route diversion. During this period, the flight crew may continue to assess the situation and carry out the appropriate checklists.

Rote following of aviation checklists cannot always guarantee success in handling an abnormal situation as indicated in the Swissair flight 111 incident (TSB 2003). The aircraft crashed in the sea of Canada after experiencing a rapidly spreading fire on board that quickly disabled all main systems. The crew chose to complete all checklists and delayed to proceed directly for landing. The captain was an instructor, one of the most experienced flight crews of Swissair, who practiced the smoke in cockpit routine many times as an emergency drill in simulator training. The Swissair 111 crew wondered whether the odd smell and the small cloud in the cockpit was an indication of fire or an innocent output from the air conditioning system. As found later in the report, neither the Swissair nor the aircraft's manufacturer checklist for "smoke/fumes of unknown origin" required crews to start immediate preparations for landing. The option to consider emergency landing was addressed at the end of the checklist, which underplayed the risk that an unknown smoke condition in the cockpit could rapidly get worse (TSB 2003). The fire spread rapidly, leaving the crew with no options for a safe handling of the situation.

This particular accident triggered a considerable debate about the issue of rote following of checklists in managing flight emergencies. Dekker (2003, 2005) argued that the flight crew of Swissair 111 faced a double-bind. People could be blamed for lack of flexibility in applying procedures without any sensitivity to the context of events but they also might be blamed for violating procedures if earlier efforts to adapt have been unsuccessful. The actions in the checklist of the Swissair 111 crew could not deal with a rapid spreading of fire. On the other hand, the crew had no clear cues that they were facing a novel fire situation and no indication that the checklists did not account for this type of event. Cases where crews face complicated situations that are only partially covered in the procedures are becoming increasingly common in aviation in recent years.

2.5 Classical Performance Models in ATC

2.5.1 Management of Occurrences

Eurocontrol has developed a standard model for handling unusual situations that was incorporated in controller training (EATMP 1999). The proposed ASSIST model (Acknowledge—Separate—Silence—Inform—Support—Time) has not been intended to replace emergency procedures but rather to act as a mnemonic used in combination with procedures (Table 2.2).

Acknowledgement is an important element because the first priority of controllers is to fully understand, correctly classify the nature of the problem, and then acknowledge it. An erroneous classification may lead to inappropriate actions that could have a negative impact. The next three steps (Separate—Silence—Inform) can be characterized as purely technical in nature and are described in detail in ATC procedures. The fifth element (Support) depends on the time

Table 2.2 The ASSIST Model for Handling Abnormal ATM Situations

ACRONYM	KEYWORDS	SHORT DESCRIPTION
A	*Acknowledge*	Make sure you understood the nature of emergency and acknowledge accordingly.
S	*Separate*	Don't forget to establish/maintain separation.
S	*Silence*	Impose silence on radio frequency if necessary. Don't disturb urgent cockpit actions by unnecessary transmissions.
I	*Inform*	Inform supervisor and other sectors/units concerned.
S	*Support*	Give maximum support to flight crews.
T	*Time*	Allow flight crews sufficient time to work on the problem.

when flight crews initiate their requests and indicate their intentions. In cases where a request for support cannot be made (e.g., crews are unable to communicate their problems or intentions), the controllers are expected to use their best judgment and reach a decision on the form of support that should be offered. The last element (Time) is directly connected to the diagnosis of the problem and the contingency planning made by the crew. ASSIST has been used in many European ATC units as a basis of operation in unusual occurrences due to its innate simplicity and usability.

2.5.2 Mental Pictures of Traffic

It has long been recognized that controller strategies are based on a mental picture or representation of current and future traffic situations (Falzon 1982; Whitfield and Jackson 1982; Shorrock and Isaac 2010). Earlier studies have indicated that controllers tend to maintain a mental picture of traffic based on a few salient features that point out potential conflicts or situations requiring continuous attention. To cope with high workload, controllers also tend to structure their mental picture by grouping aircraft into meaningful units. In this way, they manage to reduce the number of aircraft in their working memory (Bainbridge 1975). Several knowledge variables are used by controllers (e.g., proximity, vertical movement, weather information) to create classifications such as aircraft requiring continuous monitoring to avoid conflicts, aircraft safely separated at a particular moment, and so on (Amaldi and Leroux 1995; Niessen et al. 1997).

A field study by Malakis and Kontogiannis (2013) has shown that mental pictures are like cognitive maps that guide the search for important data and direct the exploration of actions that can change the environment. Cognitive maps direct action and enable practitioners to build imaginary connections between events, objects, and situations in their environment so that they become meaningful (Henneberg et al. 2006; Colville and Pye 2010). Malakis and Kontogiannis (2013) made an effort to understand how controllers build and reframe their mental pictures in response to their work progress. To achieve this objective, controllers were asked to draw cognitive maps and freely annotate what they thought to be significant in handling a number of traffic scenarios. Verbal reports from participants were used to gain insights into the processes of building and reframing mental pictures.

The results of this study showed that controllers modified their mental pictures in order to achieve a series of good criteria of performance (Table 2.3).

Cognitive maps are annotated with headings and level changes issued to the aircraft as well as the selected approach sequences of the inbound traffic. Differences between experts and novices can be illustrated in the drawings of cognitive maps for handling traffic in several scenarios. In Malakis and Kontogiannis (2013), it appeared that expert and novice controllers assigned similar approach sequences to the arriving aircraft and made similar attempts to de-conflict the traffic patterns. However, novices created tighter plans than experts that were fragile to possible requests for flight

Table 2.3 Criteria Used to Frame and Reframe Mental Pictures in Air Traffic Control

PERFORMANCE CRITERIA	BRIEF DESCRIPTION
Avoids potential conflicts	Creates patterns that do not give rise to potential conflicts
Creates open and inspectable patterns	Creates an open traffic flow pattern that minimizes the need to monitor horizontal separation distances; this creates a slack that allows controllers to engage with additional tasks
Provides more options to crews	Provides more options to crews to adapt to unexpected situations (e.g., changes in weather, turbulence, and so on). In this way, crews are given more opportunities to detect problems and recover from them
Minimizes chances for go-around	Minimizes chances of go-around by placing aircraft at the correct altitude, speed, course and distance on the final approach for landing
Makes subtle changes to flight paths	Performs small changes so that aircraft are not far from the standard instrument approaches and standard routings; this makes it easier to bring flights back to original paths should the need arise
Takes into account terrain features	Takes into account terrain factors and obstacles (e.g., mountains and obstacles on airport) that may reduce acceptability in terms of safety and quality of flight
Takes into account crews preferences	Takes into account the preferences of crews (e.g., continuous descent profiles, optimal speed profiles, preferred rates of climb /descent, direct routings)
Avoids stormy/turbulent zones	Avoids routing aircraft near stormy zones (e.g., significant weather) or turbulence zones that can make flight crews and passengers feel uncomfortable

Source: Malakis, S. and Kontogiannis, T., *Applied Ergonomics,* 44, 327–339, 2013.

deviations due to bad weather. A tight plan also implies that deviations cannot be easily accommodated as they may disrupt overall traffic planning. In addition, experts made considerable efforts to create open patterns, avoid stormy/turbulence areas, and minimize chances for go around. Novices, on the other hand, did not seem to take into account the terrain features and the crews' preferences; moreover, the choices offered to flight crews were more restrictive as traffic planning was rather tight. It is important to note that novices used the same approach sequences as experts did in most cases; their differences regarded the structuring of the aircraft routes and the overall planning of traffic.

2.6 Aspects of Complexity and Coupling in the ATM Environment

The earlier presentation of work demands and stressors has taken a general approach that is applicable to many work domains such as aviation, process control industries, manufacturing, and so on. Researchers and organizations of ATM systems have expressed a more specific interest on the complexity of air traffic and its effects on human performance. In this sense, it is more useful to define the work demands and stressors in relation to the complexity of traffic situations in ways that it should be possible to measure complexity and take remedial actions. Examples of indicators of traffic complexity may include: the number of aircraft on the airspace, their performance characteristics, their conflict geometry, their space configurations, the airspace restrictions imposed by weather, and so on. The paradigm of complexity looks at more specific work demands that provide a means of measuring complexity and devising ways of reducing it to manageable levels.

The complexity of modern ATM systems has always attracted a lot of attention from researchers, practitioners, and lay people concerned with aviation safety. However, "complexity" gets different meanings depending on the person who talks about it. The public focuses on the safety repercussions of complexity, while researchers try to define and measure complexity so that its contribution to risk can be modeled, and finally, practitioners focus on the strategies and adaptations that are necessary to manage more traffic. Traditionally, complexity has been defined in terms of the ratio of current traffic load to airspace capacity and also in terms of the "conflict geometry" of the involved aircraft. What remains often unstated is that complexity is not an abstract property of the situation but reflects the relationships between objects, people, and processes in action.

Perrow (1984) has coined the term "interactive complexity" to refer to the relationships between parts of the system and their familiarity or observability in the system. Hence, interactions that are unusual, unexpected, hidden, or not immediately comprehensible are described as "complex" or "nonlinear." Another property of systems that makes them vulnerable to risk is what Perrow calls "coupling"—that is, the extent that any slack, buffer, or alternative means exist between the parts to absorb disturbances, modify plans, and recover problems. According to the normal accident theory, accidents are "normal" in systems that are high on complexity and coupling because control can be lost in critical situations. This section looks at the factors that increase the complexity and coupling of the ATM system as they have safety repercussions.

Without doubt, everybody seems to be concerned with the increasing levels of traffic and the associated increase in complexity, especially where the conflict geometry becomes more complicated. The complexity is "interactive" because earlier efforts to resolve a conflict or reroute a plane usually come at many costs (e.g., absorption of attention, creation of another conflict in future, complaints of the aircrew that is rerouted, etc.). Complexity can also be increased when faults in computers have unexpected side effects when automated systems gradually lose control unknowingly to the practitioners, and when problems cause "common mode" failures. A classic example of unexpected system interactions is the operation of ATM in a degraded mode (e.g., equipment disabled by maintenance work) or the operation with fewer controllers on shift.

Existing measures of complexity have focused on physical interactions that can be externally observed or reported. This approach fails to address the complexity in relation to the capabilities and strategies of controllers. It is rather the controller's perception of the particular conflict and the appropriate control actions that signify the importance of a factor. The study of complexity requires an analysis of traffic demands and procedures as well as an analysis of the mental pictures and cognitive strategies of controllers. This issue is further discussed in Chapter 9, where complexity is examined in relation to the capabilities and strategies of controllers.

The configuration of flight crews, controllers, and automated agents can become tightly coupled when more constraints are imposed on the choice of acceptable actions and the time to act. As the system becomes tightly coupled, interactions increase and available time decreases. Tight coupling implies that a disturbance in one part spreads quickly at other parts because there is no slack to delay the problem, no alternative means to substitute one method for another, and no buffers to stop an escalation of the problem. It is important, therefore, to examine the factors that affect the degree of coupling in the ATM system.

Perrow (1984), Rochlin (1997) and Weick (2007) have argued that the ATM system has a moderate degree of coupling that allows people to cope better with increasing levels of complexity. For instance, controllers can provide less efficient services to flight crews when workload gets heavier while operations can continue in a degraded mode without a sudden breakdown. Another important factor is redundancy in resources and skills since controllers have a broad range of skills that makes them fill in for each other, understand other's work, and provide help even when not asked explicitly for it. In addition, redundancy in safety equipment and barriers provides an additional layer of protection in air traffic control. There are also many degrees of freedom in controlling airspace, such as rerouting aircraft, keeping aircraft on the ground, refusing early transfer of aircraft, and so on.

External coupling refers to the extent that a plan is coupled to other work activities that occur at the same time in the environment. For instance, some airlines require that the takeoff checklist be accomplished on the active runway or just prior to the entry onto the runway. In this case, the takeoff checklist is tightly coupled with other tasks (e.g., monitoring traffic communication or sequencing with other aircraft on the final approach) as well as with the pilot's mental representation of takeoff (Degani and Wiener 1994). External coupling forces plan execution to keep pace with several external activities that may result in high workload and memory errors. Dekker (2006) discusses how flight crews invent new, interesting ways to prevent forgetting items on the checklist that are coupled to other work activities in the environment.

2.7 Aspects of Uncertainty in Making Sense of Information

Perrow has referred to interactive complexity that can be unexpected or hidden, implying that there is always some degree of uncertainty in the direction of interaction. However, uncertainty deserves special attention because it affects the way that people seek and interpret information that is not handy or complete. The systems used by controllers are designed to provide the necessary information for reducing uncertainty to acceptable levels to achieve optimum decision-making. In some situations, however, the information may be incomplete, unreliable, or difficult to interpret, while controllers still have to make good decisions to regulate traffic safely.

Information uncertainty is a characteristic stressor of the ATC domain that can be considered separately from interactive complexity and coupling. For instance, flight crews in general aviation aircraft may delay to respond to controller instructions or may provide unreliable information that increases the complexity of the situation. In other cases, the ATM communication and radar systems may provide an avalanche of information that requires additional effort to process and comprehend. Any situations that produce irrelevant or inconsistent information create more uncertainty which further increases the size of the problem to solve.

Information uncertainly is always likely to be present in the ATM environment, even in normal operations. Hence, a discernible level of uncertainty must be tolerated in certain situations and controllers should be able to adapt their strategies to cope with it. It is likely that different types of uncertainty impose their own demands on performance and require particular cognitive strategies. For this reason, it is important to consider different types of uncertainty and their implications for coping strategies.

Klein (2004) has distinguished five types of information uncertainty that are discussed below in the context of ATM:

1. *Missing information* that is unavailable or cannot be localized when required
2. *Unreliable information* that may be wrong or may take additional time to verify
3. *Inconsistent information* that may be in conflict with other information or with the initial understanding of the problem

4. *Information noise* that is part of an "avalanche" of data that increases efforts at identifying critical cues
5. *Hard to interpret information* that makes it difficult to construct a coherent story of events, or an explanation of the situation.

Situations where controllers encounter more than one form of uncertainty are considered the most demanding and challenging. In the following sections, the five types of uncertainty are illustrated in the ATM context.

2.7.1 Missing Information

To perform their assigned tasks, controllers need information that is relevant to the context of work and their roles. For example, approach controllers may rely on different information than tower controllers to perform their tasks. When information is incomplete or missing, controllers have to find ways to recover such information or tolerate this event a little longer and rely on tentative assumptions. The difficulties in handling incomplete information also relate to the sources of information and the tasks at hand. There are four main information sources for controllers, namely:

- *Flight plans (FP)*: They provide basic information for most flights and originate from the departure airport; in some cases, it is very difficult to obtain missing FP information from this source.
- *Flight crew*: In the onset of an emergency, flight crews are very selective in the amount and sort of information to communicate to ATC units. As discussed earlier in this chapter, communication becomes a main priority. Controllers have to trade-off several options, such as pressing on for information, which increases crew workload, tolerating some uncertainty by utilizing other information resources, or waiting until the situation is stabilized.
- *Other controllers units or agencies*: Critical information about flights may be available from other controllers in the same unit or another agency. Asking for more information, however, comes at the cost of task interruption, attention diversion, and more delays. Since there is no guarantee that extra

information will reduce uncertainty, knowing when to interrupt others and what information to inquire for becomes an important skill in teamwork.

- *CNS systems*: Radar is the fundamental surveillance system that provides a set of useful information for aircraft identity, track, speed, and altitude. Missing parts of this information can increase the handling requirements of the situation.

2.7.2 Unreliable Information

Sometimes the information is accessible but its reliability may be low, as controllers may suspect that it is erroneous or outdated. Unreliable flight plan information is rather common in general aviation (GA) and the military. GA crews may fly on a weekend basis and may not fully comprehend the intricacies of the ATM system. In many cases, they may not know how to accurately complete a FP or conform to the required data conventions. In addition, military flights cannot be fully revealed in an FP, hence creating misleading information. Private flight crews flying in their spare time may be a source of unreliable information due to their incomplete knowledge of their onboard systems or the ATC system in general. Very early in their professional careers, controllers learn not to fully trust reports of private flight crews.

Other units or agencies can also produce unreliable information for several reasons such as rivalries, incompetence, or human errors. For instance, an inaccurate time estimate over an entry point between two neighboring sectors might be attributable to lack of competence or malfunction of the flight data processing system (FDPS). It is also likely that a controller may communicate an unreliable routing to a military flight due to operational planning limitations in the dissemination of information. Controllers may be able to recognize unreliable data from the CNS system by utilizing their training and experience. Unreliable information may take the form of false targets, omissions in presentation of targets, and distortion of weather reports. In most cases, controllers can identify unreliable data from CNS systems but this may become difficult when systems operate in degraded modes, or when maintenance work has disabled some equipment.

2.7.3 Inconsistent Information

When all necessary information has been obtained, and its reliability has been verified, controllers may find out more conflicts with other external information or their own expectations. An illustrative story concerns a private pilot who received an instruction by a controller to proceed to point ALPHA in a busy terminal maneuvering area (TMA) sector. As the radar screen showed that the aircraft was actually proceeding to another point BRAVO, the controller requested the pilot to report his course. It came as a surprise that the pilot replied that he was proceeding to point ALPHA. The experienced controller compared the reported position with other radar data and thought of possible errors made by general aviation pilots (e.g., forgetting to change the route or activating the wrong point in the cockpit GPS). Using other reliable sources and his own expectations about general aviation pilots, the controller became confident in his assessment that the pilot had not fully complied with the earlier instruction and was actually in the wrong place.

2.7.4 Information Noise

Although the increased use of automation created an information-rich ATC environment, this was done at the cost of producing more noise or unrelated data that should be filtered out by controllers. Displays are usually designed in a linear fashion assuming that roughly the same set of information is always needed, although with some minor variations. This is not entirely true because controllers may wish to filter out some information to improve observability and make data easier to interpret. The following story shows that the system may not allow controllers to reduce noise and select what information to hide or what information to display in cases of emergencies.

In an Approach sector, the executive controller (EC) set his radar display to depict aircrafts tracks from mean sea level (MSL) to FL200 (20,000 ft). Suddenly, urgent information was received from the area sector, showing an aircraft cruising at FL410 (41,000 ft) that experienced an explosive decompression; the aircraft was descending rapidly to the airport below for an emergency landing. As the controller removed the altitude filter from the radar to identify the aircraft of concern, the radar screen was cluttered with dozens of flight tracks that were difficult to identify. Altitude filters are useful tools but they do not allow controllers to tailor them to their search patterns and choose what to hide and what to display on the radar screen. Therefore, automated assistance tools may produce noise and clutter the display with unrelated data to such an extent that searching for a target may become impossible.

2.7.5 Hard to Interpret Information

Even when all critical information has been collected and verified, in some cases, controllers may find it hard to interpret the information and make sense of the situation, or they may be confronted with many stories that are all equally plausible. Many difficulties in making sense of the problem have been illustrated in the Helios flight HCY-522 crash near Athens (AAIASB 2006).

Departing from Larnaka, flight HCY-522 contacted the company operations center at 16,000 ft. and reported a takeoff "configuration" warning and "equipment cooling system" problem (AAIASB 2006). Communications between the captain and the operations center ended when the aircraft was climbing through 28,900 ft. From that moment, no further communications were established with the aircraft climbing and leveling off at 34,000 ft. The aircraft cruised at 34,000 ft and followed its course to its destination (i.e., International Airport of Athens) since the crew had programmed the FMS to follow this route. Repeated attempts to establish communication with the aircraft failed and two Greek F-16s fighters were ordered to provide close inspection of the aircraft. One of the F-16 flight crews reported that the captain's seat was vacant and the first officer's seat was occupied by someone who was slumped over the controls. After a few minutes, the F-16 pilot reported that a person, not wearing an oxygen mask, entered the cockpit and occupied the captain's seat. Later on, the left engine flamed out and the aircraft started to descend into the holding pattern until fuel was exhausted, and the aircraft crashed killing 121 passengers and the flight crew onboard.

The aircraft never deviated from its FP route and there was indication that something abnormal was happening. Initially, the Athens ACC declared an alert phase to the joint rescues coordination center (JRCC) and 40 minutes later declared a distress phase. The initiation of holding procedures over Athens left the ACC controllers and the Hellenic government puzzled as to what exactly they were encountering. Controllers and government officials could not provide a solid interpretation of the unfolding situation; the F-16 pilot's reports exacerbated the puzzle and further increased uncertainty. In the end, the aircraft was classified as "rogue" and the F-16 flight crews were instructed to shoot it down if there were clear indications that the aircraft was heading for populated areas. However, the fuel was exhausted and the aircraft crashed without any casualties to the population on the ground. This incident illustrates one of the worst forms of uncertainty that the ATC system may encounter in an emergency with regard to team collaboration with several external agencies.

2.8 Concluding Remarks

This chapter has provided a framework of the work demands and challenges faced by controllers in abnormal situations. Even in everyday situations, controllers have to manage several threats (e.g., adverse weather, degraded equipment, heavy traffic, communication failures) that have the potential to combine together and increase traffic complexity. Although work demands in the ATC domain have been viewed from three paradigms (i.e., stress, complexity, and uncertainty), all of them have emphasized the need to pay closer attention to the controller's perception of the imbalance between work demands and cognitive capabilities. The identification of work stressors, complexity factors, and uncertainties is very important in assessing the context of work in which human performance takes place. It is equally important, however, to examine the cognitive strategies used by controllers to make sense of the situation and to choose candidate solutions. This interplay between work demands and cognitive strategies is more thoroughly addressed in Chapters 4 and 9.

The three paradigms have illustrated several views of the context of work, which presents different opportunities and constrains in human performance. To respond to their work demands, controllers employ several cognitive strategies—such as decision-making, sensemaking, replanning and adaptation—that are the building blocks of cognition. Since the cognitive strategies are elaborated in later chapters, some examples have been provided in classical models of performance used by many aviation organizations.

The implication is that cognitive strategies should be presented in practical terms so that they can be easily applied by practitioners. Earlier approaches have used simple models of performance cast as mnemonics. In later chapters, the cognitive strategies are illustrated with behavioral markers or exemplars of good and poor practices.

3
Safety Organization and Risk Management

3.1 Introduction

In the past, aviation safety has taken a reactive approach that focused on the analysis of incidents and the consideration of corrective measures to prevent similar occurrences. Although this approach has provided the basis for a high safety record in aviation, it is now increasingly difficult to achieve further safety improvements with this approach. As a result, aviation authorities have realized the need for a proactive approach to managing safety that concentrates on organizational processes rather than investigation and remedial actions. This attempt has been realized as a safety management system (SMS) that sees safety efforts as an integral part of business activities rather than as an additional layer of oversight of organizations.

The purpose of a SMS is to provide a systematic way to control risks and obtain assurance that risk controls are effective. The SMS provides certificate holders with means of meeting statutory safety requirements and evaluating management capabilities. Safety management requires continuous monitoring of safety objectives, organizational processes, and accountabilities in relation to different safety areas. Under International Civil Aviation Organization (ICAO) Annexes 6, 11 and 14 (2010b, 2001, 2013a), SMSs are required in civil aviation for the operation of aircraft, air traffic services, and airports respectively. Annex 19 (ICAO 2013b) and document 9859 (ICAO 2013c) elaborate the framework and processes of SMS in aviation. At an international level, ICAO provides general guidance for, airport operators, ANSPs, and airlines. At the EU level, the implementation of SMS by air navigation service providers is mandated by regulation (EU) No 1035/2011 that will be replaced by regulation No 1377/2016 from 1/1/2019.

ICAO specifies that all states shall require that organizations in aviation implement a SMS acceptable to the state that includes the following functions:

- Identify safety hazards
- Develop risk assessment procedures and risk matrices
- Ensure the implementation of remedial actions necessary to maintain safety
- Provide for continuous monitoring and regular assessment of safety performance
- Aim at continuous improvement of safety performance

Further, safety roles and accountabilities should be specified in appropriate organizational charts, detailing responsibilities and communication requirements for role functions.

ICAO has produced an Annex and a document on Safety Management Manual (ICAO 2013b, c) in order to provide states with guidance on how to develop a regulatory framework and how to implement the SMS. The ICAO manuals will be the basis for European safety aviation agency (EASA) in formulating common SMS requirements in Europe and for national Civil Aviation Authorities (CAAs) in their function as national regulators. Ulfvengren et al. (2013) identified two important elements in the new SMS approach, namely: (1) integration of safety management with other business processes in order to deliver aviation services and (2) performance-based regulation that is capable of demonstrating effectiveness in terms of measurable outcomes related to safety. Earlier SMS generations focused on the importance of having a safety system independent from the production department in order to achieve a balance between protection and production. Modern SMS approaches seek to implement a performance-based framework that emphasizes an integration of safety management with other business processes in order to achieve commercial, quality and safety requirements. Therefore, we need a model that goes a long way beyond the borders of what has traditionally been included under the rubric of safety management.

This chapter presents some key concepts for the safety of air traffic management, ICAO's SMSs, aspects of performance management, existing methodologies for risk management, and aspects of monitoring and evaluation of safety initiatives. During the design and implementation of safety management, ANSPs should be able to cope with many challenges in overcoming conflicts (e.g., protection vs. production), methodological weaknesses, problems in managing large quantities of data, hindrances in safety communication, and drifts of performance toward the margins of safety. These challenges to safety management and risk assessment are presented in this chapter and a reference table is used to relevant chapters in this book.

3.2 Basic Safety Concepts

In order to understand and build the justification of SMS, ICAO has reviewed the strengths and weaknesses of many established approaches to safety. A contemporary approach to safety, according to ICAO, should take the view that (ICAO, 2013b, 2013c):

> Safety is the state in which the possibility of harm to persons or property damage is reduced to, and maintained at or below, an acceptable level through a continuing process of hazard identification and safety risk management.

Traditional approaches to safety have focused on active failures and tended to neglect the role of latent work conditions created by organizations. Their focus has been on the outcome of safety management rather than on the organizational processes that manage safety. Organizational processes that create latent conditions of failure are under the direct control of senior management and include: policy making, planning and communication, allocation of resources, supervision, and so forth. Safety thinking has also been expanded from studying technical factors to include human factors and organizational factors in safety.

In modern systems, practitioners, tools, and technologies interact in complex ways and present challenges that often exceed human capabilities. Understanding how system complexity affects humans at work is fundamental to safety management. Therefore, safety is not a matter of error-free performance but rather it is a question of effective error management (McDonald 2006). This implies that errors should be studied together with cases of successful action in certain working conditions. In this sense, ICAO considers operational errors as normal outcomes of complex systems where people and technology interact to achieve production goals.

In the ICAO documents, safety management is just another organizational function that must be considered with the same importance as other core business functions. Although safety may not be the first priority of organizations, the management of safety allows them to achieve their business objectives and deliver their services. In this sense, safety management should examine the organization's goals and allow for a balanced allocation of resources between production and protection.

Stolzer et al. (2008) have proposed that safety management can be seen as a system that allows organizational processes relevant to safety to be identified, measured, monitored, and finally improved. Indeed, safety should have a high priority in the organizational structure and not be an issue that is dealt within the safety department. In the same way that quality management systems cut across departments so should safety cut across common organizational silos (Ulfvengren 2010). Safety management has built on quality management principles and moved away from measuring safety outcomes in terms of undesired events. Especially for organizational change and innovation, there is a growing demand for integrating the management of quality, safety, and productivity.

ICAO argues that aviation service providers should apply their business practices to aviation safety and collect operational data in order to develop their safety space.

"Within a safety space, the organization can freely roam while delivering its services, with the assurance that it is within a space of maximum resistance to the safety hazards which exist in the context in which it must operate to deliver its service." (ICAO 2013c).

In the past, many organizations relied on reactive data collection triggered by incidents and accidents. ICAO has advocated the use of proactive data collection using safety surveys, safety audits, mandatory and voluntary reporting. More safety matured organizations adopt advanced data collection systems by making use of confidential reporting, flight data analysis, and normal operations monitoring (Rignér et al. 2009). A statistical analysis of this information may be indicative of emerging risks from a variety of sources. When combining reactive, proactive, and predictive strategies, a safety intelligence function is developed which shows the level of maturity in safety management (Kirwan 2013).

3.3 The Safety Envelope of Aviation Systems

A system of work operates within a dynamic environment that exerts pressure and makes the system modify its structure and behavior over time. Financial pressures create a gradient toward efficiency that constrains practitioners into plans that cater for efficiency and economic survival. Furthermore, demand-capacity mismatches and workload create a gradient toward economic effort that forces practitioners to find "easier ways" to do the job or take more responsibilities with fewer resources. Finally, work is constrained by safety requirements that create a third gradient toward safe performance and away from the safety failure boundary. Overall, the three boundary conditions create a safety envelope within which organizations should work (Rasmussen 1997). As the performance of organizations varies over time, some variability should be expected that is represented as an operating point or a cycle inside the safety envelope (Figure 3.1).

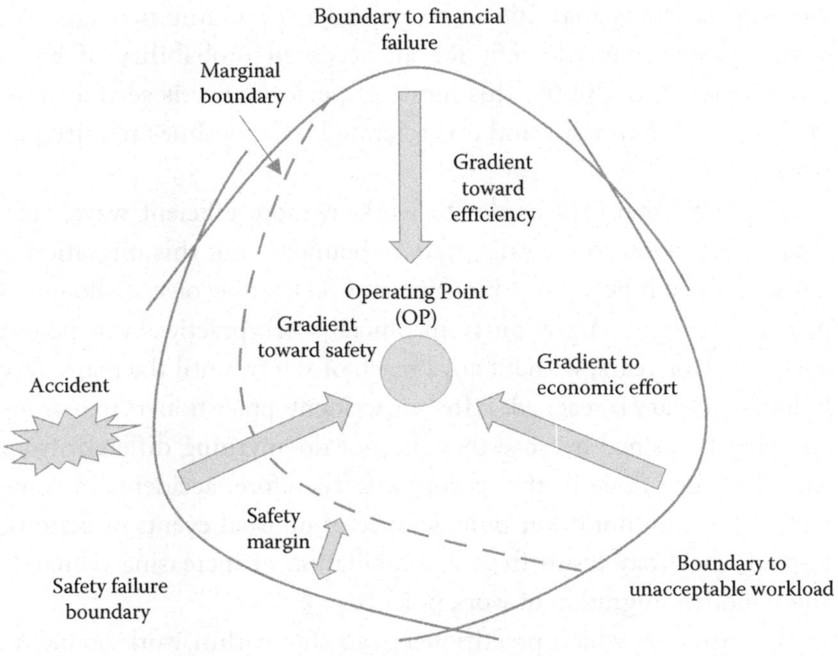

Figure 3.1 A safety envelope created by the boundaries of financial failure, high workload and safety failure. (From Rasmussen, J., *Safety Science*, 27, 183–213, 1997.)

The size of the safety envelope is a function of the constraints imposed by the boundaries of operation. By increasing competition, for instance, the financial failure boundary creates a smaller performance envelope by limiting the possible options of operating staff. The size of performance envelope may also change over time from the initial design of the system to its operation and evolution. According to Amalberti (2001), at the design stage, the system is designed to operate according to a set of rules and procedures with some regard for the likely financial pressures; procedures act as defenses against errors and constrain variability in the plans of practitioners. As the system commences operation, it must adapt continuously to new social and technical demands. The pressure to increase system output with constrained resources (that is, doing the same amount of work with less staff and tools) can make practitioners act more quickly and bypass procedures. This adaptation of work to increasing system demands may cause a migration toward the safety failure boundary. However, system performance may appear stable since there is a buffer zone or safety margin that keeps the system away from the safety failure boundary. Operating close to the safety margin can be viewed as providing management with the maximum benefit for an accepted probability of harm (Amalberti et al. 2006). This mode of performance is seen as beneficial rather than risky and it is tolerated or sometimes required by management.

As practitioners try harder to work in more efficient ways, they are coming closer to the safety failure boundary but this migration is invisible since it becomes so routine and seems to evolve without any breach of safety. Migration from official work practices can persist and evolve for years, without any breach of safety, until the real safety failure boundary is reached. After an accident, practitioners may wonder what happened because they did not do anything different from what they had done in the recent past. Therefore, accidents in complex systems do not occur only because of unusual events or actions; instead, they may result from a combination of increasing demands and a hidden migration of work practices.

The extent to which practitioners can stay within work boundaries determines how much drift the organization can tolerate without failure. Safety can be improved by three means:

1. Increasing the size of the safety space by relaxing constraints and boundaries
2. Reducing the circle of the operating point of the system by reducing variability of performance within the operating teams
3. Operating the system away from the safety failure boundary

The third characteristic of performance may be exploited by safety critical organizations in different ways. For instance, low-risk organizations may choose to stay well away from the safety failure boundary. Others may choose an operating point much closer to the safety failure boundary but where safety is achieved by knowing its location and ensuring small migrations. Cook and Rasmussen (2005) found that high reliability organizations (HROs) manage small transgressions inside the margin of safety without losing sight of the work boundaries (see Figure 3.2).

Systems may become unstable as they become more tightly coupled and attempt larger movements away from the safety failure boundary (in other words, the operating point moves closer to the safety failure boundary and also its circle size may increase). Losing sight of the safety failure boundary and attempting large migrations usually

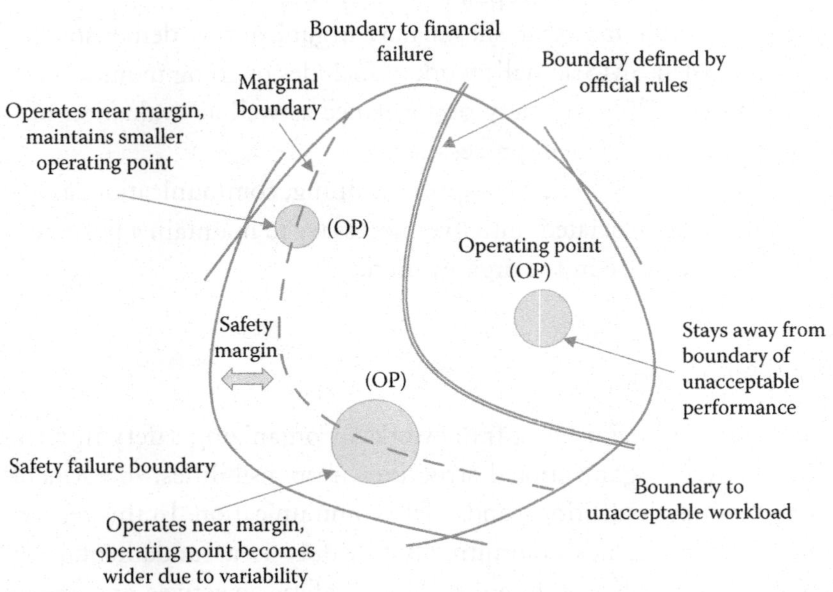

Figure 3.2 Mapping high and low reliability organizations into the safety envelope. (From Cook, R.I. and Rasmussen, J., *Quality and Safety in Health Care*, 14, 2, 130–134, 2005.)

characterizes low reliability organizations (Cook and Rasmussen 2005). Hence, organizations that come closer to the margin of safety should try to recognize such cases and institute actions to reestablish operations inside the performance envelope. On some occasions, it must be also recognized that a boundary may be crossed intentionally in execution of appropriate safety interventions (e.g., in case of unexpected emergencies due to unrecognized events and conditions).

3.4 The Four Quadrants or Pillars of Safety Management

ICAO provides a framework for the implementation and maintenance of safety management, which includes the following four components (referred to as "pillars" or "building blocks") as a minimum requirement:

1. *Safety policy* that establishes senior management's commitment to safety. The policy defines the methods, processes, and organizational structures needed to meet safety goals.
2. *Safety risk management* that determines the selection of appropriate risk analysis methods based upon what the organization considers an acceptable level of risk.
3. *Safety assurance* that shows how organizations demonstrate that their SMS actually work; it includes ongoing monitoring and recording of safety performance as well as evaluation of safety management processes.
4. *Safety promotion* that comprises training, communication, and all other associated initiatives necessary to maintain a positive safety culture in the organization.

3.4.1 Safety Policy

Safety policy establishes a framework for organizing safety in terms of objectives, organizational structures, responsibilities, investments on risk mitigation actions, and safety communication. In this respect, senior management's commitment to safety is a fundamental element. Safety policy is foremost about safety objectives and means of achievement which should be visible to practitioners at all levels. Organizational charts assign safety responsibilities to managers, supervisors, and controllers. High-level management specifies safety objectives and procedures that are monitored by supervisors on a daily basis and put in practice by practitioners.

Top-down safety management provides the means of safety, while bottom-up feedback provides early warnings about residual hazards. Problems in top-down enforcement of safety rules may include inadequate procedures, late modifications after changes, unclear procedures, and violation of procedures. This latter problem is also known as the gap between "work as done" and "work as planned" and it has been thoroughly discussed in the human factors literature (Dekker 2006; Woods and Hollnagel 2006). Problems in bottom-up transmission of feedback may include delayed feedback, distorted feedback, and selective feedback (i.e., certain events are not reported).

Safety culture is a "state of mind" of the organization that encourages safety communication in both ways. In top-down communication, management is aware of the role of practitioners in controlling situations beyond procedures so that a blame-free culture can be created. In bottom-up communication, practitioners provide early warnings about dangerous events. In this sense, the safety management loop becomes more efficient and faster.

3.4.2 Hazards and Risks

Hazard identification is dependent upon the organization's ability to identify operational conditions that may unleash the damaging potential of hazards. A hazard is defined as "a condition, or an event, with the potential of causing injury to practitioners, damage to equipment, loss of material, or reduction of ability to perform a prescribed function." ICAO (2013c) has grouped hazards into three areas: (1) natural hazards, geophysical events, environmental events, and public health events, (2) technical hazards, and (3) economic hazards.

Hazards can be identified by using many types of information sources. For instance, internal sources of hazard information may include flight data analysis (FDA), reports, safety surveys, safety audits, monitoring of normal operations, trend analysis, feedback from training, and finally, investigation of near misses. Examples of external sources of hazard information include: accident reports, state mandatory occurrence reports, state voluntary reports, state oversight audits, and information exchange systems. A high-level description of a typical occurrence investigation procedure is depicted in the flowchart shown in Figure 3.3. It is essential that the likelihood of hazards and their consequences are evaluated to calculate their risk level before any efforts are made to allocate resources to risk mitigation strategies. It is considered a common pitfall to do hazard identification only and

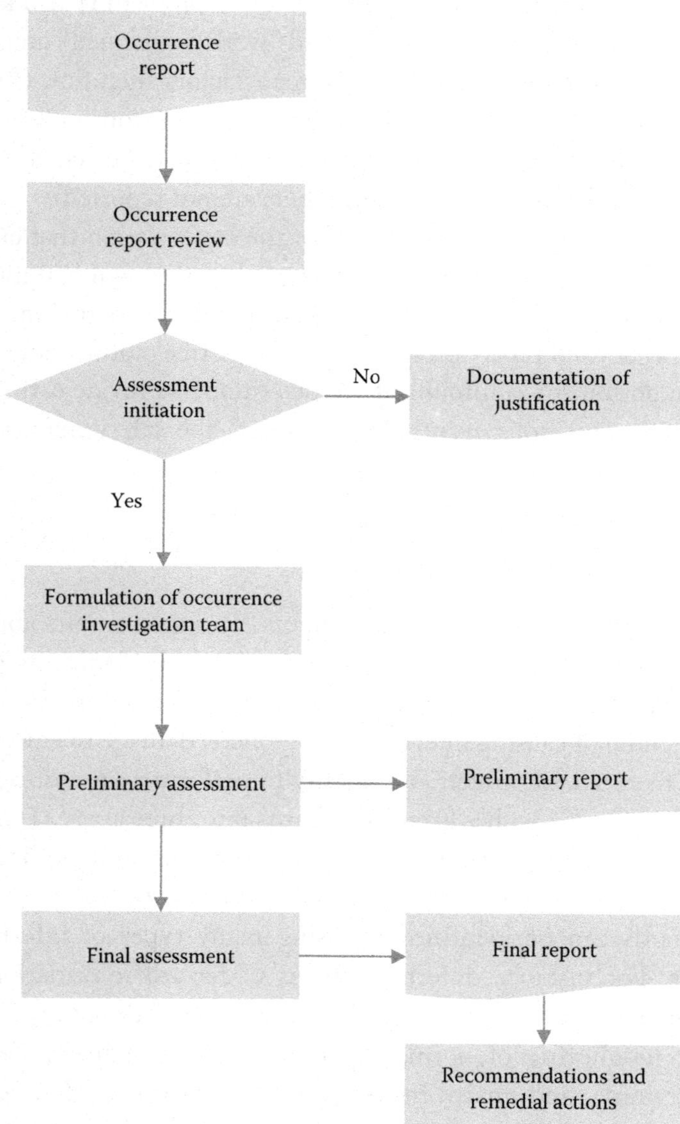

Figure 3.3 Flowchart of a typical occurrence investigation procedure.

then engage directly in risk mitigation because this would have been too resource intensive and expensive (Ulfvengren 2010).

3.4.3 Safety Assurance

Safety assurance can simply be defined as something that gives confidence. The safety assurance function ensures that corrective actions are taken in response to reports, studies, surveys, audits, and procedures for verifying effective SMS implementation. It also provides for the identification of new risk controls in response to changes in the operational environment. In this sense, safety assurance provides feedback on SMS performance and valuable input for any necessary changes.

Safety assurance is also about performance management, which includes: safety requirements, safety performance targets, and safety performance indicators (e.g., a 10% reduction in the number of runway incursions per year). Several information sources can be used for performance monitoring and measurement such as, hazard reporting, safety studies, safety reviews of changes, audits, safety surveys, and internal safety investigations.

Change management also falls within the scope of safety assurance because changes may introduce unexpected events for which existing safety barriers may be inadequate. Changes may be the result of programmed modifications (e.g., new procedures or new technology) or high-level organizational interventions (e.g., unforeseen company growth or operation to new destinations with minimal airport facilities). Hazards due to system changes should be identified and quantified in a process similar to risk assessment. A high-level description of a typical safety assessment procedure following a system change is depicted in the flowchart in Figure 3.4. Risk-based change management has also other benefits, such as prioritizing interventions and choosing efficient risk control measures (McDonald et al. 2012).

Finally, safety assurance includes a process of continuous monitoring of performance after the implementation of safety plans or system changes. Safety monitoring is based on the targets and safety indicators set as part of performance management. Therefore, safety performance indicators and other informal measures can be used in this process.

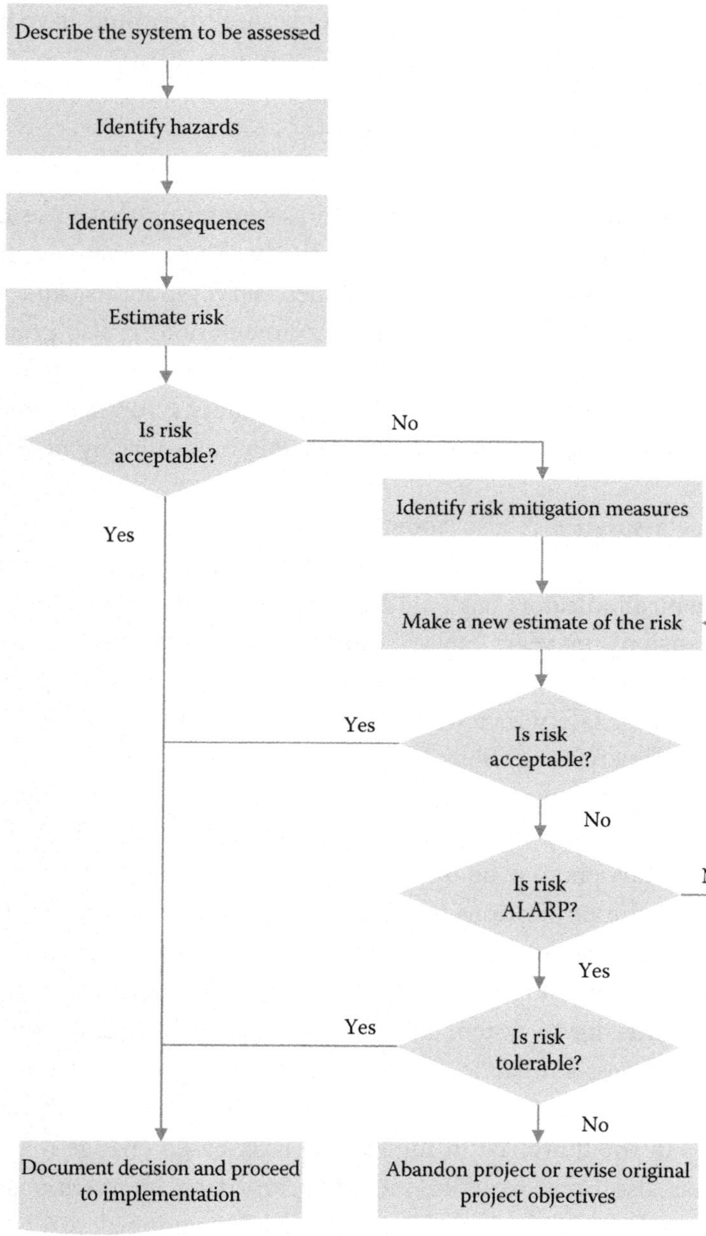

Figure 3.4. Flowchart of a safety assessment procedure following a system change.

3.4.4 *Safety Promotion*

Safety promotion includes training and education, safety competencies, and safety communication. Safety training may be provided to practitioners at all levels in the organization. For instance, for senior managers, safety training may include compliance with national safety requirements, allocation of resources, and effective communication across departments as well as active promotion of the SMS.

Safety training for managers and supervisors should address safety promotion and encouragement of operational practitioners in hazard reporting. In addition, training should include a thorough knowledge of safety processes, hazard identification, and risk assessment techniques. Emphasis should also be given to training for managing changes in technology, workplace, and operating resources. Finally, training for operational practitioners should address safety procedures and introduction to SMS fundamentals.

Safety communication refers to the delivery of safety instructions to the operating lines and the setting of feedback channels. Organizations should communicate the SMS policies and procedures to all practitioners as well as receive feedback about early warnings. The SMS should be visible in all aspects of the organization's policy so that supervisors are accountable for their job roles and practitioners are clear about the job objectives and their degree of autonomy. In a way, safety promotion fills in the blank spaces in the organization's policies, procedures, and processes, hence providing a sense of purpose for safety initiatives.

3.5 A Control Framework Linking the Four SMS Pillars

Within the aviation community, the four pillars are generally accepted as a means of compliance to satisfy requirements. Organizations tend to excel in each pillar, however, they rarely see how the pillars can be connected to produce a workable SMS. To fully appreciate how the SMS parts should interlock, it is beneficial to regard safety management as a control process which takes place at three levels of functioning (Figure 3.5), namely:

1. *At the policy level*, the manner in which managers and supervisors handle conflicts and prioritize goals is important for safety management. This brings to the fore the role of organizational

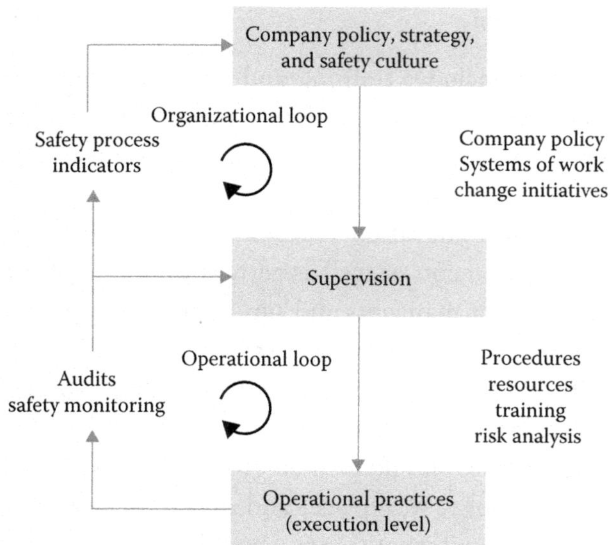

Figure 3.5 System safety as a control process between organizational levels.

knowledge and culture that constitute the deepest set of beliefs about how the world works, about potential hazards, and about perceptions of organizational capabilities. This mindset can remain resistant to change because social beliefs and assumptions remain largely unstated (Meadows 1999).

2. *At the supervisory level*, safety goals are passed onto supervisors and are transformed into specific plans for action that are assigned to different practitioners. This level should specify a set of explicit plans and procedures for risk control in order to guide and co-ordinate the execution level. New hazards reported from the execution level should result in an update of the plans and procedures.
3. *At the execution level*, sharp-end practitioners (e.g., air traffic controllers, flight crew, and airport staff) translate policies, procedures, and standards into specific practices in order to adapt to variations in the environment. The primary control of hazards takes place through the actions of practitioners directly in contact with the system. To assess the adequacy of safety plans and update the mental models, a feedback loop is established to the higher levels of supervision and management.

A periodic review of the SMS can be triggered by trends in safety performance, benchmarking with other competitors, unexpected events, and finally, changes in the company situation or policy. This review can result in rethinking the framework of risk control, in better adjustments to changes of the organizational structure, or in challenging new performance targets.

The SMS translates the theoretical ICAO SMS framework components into a working process. Safety policy sets the framework for designing plans and procedures at the supervision level. The execution level translates plans and procedures into specific actions for managing operations in an efficient and safe manner. At this level, practitioners interact with the technical system and manage risks by making use of their safety procedures and resources. The SMS elements cannot be maintained without good training, proactive communication, and a positive safety culture. Although the integration and deployment of SMS elements is critical to their success, safety risk management remains a keystone of SMS, which is discussed in the following section.

A systemic control view of safety allows analysts to pay closer attention to its dynamics and modes of control. With regard to system dynamics, the organizational and operational loops may have different timescales of change, which affects the flow of control actions and feedback and hence, the way that control loops work in enforcing safety. For instance, safety issues at the operational loop (e.g., incident reports, warnings, and change requests) have a timescale measured in hours, days, hours, or weeks. In contrast, safety policies at the organizational loop may take months to develop or change, which may keep this level behind current technologies and practices. Such time lags may result in asynchronous evolution of the control structure (Leveson 2012). Risk analysis must include the influence of these time lags and potential changes over time (McDonald et al. 2011). A common way to deal with time lags and delays is to delegate authority to lower levels that are faster in obtaining operational feedback and in making timely decisions.

With regard to modes of control, authority and coordination may be enforced either in a prescriptive mode (i.e., feedforward control or coordination by rules) or in a loosely implemented mode as performance objectives with many degrees of freedom to match the local context (feedback control or coordination by mutual adjustment). Two-way communication channels are used to exchange formal and informal information between practitioners or between organizational levels. External communication of the organization with regulatory institutions and public interest groups could also be considered in order to examine the organizational interface with the environment. In addition, coordination can be viewed at the management and supervisory levels required to synchronize management of change and organizational reforms.

3.6 Challenges to Safety Management

ICAO's SMS documents provide only general directions for setting safety objectives, applying risk assessment tools, and evaluating safety programs. ANSPs should use their own judgment how to cope with many challenges in overcoming resource problems, conflicts (e.g., protection vs. production, work as planned vs. work as done), methodological weaknesses, large quantities of data, performance drifts, and hindrances in safety communication. It is important, therefore, to present SMS challenges briefly here, with further discussions in relevant chapters of this book (see Table 3.1 for an overview presentation).

3.6.1 Safety Policy

In the ICAO documents, safety policy refers to management's commitment to safety and regulation of lines of authority to meet safety goals. One of the challenges highlighted by ICAO concerns the development of a business case for safety to demonstrate that safety investments have returns in productivity and product quality. Many efforts have been made recently (Hopkins 2009; HSE 2006) to examine how to define safety performance indicators (SPIs) and how to integrate them with performance indicators for other business processes. Strategic management of safety requires internal intelligence on how the organization is performing, how to plan for changes, and how to improve productivity and safety (Kirwan 2013). The other side of strategic management includes monitoring the commercial environment and benchmarking with other business competitors. This requires some sort of external intelligence in order to understand

Table 3.1 Challenges to Safety Management Addressed in the Book Chapters

SMS ELEMENTS	CHALLENGES	CHAPTERS
	Safety Policy	
Management commitment and responsibility	Safety objectives and planning Common picture of risks The business case for safety	Organizational models (Chapters 11, 12)
	Change initiatives Intelligence on how to change the system	New challenges in ATM (Chapter 10)
Safety accountabilities Appointment of safety practitioners	Communication of safety policies Safety role structures Direct/indirect communication channels; work procedures and safety compliance	The ATM system (Chapters 1, 2)
Emergency response plan	Change of authority in abnormal situations Competence in handling abnormal events	Performance models in abnormal situations (Chapters 4, 5, 14)
	Risk Management	
Hazard identification	System modeling (e.g., time lags, feedback loops, nonlinear effects) for hazard identification	System modeling (Chapters 11, 12)
System risk assessment	Risk acceptance criteria Assessment of safety barriers	Factors affecting ATM (Chapter 2)
	Work as done vs. work as planned Systemic vs. operational risk assessment Safety as part of everyday activities	Error modeling and work practices (Chapters 6, 7)
	Safety Assurance	
Safety performance	Performance management and safety indicators Lagging vs. leading indicators	Organizational models (Chapters 11, 12, 13, 14)
Continuous improvement	Safety indicators drifting slowly away from safety standards Recording and prioritizing risk information	
Internal safety investigation	Risk information from the operations room Recording of active and latent failures Early assessments of incidents	Managing workload and complexity (Chapter 9)
Change management	Assessing effectiveness of interventions and change	New challenges in ATM (Chapter 10)
	Safety Training and Promotion	
Training and education	Training needs analysis and job competence	Refresher training (Chapter 8)
Safety communication	Job design and autonomy Hindrances in risk communication Capturing and communicating efficient work practices	Organizational models (Chapters 11, 12) work practices (Chapter 7)

competitive challenges and new business objectives. A discussion of the role of organizational policy and intelligence on safety appears in Chapters 11 and 12.

The tactical management of safety refers to the safety roles of managers and employees, the assignment of key roles in the organization, the communication of safety policies and rules, and the processing of operational feedback. Although the SMS manual specifies the safety processes of tactical management, several challenges face the reality of aviation systems where multiple stakeholders may be involved in the running of everyday business (e.g., ANSPs, airlines, airport, maintenance and so on). Local optimization of individual stakeholders may create side effects on others and this could create latent conditions of system failures. One challenge in managing system safety regards the development of a common risk picture where decision trade-offs are made by examining both internal and external threats to all stakeholders. Chapters 1 and 2 on the ATM system go beyond the immediate needs of air traffic control and provide a succinct description of the interaction between stakeholders.

Under the rubric of safety policy also comes the management of abnormal situations and emergencies. The SMS manual specifies the need for a proactive approach to emergency management where hazards are identified in advance and risks are mitigated with technological supports or additional safety training. The conduct of regular emergency exercises is very important for maintaining organizational readiness and practitioners' skills in managing systems under time pressure. At the organizational level, a major challenge in emergency management regards the adaptation of authority roles from normal everyday situations. Since unexpected and abnormal situations have different event dynamics, time constraints, task allocations, and repercussions from everyday situations, the organizational mode of control may have to be adapted accordingly. Whereas a feed-forward mode of control and a hierarchical structure may be efficient for normal operations, the changing demands of an emergency may require a greater reliance on feedback control and a flatter structure where operational staff obtain a higher degree of autonomy (Kontogiannis 2010a,b). In this sense, emergency management goes beyond risk analysis and requires a better understanding of organizational and situational demands (Chapter 2), knowledge of task work and teamwork adaptations (Chapters 4 and 5), as well as a consideration of alternative operating modes (Chapter 14).

3.6.2 Risk Management

Risk management includes the identification of hazards, the calculation of their risk potential and the design of risk mitigation measures. Effective risk management requires not only the collection of historical data on system operation but also the development of models that describe "how-the-system-works" and "how-the-operations-function" to achieve safety. In this sense, risk analysts have been using several models of system functioning (e.g., Functional Block Diagram, Structured Analysis and Design Technique) and task analysis. Recently there has been a wider recognition of modeling the non-linear relationships between functions, their time lags, and feedback loops so that the level of risk is evaluated along different time spans. For this reason, the systems-theoretic accident model and processes (STAMP) has been extensively used in studying the dynamics of the aviation system (see Chapters 11 and 12).

Traditionally, the aviation industry has placed a lot of importance on failure prevention by creating standards for policies, programs and procedures. This work standardization was beneficial in reducing incidents in the past but it is no longer suffices to increase safety levels. ICAO has recognized the need for making the next step to error management where errors and failures that can cripple inside the system could be recovered in a timely fashion. For this reason, Chapter 6 is included on error detection and recovery while implications have been made for error management training.

A usual pitfall in risk management is that deviations from standard procedures are considered hazards that may lead to adverse events. In other words, the gap between work as planned and work as done has been considered a hazard. In traditional approaches, standard procedures have been used as reference material for conducting risk analysis. However, there has been ample evidence that work practices that deviate from procedures could be a source of resilience in unforeseen circumstances. To some extent, everyday learning in operational rooms implies situations where controllers experiment with procedures to find more efficient ways of doing their jobs. Some new practices may receive wider recognition and become formal procedures themselves. In hindsight, modifications or workarounds may be seen as violations in cases where a problem is not managed properly. Risk management should rely not only on written procedures but also on actual descriptions of work practices. Therefore, a better understanding is required of the reality of work as done in the aviation industry. This is the aim of Chapter 7, which presents factors affecting modifications of work practices, practitioner methods for optimizing performance, and organizational approaches to the communication of best practices invented by practitioners.

Risk management has taken a systems view of operations and looked into the role of workplace, technological, and organizational factors in accident causation. Another type of risk management can be specified for operational practitioners that is practiced on a daily basis for previewing risks in everyday operations. Operational risk management (ORM) is a simplified version of systemic risk management that focuses on daily hazards and has been part of safety briefings and safety previews.

3.6.3 Safety Assurance

Safety assurance refers to the degree of confidence that the SMS can work in practice. It includes ongoing monitoring of safety performance and periodic evaluation of safety management practices. The ICAO SMS manual presents several sources of ongoing and periodic monitoring. such as event reporting, safety reviews and surveys, safety studies, and internal safety investigations. The challenges in safety assurance mainly have to do with performance management and include: defining lagging and leading indicators, encouraging voluntary feedback of operations, identifying early warning signals, and analyzing statistics of risk information. In accident investigation, the challenges relate to considerations of latent organizational failures that set the conditions for the recurrence of similar events. As formal investigations take a long time to get published, it is essential that organizations are able to draw preliminary conclusions from internal investigations so that organizational weaknesses are corrected and similar accidents are avoided (see Chapters 11 and 12).

Continuous reporting of safety-related events is emphasized in the ICAO SMS manual because organizations may slowly drift away from their safety standards without noticing it. To a certain extent, most organizations operate within this practical drift. Some organizations may stray from their standards and then oscillate a short distance from them. Yet there are organizations that begin to deviate very slowly, almost insidiously at first, then accelerate quickly away from the safety boundary. Monitoring operations to identify drift is part of a mature SMS. The challenge here is in capturing the right data at the right time. ICAO makes reference to navigation aids within the practical drift that help organizations navigate the currents and obstacles.

Safety assurance makes feedback on SMS performance possible but also provides valuable input to many system changes. The management of change has been an essential part of the SMS manual because ANSPs have to adapt continually to increasing system demands and competition. This adaptation requires a policy and a program of introducing changes and managing risks. The challenge for managers and designers here is to imagine how the new system of work will operate in the near future and predict likely hazards in new operations. Although some hazards may be identified and prevented prior to design, others may make their way through to future operations; their effects can be unmasked by a safety assurance program. New ATM initiatives such as SESAR and NextGen can be considered as change management initiatives and are considered separately in Chapter 10.

From another perspective, change management may be seen as a separate safety case, which identifies change objectives (safety policy), controls risks in the design stage (risk management), assures that "residual risks" are captured (safety assurance policies), and equips practitioners with necessary skills for transferring their skills into the new context of work (safety promotion). In other words, change management may require all four pillars of safety management.

Since new change initiatives are made in response to increasing task demands and traffic patterns, the issue of coping with workload and complexity becomes fundamental. In this respect, Chapter 9 deals with the issue of complexity at the operational level where controllers have to manage heavier traffic levels. Chapter 10 looks into the strategies that controllers use to reduce complexity or manage high complexity in situations of high traffic or new situations in SESAR and NextGen scenarios.

3.6.4 Safety Promotion

ICAO's emphasis on safety training focuses on how to make practitioners at all organizational levels familiar with the SMS and provide training in the identification of hazards, their prioritization and their means of mitigation; a separate issue here is safety communication that should be sensitive to safety warnings. Although, safety training is a legitimate aspect of improving human performance, it appears that safety has been treated as an important but isolated aspect of total performance management (e.g., productivity, quality, and maintenance). A challenge for organizations would be to consider new approaches that make a business case for safety by integrating safety and productivity aspects of performance (McDonald et al. 2012).

Existing risk management approaches emphasize this division of mind between safety and productivity as illustrated in fault-tree analysis. Most probably, controllers do not perceive of conflict detection as a task separate from other tasks, such as putting arriving aircraft in sequence, establishing departure flows, coordinating with other sectors, and so on. The challenge for controllers is to manage these tasks efficiently and safely. Therefore, the focus should be on the cognitive functions of controllers that usually address both efficiency and safety, although with different priorities depending on the circumstances. In this respect, several methods of cognitive tasks analysis (CTA) are presented in Chapter 8 as a basis for controller training. CTA methods are based on models of human performance and behavioral markers that exemplify aspects of poor and excellent performance.

Safety promotion also includes the collection and dissemination of risk information across all organizational levels. Safety communication is very important for remaining alert to early warnings of danger as well as learning lessons from recorded near misses and incidents. Experienced controllers manage to fix problems quickly, hence overcoming several systemic problems that may persist for long periods as latent failures. Fixing a problem usually is not followed by early warnings to upper management, which prevents a systemic solution. It is important, therefore, to examine how practitioners develop their quick-fix practices on the job and how organizations can design systems to capture knowledge about work practices that affect safety (see Chapter 7).

Safety communication may encounter many filters and hindrances as it travels upward through organizational levels. In this sense, organizational communication and safety culture are important aspects of managing safety communication. Organizational communication can be considered together with other managerial processes within the framework of system theoretical models such as a STAMP and Viable System Model (see Chapters 11 and 12).

3.7 Revisiting the Safety Envelope and Applying Resilience Engineering

Resilience engineering looks at how organizations adapt to unanticipated situations by moving closer to the safety margin. Adaptation involves managing shifts in strategies, organizational processes and coordination patterns. Resilient organizations usually exhibit the following characteristics (Woods 2006):

- Develop a compensation capacity to system disturbances without a fundamental breakdown in performance
- Monitor how closely or precariously the system is operating relative to the boundary conditions
- Manage to degrade gracefully as pressure increases but do not collapse abruptly
- Reframe their "model of safety" and restructure their processes in response to changes in the environment. This reframing process involves noticing weak signs of unacceptable performance, calling into question ongoing models of safety and considering potential revisions (Klein et al. 2007)

The first aspect of resilience refers to the capacity of organizations to recognize situations closer to the safety margin and compensate without a fundamental breakdown in their performance. The ability of organizations to adapt to minor disturbances and normal variability of conditions is referred to as *adaptability*. Woods (2006) used the term *resilience* to refer to a broader capacity to adapt to situations close to the safety margin, which challenges ongoing models of safety and procedures. This leads to the second aspect of resilience that requires organizations to know how close the operating point of the system has come relative to boundary conditions (see also Figures 3.1 and 3.2).

However, assessing the safety margin involves more than knowing the distance from the margin since the adaptive capacity of the system to remain within this margin may be exhausting. Sarter et al. (1997), for instance, have found that automated aircraft systems may be working very hard to maintain control in the face of disturbances but their stretching may be hidden from flight crews. Their successful compensation partially masks the presence of disturbances and their stretching of compensating capacity. In a second phase, after the capacity is exhausted, automated control may collapse as the disturbance may persist or grow. Unfortunately, there are no direct cues that could alert flight crews of the trouble that automation has experienced in maintaining control of the situation. This decompensation pattern can be difficult to detect because it develops slowly over time but eventually collapses in a dramatic manner leaving little time for transfer of control to human controllers (Woods and Cook 2006). Hence, the third aspect of resilience involves a graceful degradation when demands exceed capacity.

While this pattern has been noted in aircraft supervisory systems, it may also apply in assessing how organizations evaluate their resilience (i.e., their adaptive capacity to maintain control inside the safety margin zone). In the first stage, safety management may be able to handle disturbances or problems; however, this adaptive capacity may be either a sign of success or a sign of incipient failure. Therefore, organizations should be able to monitor their adaptive capacity and assess its stretch to various sorts of and sizes of disruptions.

The fourth aspect of resilience regards how organizations reframe their model of how safety is created before an adverse event occurs. Therefore, understanding the reframing process of organizations is very important in order to develop resilience indicators and supportive means (see Chapter 14 for a more thorough discussion).

3.8 Risk Assessment Approaches

The most widely recognized element of safety organization has been the management of safety-related risks. Existing methodologies of risk assessment in the aviation domain have been based on a typical framework that includes the identification of hazards, their screening according to a risk matrix, the quantification of risks, their prioritization, and finally, the recommendation of risk mitigation measures or corrective solutions. There is also a change management policy so that the effectiveness of changes and risk mitigation measures is monitored according to a formal plan (i.e., the risk monitoring stage).

The ARMS methodology (EASA 2011) and the risk assessment tool (Eurocontrol 2009) both comply with this typical framework of analysis.

However, many aviation practitioners and organizations have indicated several difficulties with the quantification and risk mitigation stages. There are many reasons for this difficulty, including high requirements for expertise, greater human and time resources, access to historical data, and cost issues pertaining to the choice of risk counter-measures. In fact, a formal survey in safety-critical industries shows that risk assessment methods have been used only in design and modification projects and not during daily operation (Andersen and Mostue 2012). As a result, risk assessment may be seldom updated and hence, important changes in safety functions are not monitored on a continuous basis. To readdress such problems and make risk assessment a practical tool for safety practitioners, this section proposes several requirements derived from the literature review, the views of several safety practitioners, and the experience of the authors. To understand potential problems and areas of improvement for existing risk assessment methods, a basic background is provided first of a typical risk-assessment framework.

3.8.1 Systemic Risk Assessment

Risk assessment is a complex process that requires a team of experts in order to identify hazards, collect historical data about component failures, construct risk models, assess the influence of workplace and organizational factors, and design risk counter-measures. Because of the large demands in human resources and data requirements, risk assessment is carried out when new designs or modifications are introduced into the system. For this reason, it is usually called systemic risk assessment to distinguish it from other types of operational risk assessment carried out daily by the practitioners themselves.

Eurocontrol has developed an integrated risk picture model (Eurocontrol 2006) that is used to estimate frequencies of several accident types (e.g. taxiway collision, midair collision, and so on). For each accident type, a separate causal model is constructed using fault trees that represent precursors to accidents and failures of barriers. Precursors are unsafe conditions that can lead to more adverse states, provided that the barriers cannot manage them successfully. The precursors are arranged in severity terms from conflicts to separation losses, air proximity events, and accidents. Barriers (or safeguards) include equipment, safety nets, procedures and processes that can be managed so as to prevent precursors from progressing in severity to accidents. The fault trees allow quantification using historical incident experience and judgmental modifications to fit specific organizations. An influence model is used to show the effects of workplace factors, environmental factors, and safety processes that contribute to accidents. The structure of the influence model is based on a system model that defines the major system elements, represents the concept of operation, and identifies interdependencies due to common resources. As a minimum requirement, the system model shows the necessary inputs and outputs for each element, the required resources, and safety constraints.

Eurocontrol has proposed the SADT model (Marca and McGowan 1987) as a basis for describing the system; other models include STAMP (Leveson 2004, 2012) and SCOPE (McDonald et al. 2011). The output of the influence model is a set of modification factors applied to the frequencies and probabilities of the base events of the fault trees. In general, the initial system model should be able to assist analysts in developing the risk influence model and tailor risk assessment to the specific problem at hand. For quantification purposes, this approach can result in an exponential growth in the tree size because managerial influences should also be seen as common causes of individual causal factors. Fault trees are based on the Swiss cheese model (Reason 1997), where causal factors are represented as holes in a series of organizational processes or barriers. A similar diagram in Figure 3.6 shows how barrier failures create unsafe states in increasing severity from flight conflicts to midair collisions. Failure of barriers includes not only technical factors, but also human failures to respond promptly and prevent the next stage precursor.

Accidents may arise in innumerable ways, hence it is not practical to identify every possible accident scenario. In building a causal model, it is only necessary to identify scenarios that involve combinations of barrier failures or scenarios that involve practitioners bypassing certain barriers. For convenience, conflicts due to pilot noncompliance and unauthorized penetration of controlled airspace (e.g., military

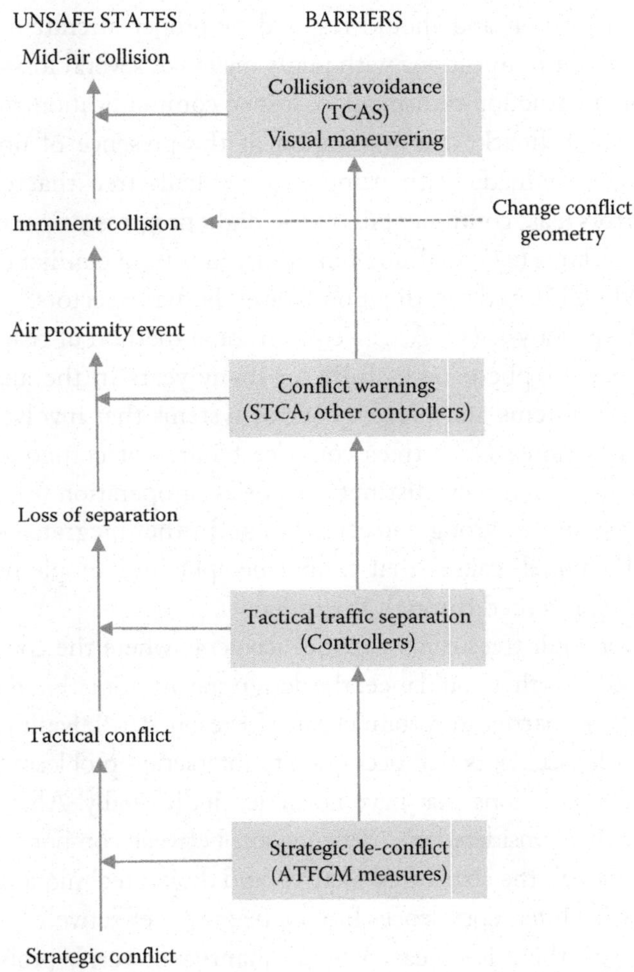

Figure 3.6 Barrier failures create precursors in increasing severity from conflicts to accidents.

flights) are referred to as unplannable conflicts, since they involve similar barriers. Tactical conflicts that could have been prevented by strategic traffic planning and synchronization are described as plannable conflicts.

An unplanned conflict can be presented due to traffic synchronization problems managed by air traffic flow controllers. The executive controller is responsible for tactical separation, which involves monitoring radar information, detecting conflicts, resolving conflicts by tactical planning, and liaising with the coordinating controller. As a result of tactical separation, flight crews should receive separation instructions in a timely fashion and should respond by proper aircraft maneuver. Inadequate communications with pilots may take several forms, incorrect or late instruction transmission, loss of communication, and inadequate pilot feedback. The estimation of the presence of unplanned conflict can be made with using another fault tree that combines human errors (e.g., controller fails to use flight progress strips correctly), technical failures (e.g., malfunction of medium term conflict detection system–MTCD), and coordination failures between sectors.

Fault tree analysis (FTA) is a conventional method of risk analysis that has been applied successfully for many years in the analysis of mechanical systems and process control systems that involve routine human operations. Fault trees consider failures of components and human actions that have distinct categories of operation (e.g., failed/operating states or wrong/correct actions). In the integrated risk picture (IRP) model, causes that cannot be split into simple operating states are represented through the influence model.

However, fault trees cannot model accidents where the components have not failed as they could meet the design specifications but the design cannot act as a barrier to certain events (Leveson 2012). Fault trees also cannot model accidents that occur due to interaction problems between components or actions that have not failed individually. Although the IRP model has considered interdependences between components in the influence model, the structured analysis and design technique (SADT) used to model interdependences has not been very effective.

In general, there is no easy way to quantify or verify probabilities that are used in fault trees because human errors depend on the work context and the dynamics of the environment. For instance, the probability of failure to detect a planned conflict has been estimated to $6\ 10^{-3}$ per flight, which is the nominal failure probability of a rule-based task. However, many could lead to undetected conflicts that have different dynamics and contexts of work. The authors have recorded at least five scenarios leading to failures of detection (see Table 3.2). These work scenarios indicate how work context and work practices can influence human detection.

The IRP model addresses many work influences in a generic manner in order to generate “modification estimates” without any reference to the particular practices and work contexts. The IRP model considers the following generic influences on performance:

Table 3.2 Scenarios Making "Unplanned" Conflicts Difficult to Detect for Controllers

SCENARIO	DESCRIPTION
Undetected conflicts in low traffic parts of radar screen	Scanning strategies enable controllers to monitor traffic especially when workload is high and identify conflicts at early stages so that sufficient time is allowed for conflict resolution. On some occasions, the traffic pattern is uneven so that heavy traffic appears in certain parts of the screen while other parts display areas with less traffic. This uneven traffic pattern may increase the chances of unrecognized conflicts in the low traffic areas when controllers have devoted their attention to the heavy traffic area.
Undetected conflicts in parts of the screen that have been filtered out	Controllers may choose to increase the screen scale in certain traffic areas where interesting events are presented and filter out traffic in areas where conflicts are least expected (e.g., traffic below 3,000 ft. may be filtered out when visibility is good and traffic is low). The risk is that filtered out traffic areas may be left unattended too long until a short-term conflict alert is displayed in the system.
Undetected conflicts due to early transfer of aircraft	A conflict may remain undetected in cases where the controllers agree to accept early an aircraft transferred from an adjacent sector but they may be late in considering the aircraft within their responsibility area. This can occur, for instance, when an early transfer aircraft is left in gray color, indicating that it is not under their control yet. Leaving the change of color until later may result in late recognition that the aircraft has already been accepted from the adjacent sector.
Converging traffic that becomes in conflict during transition to another sector	Traffic may become in conflict while aircraft are in transition from one sector to another with different characteristics. For instance, two aircraft descending at different speeds may be safely separated in one sector but may become in conflict as they cross sectors; this is more likely to occur when the next sector has higher separation minima.
Differences in aircraft performance may result in unexpected conflicts	Projecting aircraft trajectories in the future requires a good knowledge of performance characteristics. Unknown to the controller, an aircraft may be heavily loaded and climbing at a slower than expected rate; this can result in conflict with lighter aircraft following the climb behind the heavy aircraft.

- *Operating environment* (i.e., traffic density, airspace design, terrain, weather, visibility conditions, and so on) and quality of equipment
- *Workplace factors* (i.e., man-machine interface, human reliability, job aids) and organizational factors (i.e., procedures, training, resource allocation, and teamwork)
- *Safety management system* (i.e., safety policy, safety communication, safety assurance, and safety promotion)

Another challenge in risk analysis concerns the modeling of interactions between human activities. It is often the case that separation planning is seen as a sequential phase that follows the identification of traffic conflicts. In fault trees, inadequate tactical separation can be due either to undetected conflicts or to inadequate separation planning. Failures in detection and planning are added through an OR gate to estimate the overall tactical separation failure. In operational practice, however, the two human activities are ongoing since controllers may detect several conflicts with different dynamics and mentally play out a couple of candidate deconflicting strategies. This may result in a decision to intervene late which makes an external observer believe that planning follows detection. On other occasions, however, an early intervention can be made to avoid imminent conflicts in the near future but this early resolution of converging traffic cannot be captured by external observers.

Separation planning seems to have two cognitive elements: (1) micromanaging where imminent conflicts are managed and (2) anticipatory planning with a longer horizon of attention where traffic projections are made in the long term and future points are decided for closer traffic monitoring. In micromanaging, controllers resolve an imminent conflict but also try to avoid side effects in other areas. In anticipatory planning, controllers try to stay ahead of traffic and maintain awareness of the traffic dynamics. Fault trees usually consider failures at micromanaging, such as delayed resolution or unsuccessful resolution.

In addition, failures at managing traffic are not static events as assumed in fault trees. Micromanaging conflicts is a dynamic process that may start with a well-thought-out plan that goes astray because of unexpected events and surprises. For instance, separation planning may be tightly coupled, leaving little scope for crew diversions or unexpected events. A traffic pattern that is tight may not be recognized by crews who may wish to change their flight to a continuous descent from a stepwise one, hence, giving rise to other conflicts later on. In a similar sense, a correct resolution plan may be interrupted by other crews blocking the radio frequency for too long (e.g., take long in initial contact formalities). That is, the plans may start well but can remain incomplete due to other interruptions.

In other cases, separation planning may be recorded officially as unsuccessful yet it may cause no harm. For instance, a resolution plan may result in two aircraft violating the separation minima but the conflict geometry may be such that the conflict is resolved soon after its recording by the system; in this case, there may be no complaints from the crews nor from the safety managers. This may happen because avoiding a temporary conflict may require tremendous effort while its tolerance may improve traffic separation in the near future.

Anticipatory planning is an important controller strategy but it is not considered in fault trees. Inadequate anticipation may lead to a tight traffic pattern that increases the chances of conflict but leaves little scope for recovering from unsuccessful resolutions or surprises later on. Anticipation involves staying ahead of traffic (i.e., having a longer attention horizon) and creating open traffic patterns that allow crews and controllers to cope with unexpected events. Anticipatory planning is an invisible strategy to the fault tree analysts because its results are seen later as undetected conflicts or unsuccessful resolutions. Many analysts consider anticipatory planning an activity for the coordinating controllers only, but recent research conducted by the authors has shown that executive controllers also actively try to stay ahead of traffic.

3.8.2 Operational Risk Management

Operational risk management (ORM) involves a fast risk analysis where a risk index is calculated and compared to a target risk level. If the estimated risk level is higher than the target then the practitioners can propose risk mitigation measures to manage the risk. ORM can be applied by the practitioners themselves on a daily basis. Several ORM methods have been proposed in the aviation domain, including threat and error management (TEM), failure likelihood indexes, and so on. In 2009, Eurocontrol developed the risk analysis tool (RAT) as a basis for rating the severity of near misses and traffic incidents but RAT can also be used in predicting and assessing risks in future scenarios. Specifically, the European Union requires all ANSPs to use the RAT tool in order to rate the severity of recorded ATM safety occurrences. The general philosophy of the RAT tool is similar to the ARMS method and can be summarized as follows:

$$R = S \times C \times R$$

Where:

- R: Risk is the product of Severity, Controllability, and Repeatability.
- S: Severity of hazardous event refers to task, traffic, or environmental consequences.
- C: Controllability refers to the actions of practitioners in order to prevent, resolve, or control and recover the hazardous event.
- R: Repeatability refers to the likelihood that a hazardous event may occur in the future, which is a function of workplace conditions and organizational factors.

Although RAT specifies a numerical calculation of risk factors, its causal model can be used as a general guidance for designing a data structure that is useful in the context of several ORM methods. In combination with bow ties, RAT can be used to generate a rich data structure that is useful for producing fault trees and event trees. A bow tie is a widely used risk analysis method that identifies hazards, their causal factors, and their likely consequences. The left side of a bow tie examines possible safety barriers that could prevent the occurrence of a hazard, while its right side examines barriers that provide hazard protection and minimize its adverse consequences. An actual example of a bow-tie analysis of a low level winds shear hazard is shown in Figures 3.7 and 3.8 (for a detailed description of LLWS phenomena see Chapter 5).

The data structure can be rich enough to enable analysts to build their risk models regardless of the specifics of their preferred techniques. The argument for designing a rich data structure driven by risk models appeals to many commercial aviation sectors as they are still free to choose the risk model they feel most comfortable with. It may be seen that this sort of data structure (Table 3.3) provides a complete description of the risk items required in conventional risk assessment but little information about the connections or gates that create the critical paths to hazardous events. In this sense, additional processing of the data may be required in order to create a risk model such as a fault tree. Although the risk data structure does not convey all the details available in risk models, it is not specific to a risk model and has wide applicability.

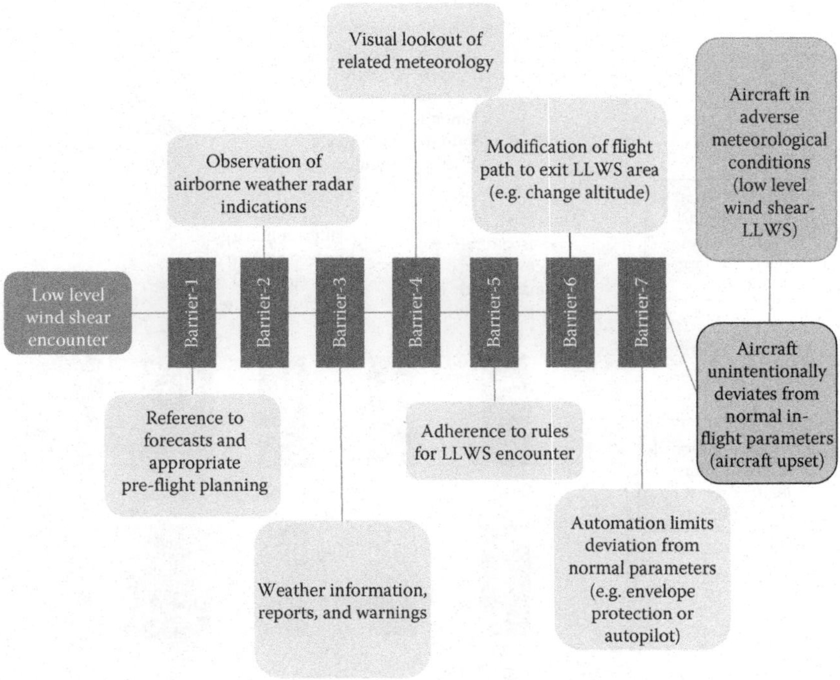

Figure 3.7 Left hand part of a bow-tie analysis.

3.9 EASA Requirements of Risk Assessment Methods

As the aviation system becomes more complex and the coupling between ANSPs, airlines, and airports becomes tighter, risk assessment methods face new challenges. For instance, EASA has proposed that risk assessment methods consider findings from separate aviation systems in an integrated fashion to overcome problems of complexity and tight coupling. The dynamics of risk also impose new challenges since an organization may be found to be safe but the drift to failure dynamics may create risks at a later phase.

In addressing these challenges, EASA has reviewed several risk analysis approaches and proposed a risk framework specified in Masson and Morier (2011):

- Define the focus and system borders of the analysis.
- Describe/model the system and its nominal operations (i.e., organizational structure, allocation of safety responsibilities, flow of safety information, human-machine interactions, existing safety barriers, work rules, and so on).

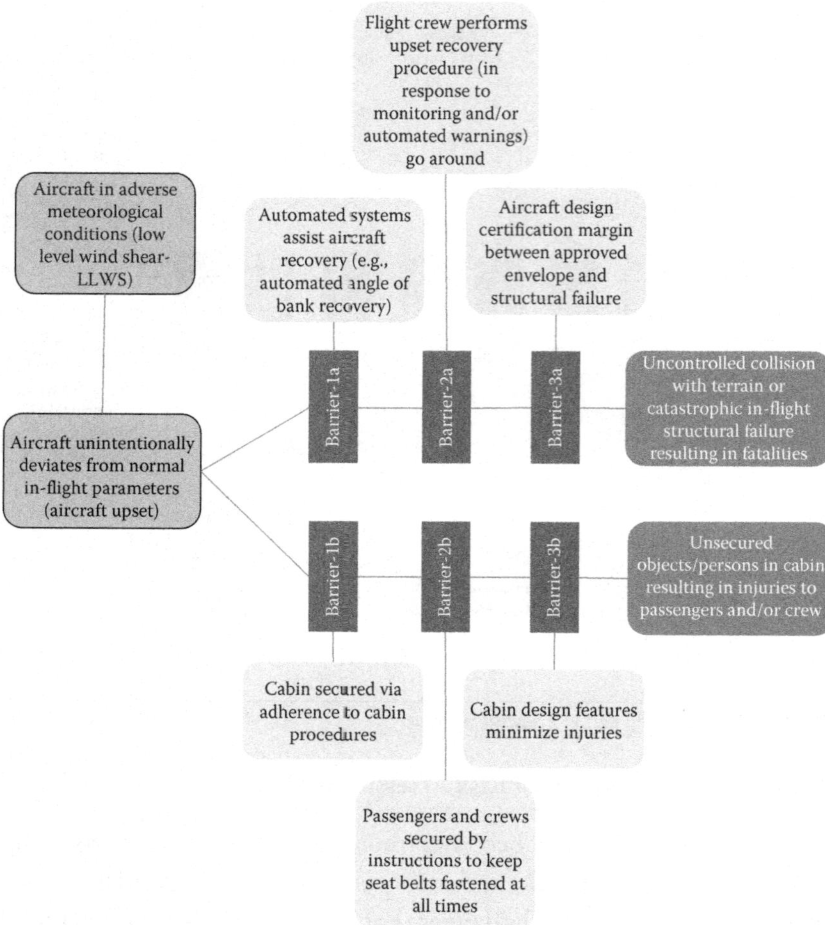

Figure 3.8 Right hand part of a bow-tie analysis.

- Identify hazards using a multidisciplinary team of analysts and ensure that the team develops a common mental model of hazards.
- Combine hazards into a risk framework and develop a risk model of the system; examine how risk levels are influenced by various contributing factors that may amplify/damp the consequences of hazards.
- Evaluate risks and assess how risks may evolve over time.
- Identify potential risk controls (barriers) and reassess the residual risk until the ALARP criterion is reached (i.e., risks should become as low as reasonably practicable).

Table 3.3 A Generic Data Structure for Conducting Risk Assessment

- SEVERITY
  - Traffic complexity
    - Rate of closure
    - Traffic density
    - Number of convergent routes
    - Number of path changes to aircraft
    - Number of aircraft around conflicts
    - Number of intersecting flight paths
  - Environment
    - Weather conditions
    - Volcanic ash
    - Visibility conditions
    - Wind shear
    - Day/night
    - Terraino
  - Safety nets/ barriers
    - Short term conflict alert (TCA)
    - Medium term conflict alert (MTCD)
    - Traffic collision avoidance system (TCAS)
  - Prevention
    - Data for traffic synchronization
  - Plannable conflict
    - Air traffic flow and capacity management (ATFCM) measures
  - Plannable conflict
    - Conflicts from traffic sequence
  - Unplannable conflict
    - Conflict from airspace penetration
    - Conflict from unmanned aerial system
    - Conflict from VFR traffic
    - Conflict from flight deviation
- CONTROLLABILITY
  - Resolution
    - Conflict detection by controllers
    - Conflict detection by pilots
  - Recovery
    - Controllers recovery the problem
    - Pilots take evasive action
  - Coordination
    - Within sector coordination
    - Inter-sector coordination
- REPEATABILITY
  - Equipment
    - Degraded modes
    - Design of equipment
  - Procedures
    - Incomplete/procedures, incomplete ambiguous, operations manual, unit training plans, unit competency schemes, and contingency plans.

- Establish a plan for monitoring the effectiveness of risk controls proposed in earlier steps (i.e., safety monitoring and verification); this may involve setting key performance indicators that show the work progress made.
- Consider the feedback loop of organizational learning and process improvement.

Because it is unusual to find a single method that satisfies all criteria, safety analysts may choose a battery of two or three methods as far as they are compatible to each other.

The literature has proposed several criteria for comparing risk assessment methods, which are beyond the scope of this chapter. However, it is worth presenting some criteria that have been proposed in Action EME 1.1 of the European Aviation Safety Plan (EASp). In assessing future risks in aviation, risk assessment methodologies should:

- Yield an integrated risk assessment
- Have sufficient power of anticipation
- Consider a range of possible hazardous scenarios in future
- Evaluate system variation during normal operations
- Consider the complexity of the system
- Have the ability to model dynamic phenomena
- Assist in identifying unanticipated uses of technology or procedures by the operational practitioners
- Provide a means of prioritizing hazards/risks
- Identify warning signals that indicate a drift to failure
- Be simple and practical to apply by knowledgeable domain experts

General risk assessment requirements can be supplemented by recent developments in complexity theory. Modern industries should pay particular attention to the emerging risks, that is, risks arising from structural changes in future aviation environments, risks emerging slowly over time without immediate symptoms, and risks arising from migration to safety boundaries. Many emerging risks have been discussed in SESAR and NextGen while others have been addressed by general studies at the societal level. The authors have done a review in order to identify contributing factors that could amplify risks as modern aviation increases its complexity. Contributing factors create a fertile ground for risks to crop up and amplify.

An indicative list of contributing factors is described below as an essential element of risk assessment methodologies (IRGC 2010; Dekker 2011; Stacey et al. 2000):

- Loss of safety margin created by the tight coupling of systems that leave little margin for recovery; the margin can be understood as the system's buffering capacity or time slack in the event of system overload or failure.
- Trade-offs between different goals and interests that may tip the balance to different directions under different work scenarios.
- Positive or reinforcing feedback loops that may strengthen the initiating event and produce nonlinear disastrous effects.
- Time dynamics where initiating events take a long time to display observable symptoms for the operating teams to detect and take action. Alternatively, the time cycle of the event may be much longer than the decision cycle of a safety manager who may focus on short-term goals.
- Tipping points or thresholds where changes or transitions occur unexpectedly as the system flips from one state to another.
- Unforeseen adaptations where workers may use procedures and barriers in ways that have not been foreseen by the designers.
- Bumpy transfer of control between automation and controllers.
- Social system dynamics and cultural issues that may amplify or dampen earlier perceptions of risk.
- Asymmetries in information from withholding of safety information, wrong delivery of information or delays that impede an understanding of risks.
- Perverse incentives or goals, such as seeking short-term productivity gains at the expense of recognizing risks that take longer periods of time to manifest themselves.

These contributing factors can be used by safety analysts to select appropriate system models that describe the complexity and coupling of modern systems, to examine factors that may influence the risk model, to see how risks may change over time, and to think of candidate risk solutions that avoid side-effects. Chapters 12 and 13 provide further discussion on how to incorporate these complexity factors in the organizational analysis of systems using system thinking approaches.

3.10 Concluding Remarks: Toward Resilient Risk Assessment Methods

Existing risk assessment methods have focused on risk prevention by specifying procedures, technology designs, and control systems that minimize the chances of encountering hazards. Civil aviation usually takes this approach to prevent flight crews from being exposed to surprising conditions or hazards that are difficult to control. For instance, airlines may have procedures that prevent crews from flying aircraft in hurricane conditions because the risks are too high. For this reason, flight crews have no formal training in such unlikely scenarios. At least, airlines should make clear to their crews of the trade-offs that have been made in risk analysis and the reasons for avoiding unfamiliar scenarios. When hazards are successfully controlled then organizations may consider a reporting system that documents resilient strategies employed by practitioners. For example, cases where teams of controllers manage to respond to ATFCM inadequacies (e.g., over-deliveries of flights in a sector) should be recorded and resilient strategies that were crafted on the spot should be documented.

The resilience approach emphasizes the need of organizations to develop a capability for dealing with unknowns. Being resilient requires that practitioners are able to improvise and adapt procedures to unfamiliar situations. There is some criticism that risk assessment methods are built using the rear-view mirror. They look at the past to generate warnings for the future. Unfortunately the world is not quite so linear; it evolves through alterations that change the assumptions of risk analysts. A case in point is the new aviation environment that will operate in different ways from the current environment. As a result, the history of failures and adverse events may not be so useful in predicting future patterns of operation in aviation. Hence, the need for risk assessment methods that look toward the future and make use of modern approaches of system thinking and complexity. In this sense, resilience engineering emphasizes the adaptive capacity of organizations to survive adverse events (e.g., making the system less tightly coupled, switching to new organizational structures, and providing more autonomy to practitioners).

Resilience engineering places a lot of emphasis in the ability of controllers to adapt procedures and recover from their own errors in a timely fashion. Error detection and error recovery should be essential elements of risk assessment methods. Unfortunately, errors that have been recovered are not recorded since they have not presented any threats or hazards to the system. Recovered tasks are considered to be normal tasks with possibly some delays. This is one of the reasons that Hollnagel (2009) emphasizes the need to collect not only reports about failures but also reports about successful operations particularly recovered operations so that we learn of the strategies that controllers use in error detection and recovery.

In some respects, real organizational life involves controllers responding to pressures at work and adapting their work practices. This informal organization of work departs from the formal procedures usually consulted in traditional risk analysis studies. As a result, production pressures and other economic issues may create a work environment that is different from the formal system model that was used as a basis of risk analysis. The implication is that new risk assessment methods should start with a formal system model but should also consider variations of the model due to production pressures. For instance, a series of tasks may be prescribed in a sequential order in the formal system model but the same series may be executed in parallel under conditions of time pressure. This implies that different sorts of errors and hazards may be associated with the organization of tasks in the real work environment. The requirement here is that risk assessment should consider several variations of the system model. In this respect, Part IV provides a further discussion on system archetypes that encapsulate past knowledge of complex systems.

PART II

Cognitive Engineering

4

Decision-Making

4.1 Introduction

Air traffic controllers (ATCOs) make decisions most of the time, from strategic flow management to tactical resolution of aircraft conflicts. Unfortunately, the decisions that get most attention by public media are those that result in adverse events, (e.g., the decision to allow a takeoff without ensuring that the runway is clear of traffic on a foggy night). While the majority of aviation incidents may be due to tactical and coordination problems, decision errors are those most likely to have serious consequences for flight safety. Hence, effective decision-making is very important for maintaining safe operations, especially under threatening conditions (Helmreich et al. 2001).

Decision-making requires a lot of cognitive effort which adds up to the existing taskload of controllers and creates the conditions for poor human performance. For this reason, Eurocontrol and the Federal Aviation Administration (FAA) have established standard procedures and checklists for a range of typical scenarios in order to simplify everyday decision-making. However, poor decisions still occur in routine operations (Orasanu 1993) due to adverse work conditions that increase risk (e.g., high workload, bad weather or heavy traffic). Controllers are usually the last line of defense as procedures and automated systems cannot always cope with all emergencies. Therefore, several studies have looked into the critical question of how flight crews and controllers make decisions, what factors make decisions difficult, and how poor decisions can be recovered from (Flin et al. 2003; Thomas 2004; Burian et al. 2005; Li and Harris 2006; Nikolic and Sarter 2007).

Earlier research in normative or analytic decision-making has tended to view decisions as choices between options made in an analytical manner so that optimal solutions are reached. Normative approaches have based their models of decision-making on experimental studies with subjects in well-formulated situations where enough time was available to make the right decisions. Recent research in naturalistic decision-making (NDM), however, has shown that making choices between options is not the only cognitive activity that underlies decision-making (Zsambok and Klein 1997; Klein 1998; Cohen et al. 1996). Experienced flight crews and controllers can spend more time assessing the situation and classifying the problem to categories for which standard procedures are available or well learned. In this sense, decision-making involves several activities, such as recognizing threats, assessing situations, choosing options, or following rules when a typical choice is encountered.

In addition, some decisions may need authorization by supervisors or coordination with other airspace sectors to avoid side effects in other areas of work. In emergency situations, decision-making may involve a dynamic interaction between controllers in separate airspace sectors in order to decide the route upon a landing diversion to an airport. Hence, team coordination may affect decision-making and especially the way that controllers share information, communicate their intentions and manage conflicting interests. Research in team decision-making has shown that coordination, shared team models, and replanning are also important cognitive functions that are part of what is called *decision-making* (Entin and Serfaty 1999; Entin and Entin 2000; Flin et al. 2003; Salas et al. 2008). Hence, decisions appear to differ in the degree to which they call upon different types of cognitive functions. For instance, a decision how to resolve a conflict may not require the team processes necessary for handling an emergency scenario (e.g., diverting to another airport due to bad weather on destination).

The nature of cognitive functions involved in decision-making depends on the structure of the decision task and the situational characteristics: How familiar is the problem? How much time is available to make a choice? Are the options described in a standard procedure? Are there any other stakeholders that may be affected by the decision? Controllers may encounter a variety of decision problems calling for different approaches. According to the cognitive continuum theory (Hammond et al. 1987), good decisions depend on the correspondence between decision strategies and situational demands. Analytical decisions applied to ambiguous and time-critical situations may be ineffective whereas other more intuitive decision strategies applied to situations where time is available may be suboptimal. Therefore, the good decision maker has to assess the situation and select a decision strategy that matches the situational demands.

The aim of this chapter has been to explore the cognitive functions involved in different decisions and propose a decision process model that considers both individual and team aspects of behavior required for a broad range of air traffic control (ATC) tasks. A review is presented first of earlier models of decision-making, ranging from normative or analytical approaches to more intuitive or naturalist ones. Subsequently, the taskwork/teamwork for effective and adaptive management (T^2EAM) model is presented for making decisions in a team environment. The T^2EAM model has been based on an experimental study and validated in the context of ATC (Malakis et al. 2010a, b).

4.2 Theoretical Foundations

According to the cognitive continuum theory (Hammond et al. 1987), decision models range from rational or normative models to intuitive or naturalistic models, with many others falling in a quasi-rational area that involves a mixture of both approaches. Multi-attribute utility theory (MAUT) has been a proponent of the rational approach (e.g., the DECIDE model in Robson 2008) while the skill-rule-knowledge model (Rasmussen 1983) and the recognition-primed-decision model (Klein 1989) have taken a naturalistic approach at the opposite end. In the middle area, choices between options can be made in a mixed manner (e.g., elimination-by-aspects, heuristics, and so on). The cognitive continuum theory (Figure 4.1) proposes that there is no one best way of making decisions but that good decisions should be matched to the type of problem at hand. For instance, problems that are ill structured (e.g., ambiguous situations, delayed cues, unclear objectives) invite a mode of cognition that is more intuitive than analytical. In contrast, problems that are well structured, with sufficient reaction time and auditability requirements, invite rational models of decision-making.

The purpose of the review of decision-making models has been twofold, that is, (1) examine decision types that are appropriate for different situations and (2) identify a range of cognitive functions that are related to decision-making (e.g., problem recognition, situation assessment, risk

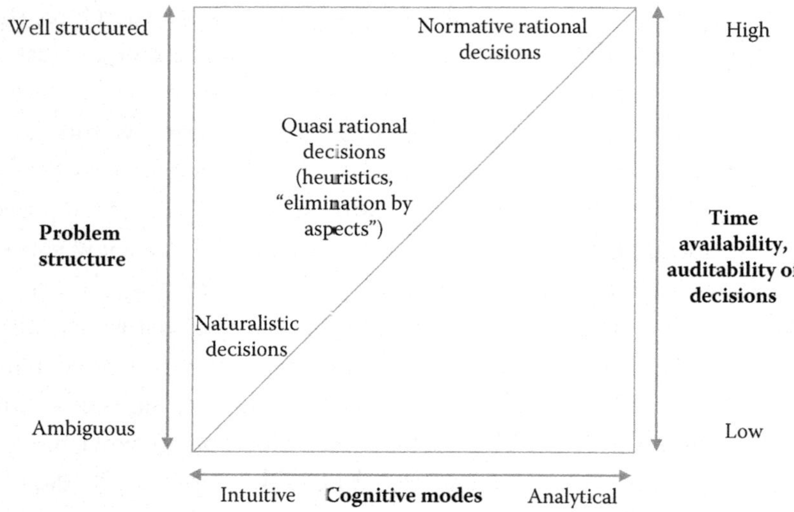

Figure 4.1 Matching decision models to different situations in a cognitive continuum.

assessment, decision-making, team coordination, and recovery of poor decisions). The identification of cognitive functions is very important for proposing a decision-making model that could be tailored to different situational demands in the air traffic management (ATM) domain.

4.3 Rational or Analytical Decision-Making

The normative approach has been based on the model of the rational man who receives all necessary information to make decisions, knows well the alternative options, and has adequate time to trade-off options and find the optimal one. Classical decision-making was based on the utility principle, combining Bayesian probability theory with multi-attribute utility theory. In general, classical decision-making follows five stages:

1. Definition of problem parameters
2. Collection of relevant information
3. Identification of available options
4. Assessment of all options based on predetermined criteria
5. Selection and implementation of the optimum option

Classical decision-making has a solid mathematical foundation and it has been published in two variations. The first variation considers only discrete options where there is very little uncertainty in the selection process. In this case, the overall value of each available option depends on the magnitude of each attribute and the utility of each attribute.

This model uses the following mathematical formula:

Utility Equation

$$U(P)=\sum_{i=1}^{n} m(i)\,u(i) \tag{4.1}$$

Where:

- $U(P)$: overall utility of option P
- $m(i)$: magnitude of ith attribute
- $u(i)$: utility of the ith attribute
- n : total number of attributes

It is obvious that the optimum option is the one that maximizes the value of function $U(P)$. This model of decision-making is appropriate for static problems where uncertainty can be reduced to zero and the selection time frame is very long.

The second variation is similar to the first one but incorporates uncertainty in terms of probabilities of outcomes. As a result, the concept of overall utility is replaced by the overall expected value. In mathematical terms, the expected value of each outcome depends on the probability of the ith outcome occurring and the value of the ith outcome:

Expected Utility Equation

$$E(P)=\sum_{i=1}^{n} v(i)\,p(i) \tag{4.2}$$

Where:

- $E(P)$: overall expected value of option P
- $v(i)$: value of the ith outcome
- $p(i)$: probability of the ith outcome
- n : number of possible outcomes for option P

It is apparent that the optimum option is the one that maximizes the function $E(P)$.

The following simplified example illustrates the application of classical decision-making (Malakis 2009). Consider, for example, an approach control unit that faces a staffing problem for a certain shift. For a time period it is uncertain whether the expected traffic will be kept at high or medium levels (e.g., public events may attract more unscheduled traffic to the airport than events expected by the approach unit). Hence, the unit manager faces the dilemma whether the shift will be staffed with four or three ATCOs. By calculating the probabilities of high and medium traffic conditions, a decision tree can be constructed as in Figure 4.2.

According to the decision tree, the two options are compared according to their expected value scores as shown in Table 4.1.

From the above example, it appears that the optimal choice is the work shift with three controllers that achieves the higher expected value score.

Classical decision-making models can be used in cases where uncertainty can easily be assessed in an approximate manner and

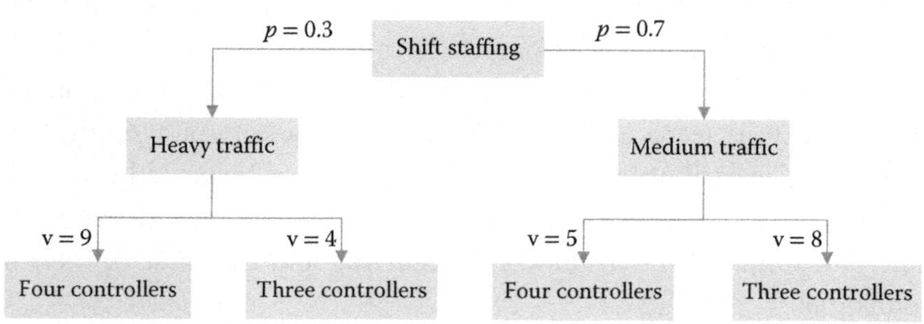

Figure 4.2 A decision tree for comparing teams of three or four controllers.

Table 4.1 Expected Values for Two Shift Options

		A SHIFT WITH FOUR CONTROLLERS		A SHIFT WITH THREE CONTROLLERS	
Options	*Probability of outcome*	*Value*	*Expected value*	*Value*	*Expected value*
	p	v	p(i)v(i)	v	*p*(i)v(i)
Heavy traffic	0.3	9	2.7	4	1.2
Medium traffic	0.7	5	3.5	8	5.6
Overall expected value			E(*P*1) = 6.2		E(*P*2) = 6.8

there is ample time for selecting and comparing options. In the aviation domain, a widely used classical decision model has been the DECIDE model (Robson 2008) which involves six stages for pilots to make good decisions in a logical manner, that is,

D : Detect that the action is necessary
E : Estimate the significance of the action
C : Choose a desirable outcome
I : Identify actions needed in order to achieve the chosen option
D : Do the necessary action to make a change
E : Evaluate the effects of the action

The DECIDE model has been extensively used in analyzing crew decisions in the aeronautical domain. It is a good example of the normative approach, which relies on the following assumptions:

- A clear definition of the problem exists that requires a good decision; there are no circumstances that may change the nature of the problem that requires setting new goals
- Decision-making is seen as trading off options while little attention is paid to situation awareness
- All relevant information is timely available, options are known, and time is sufficient to do the work; information that is unreliable or delayed may create problems in the estimation of the optimal solution and so does the lack of adequate time
- The outcome should be an optimal decision and not a good enough or viable decision
- A formal method (e.g., MAUT) should be applied to find the optimal solution; other informal methods are not acceptable (for example, heuristics)

For decades, classical decision-making has been fully accepted by the majority of researchers and practitioners. The rational approach to decision-making suggested that problems could be formally represented in a mathematical form, allowing optimization of resources. As a result, failures in problem-solving were attributed to the practitioners rather than the adopted method of work. In the period 1965–1985, decision researchers produced evidence that humans do not usually follow rational models of decision-making. New studies have built on the concept of bounded rationality (Simon 1957), that is, rationality limited by the tractability of the decision problem, the cognitive limitations of the human mind, and the time available to make the decision. In this view, decision makers act as satisfiers, seeking a satisfactory solution rather than an optimal one.

Simon's findings triggered new studies that looked into rational and other informal models of how experienced people make decisions. Kahneman et al. (1982) presented research indicating that humans use a range of heuristics that involve less cognitive effort than rational decisions, although the final decision may not be optimal. Among the well-known heuristics have been the availability heuristic (i.e., a reliance on a good solution that is readily available and worked well in the past) and the confirmation bias (i.e., a selective attention to information that confirms a preferred option). Heuristics are powerful tools in making decisions under time pressure because they do not entail the same cognitive effort that rational decisions do. Another quasi-rational model of decision-making proposed by the same researchers was elimination by aspects. According to this model, people often do not have time to consider and weigh all attributes of different options. In this situation, they would start by establishing a minimal criterion and eliminating the options that fail to satisfy it. Subsequently, another criterion can be selected to eliminate more options until a stage is reached where the last option satisfies all the criteria that remain. However, as options get eliminated in a serial fashion, people may miss an option that has a low score in the first few criteria but compensates with a highest aggregated score for all criteria.

4.4 Naturalistic Decision-Making

In middle 1980s, a new naturalistic approach to decision-making emerged that shifted research from laboratories to natural work settings (Klein et al. 1986). Prominent examples of the naturalistic approach include the skill-rule-knowledge model (Rasmussen 1986), the recognition primed decision model (Zsambok and Klein 1997), and the critical thinking model (Cohen, Freeman, and Wolf 1996). Classical decision-making theory has been criticized as cognitively incompatible with the decisions made by experienced practitioners in complex and dynamic organizations (Cohen et al. 1998). The emergence of the new paradigm boosted research in decision-making in complex organizations as it paid attention to several work constraints in the real environment, such as:

- Multiple forms of uncertainty
- Environments where the situation is constantly changing
- Information rich but noisy environments which increase the monitoring workload
- Goals that may evolve in time and may compete with other goals elsewhere in the organization
- Severe time constraints that force practitioners to act quickly
- High consequences arising from decision errors
- Decisions dependent on multiple practitioners working in distant locations

The ATM system represents a typical domain for the application of the naturalistic decision-making (NDM) paradigm since most of these constraints are brought into play, especially in emergencies and abnormal situations.

A proponent of the NDM approach has been the Skill-Rule-Knowledge (SRK) model which describes the decision-making processes that people use, depending on their level of expertise and the context of the situation (Rasmussen 1986, 1993). The SRK model (or decision ladder) has been very popular because it emphasizes the role of other cognitive functions (e.g., situation assessment and planning) in decision-making and identifies several shortcuts for speeding up decisions. The SRK model (Figure 4.3) postulates that practitioners can make decisions at three levels, depending on the competence levels and the situational characteristics (i.e., familiarity, response time, and availability of rules).

Most situations in the ATM domain have been previously encountered by experienced controllers and flight crews while supportive checklists and procedures exist for making suitable choices, especially under time pressure. Experts make decisions based on rules that are externally stored (e.g., procedures) or internally stored in their memory (e.g., acquired through experience and training). Rule-based decisions entail recognition of the problem symptom that helps experts classify the problem and identify appropriate rules to make a choice; usually a standard procedure may present practitioners with two or three

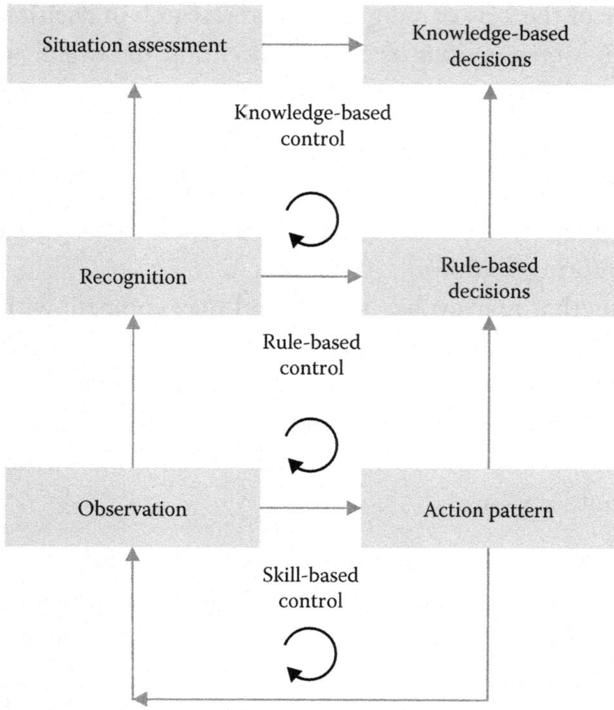

Figure 4.3 An adaptation of the Skill-Rule-Knowledge model.

options and a method to make a quick choice. With extensive practice, however, practitioners are able to associate an observed symptom with a preferred action pattern in an effortless way. Hence, behavior at the skill-level allows practitioners to time-share many tasks and speed up their performance. However, "perception–action" patterns may produce certain lapses and omissions that are difficult to capture since attention is shared among several tasks.

The most challenging decisions are made at the knowledge-based level where controllers are faced with unfamiliar situations and have to rely on their knowledge to identify candidate options and select the most appropriate one. Knowledge-based decisions may involve analytical comparison of options or may require the creation of a new option. The SRK model is not dogmatic about the decision criteria that experts use. It does not require people to find an optimal decision through a formal comparison of options. Practitioners may use their own knowledge and judgment to make decisions using quasi-rational approaches (e.g., "elimination-by-aspects"). The important thing about knowledge-based decisions is that practitioners have to spend time in assessing the situation before making a decision. Situation assessment may involve thinking about the causes of the problem, projecting how the situation may evolve when an option is selected or even thinking how to share knowledge between team members to avoid side-effects.

Human cognitive control can be shared between two levels, such as when practitioners apply a set of rules but also try to interrogate them using deeper knowledge about the system. In this sense, knowledge can take a supervisory role in the application of rules and even provide a ground for learning from experience. Human heuristics in decision-making can be seen as shortcuts at the knowledge level where practitioners resort to rule-based decisions to speed up their cognitive processes; however, the risk remains that short-cuts prevent practitioners from introspecting their rules. Effective decision-making in safety critical domains depends on all three levels of cognitive control of the SRK model.

Another decision model that has been very popular in the NDM paradigm regards the recognition primed decision (RPD) model (Klein et al. 1993, 2004). The main tenant of the RPD model is that people make decisions by drawing analogies from their experience once they recognize that have encountered similar situations in the past; this pattern-matching process allows practitioners to identify suitable options and courses of action that may still apply to the current situation (see top loop in Figure 4.4).

In some cases, however, practitioners may wish to explore alternative options either because they are not certain about the situation or because the cost of errors are very high. In the RPD model, options are evaluated by imaging how their consequences might unfold in the future according to a mental model of the system. All options are evaluated in a serial fashion until a suitable one is found that is fit for purpose. Hence, the middle-loop of evaluation (Figure 4.4) relies on a process of mental simulation that requires a good knowledge or model of the system.

The most complex case is where the situation looks unfamiliar, which requires a thorough assessment of the situation or fault diagnosis. The situation assessment loop (at the bottom of Figure 4.4) can be supported by a good mental model of the system that guides the search for further information and helps practitioners identify possible

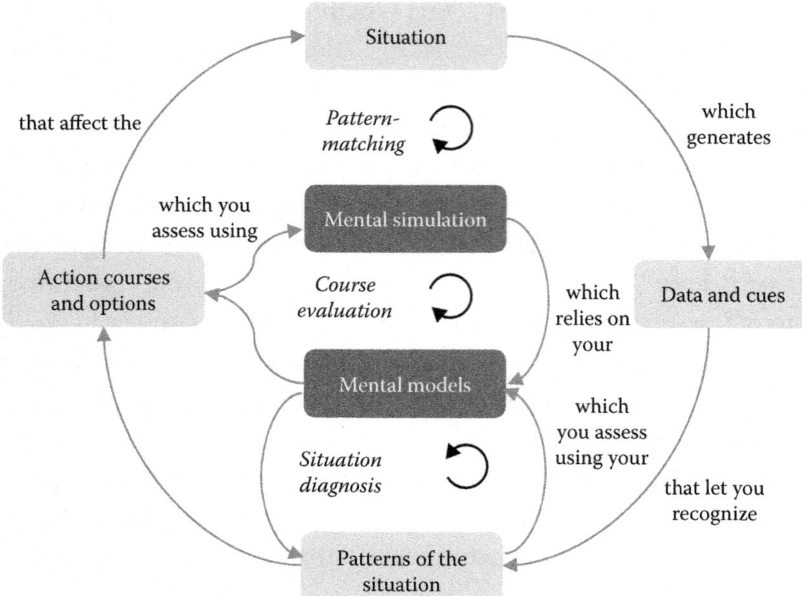

Figure 4.4 The Recognition Primed Decision (RPD) model showing the functions of pattern-matching, situation diagnosis, and course evaluation. (From Klein, G.A., *The Power of Intuition,* Currency Books, New York, NY, 2004.)

causal factors of the problem. Very often, the diagnosis of the situation is likely to be followed by an exploration of new options that are evaluated by mentally simulating their effects.

In contrast to the analytical decision theories, the RPD model includes consideration of situation assessment that interacts with option evaluation in several ways. For instance, situation assessment can make option evaluation easier by reducing the set of candidate options to choose from. Hence, the RPD model provides an integration of situation assessment and decision-making that varies according to the characteristics of the context of work.

In novel situations, where no familiar patterns exist, proficient practitioners supplement situation assessment with a supervisory process that verifies the results of mental simulation and corrects any problems; this supervisory process has been referred to as metacognition. This higher order cognitive function has been addressed by the recognition/metacognition (R/M) model (Cohen et al. 1996). The R/M model describes a set of critical thinking strategies that supplement recognition processes in rapid decision-making. Metacognition involves a number of cognitive strategies regarding whether it is worthwhile to think more about a problem, how to critique a situation model for incompleteness, conflict or unreliability, and how to improve it by collecting new information or revising assumptions.

The origins of recognition/metacognition model are traced into a United States Navy (USN) research program known as tactical decision-making under stress. TADMUS was launched after a tragic accident where the US Aegis cruiser Vincennes shot down a commercial Iranian airliner, killing 290 passengers (1989). The accident was attributed to many work factors with the most important being flaws in the decision-making process of practitioners. Among other approaches, the R/M model has been developed and validated in the context of the TADMUS program.

The recognition/metacognition model claims that practitioners build a mental model of the situation and of suitable plans that are subject to critique and correction. These metacognitive functions depend on the characteristics of the situation (i.e., time availability, high stakes, and uncertainty). Figure 4.5 shows how the functions of model building, critiquing, and correcting are adapted to the work environment as follows:

1. *Carrying out a quick test,* which rapidly assesses whether it is worth taking more time for critical thinking rather than acting immediately on the current recognition of the situation
2. *Critiquing the results of recognition* in order to handle three kinds of uncertainty:
    a. incompleteness in understanding the situation or in formulating response plans
    b. conflicting evidence or goals
    c. explicit or implicit assumptions made to simplify the problem
3. *Correcting flaws* by shifting attention to other evidence or making other assumptions.

Metacognition occurs when the benefits associated with critical thinking outweigh its costs. This is likely to be the case when the situation is novel (i.e., the uncertainty is high), the cost of errors are considerable but there is sufficient time for critical thinking. The quick test considers these factors and, if conditions are appropriate,

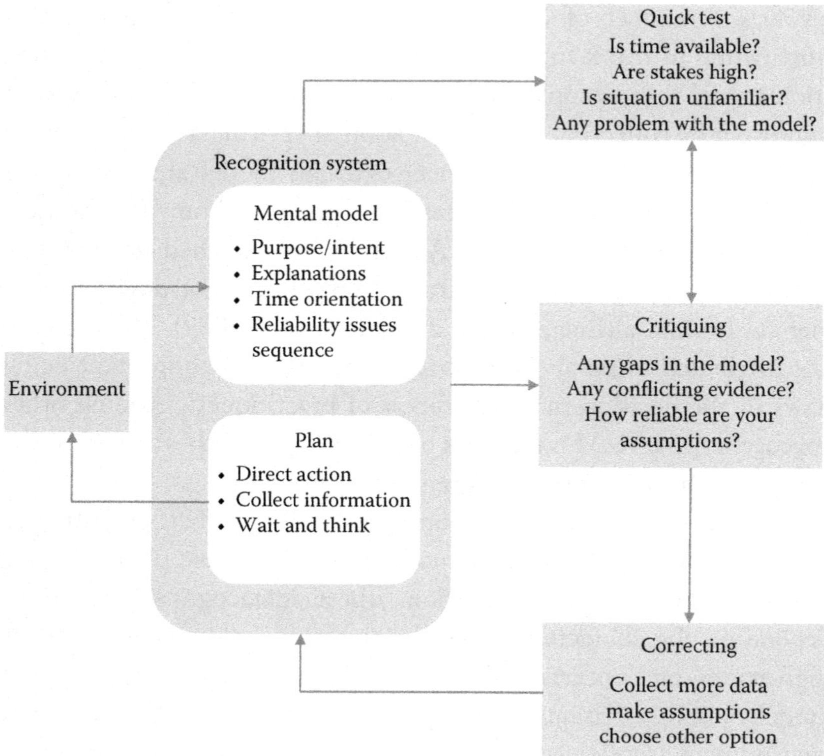

Figure 4.5 The Recognition/Metacognition Model. (From Cohen et al., *Human Factors*, 38, 206–219, 1996.)

interposes a process of critical thinking. The cornerstone of the R/M model is a critique of our current understanding of the situation and our earlier decisions. Critiquing models of the situations or decisions is a means of making decisions when uncertainty is high but there are expectations that additional information will be available for later improvements.

4.5 Toward a Decision-Making Model in ATC

This section presents a decision-making model based on an experimental study undertaken by the authors that recorded strategies in making decisions in unfamiliar situations (Malakis et al. 2010a, b). NDM approaches have pointed out that decision makers are not solely concerned with selecting and comparing options; they are also likely to be involved in situation assessment. In critiquing their earlier decisions, practitioners should comply with the decision requirements and constraints set in operating procedures and organizational rules of conduct. In our effort to develop a comprehensive decision model, we have gleaned from the literature review several cognitive functions that may be related to decision-making such as the following:

- Recognizing and anticipating threats
- Building a model of the situation and defining the problem
- Identifying and evaluating options—either by direct comparison or by mental simulation
- Complying with procedures organizational rules
- Critiquing mental models and earlier decisions, especially when uncertainty is high

It is apparent that emergencies present controllers with many challenging issues. Threats and concerns must be detected promptly because time is limited and original plans must be modified as the situation evolves over time and new threats may appear. There is a need for assessing the situation continuously in order to fill in gaps, correct explanations, and clarify assumptions. This calls for strategies in anticipation, situation assessment, coping with uncertainty, and managing workload. These cognitive functions in decision-making are important sources of resilience that provide a good basis for debriefing controllers after critical events, developing training programs, and comparing alternative automation designs.

In order to develop a decision-making framework of ATC operations, a field study was conducted involving *ab initio* training of novice controllers and refresher training of experienced controllers in two major European sites (Malakis et al. 2010a, b). The approach was based on ergonomic research methods widely adopted in aviation (Seamster et al. 1993; Flin et al. 2003), the military (Cohen et al. 1996) and process control (Kontogiannis 1996; Woods and Hollnagel 2006).

4.6 Taskwork Functions and Strategies

In the early 1990s, the NDM approach produced new and refined decision-making models including, the recognition-primed decision model (Klein 1989; 1998) and the recognition/meta-recognition (R/M) model (Cohen et al. 1996). To develop an inventory of cognitive strategies for ATCOs, the two models were integrated with the contingent operator stress model (COSMO) (Kontogiannis 1996, 1999a) and the anomaly response model (Woods 1994; Woods and Hollnagel 2006). Specific models of performance in ATC (Reynolds et al. 2002; Oprins, Burggraff and Weerdenburg 2006) and cognitive analysis of ATC tasks (Seamster et al. 1993; Kallus et al. 1999) have also been used to tailor the generic models of decision-making into the requirements of the ATC operational context.

Our model was initially termed taskwork and teamwork strategies in emergency air traffic management (T^2EAM) and was based on a core set of five cognitive functions: anticipation, recognition, uncertainty management, planning, and workload management. The initial model has been improved using data from later studies conducted by the authors and was changed into taskwork/teamwork for effective and adaptive management (T^2EAM). To establish a suitable structure for the taxonomy of taskwork skills, we adopted the format of the European behavioral marker system for rating pilot's nontechnical skills—NOTECHS (Flin et al. 2003). This scheme has a three-level hierarchical structure of functions (e.g., recognition), strategies (e.g., noticing distinguishing cues), and behavioral markers (e.g., identifying military aircraft as a threat). Performance can be rated at both the function and strategy levels, depending on the purpose of the assessment and the amount of feedback required. The behavioral markers were intended to help external raters to recognize the types of behavior associated with the performance of each strategy.

The five cognitive functions and accomplishing strategies for dyadic teams in ATC are presented in Table 4.2. The behavioral markers can also increase the reliability of raters in the assessment of approach and en-route controllers during simulated and real-life scenarios (Malakis et al. 2010a, b).

The following sections provide a detailed description of all taskwork functions and strategies.

4.6.1 Recognition

Recognition is a cognitive function that enables controllers to recognize signs of impending emergencies and build a model of the

Table 4.2 Taskwork Functions and Strategies (T^2EAM)

COGNITIVE FUNCTIONS	COGNITIVE STRATEGIES	BEHAVIORAL MARKERS
Recognition	Noticing cues and recognizing states	• Detecting unauthorized deviations in altitude, vertical/horizontal speed and route • Detecting cues of CNS malfunction/degradation
	Projecting and estimating states	• Projecting the most probable flight path of an "emergency descent" and the aircraft that will be affected during the descent • Projecting the aircraft that will be affected by a Non-RVSM (reduced vertical separation minima) flight
Modeling and critiquing	Critiquing models of the situation	• Requesting route and/or altitude information for military traffic that has not specified their intentions • Requesting information about the extent and expected duration of a system degradation • Requesting altitude verification for aircraft that appears to experience a mode C malfunction
	Critiquing goals and responses	• Testing time parameters related to the unfolding situation (e.g., judging the available time for planning/acting of an emergency) • Deciding on immediate actions or questioning the nature of an observed altitude and/or route deviation • Deciding on acting or allowing traffic alert and collision avoidance system (TCAS) to resolve a conflict situation
Anticipation	Acknowledging threats	• Increasing the lateral and/or vertical distance between military and civilian traffic • Avoid vectoring close to areas of known or observed military activity • Placing a visual reminder on a suspected threat (e.g., military traffic aircraft with malfunctioning transponder)
	Exploiting less busy periods to plan	• Utilizing velocity leaders of more than 1 minute to perform planning • Timing the initiation of an altitude change of aircraft experiencing radio communication failure (RCF) • Estimating time horizons of emerging events due to the unfolding situation
Planning	Standard planning	• Providing increased vertical and lateral separation between aircraft experiencing an emergency and the other traffic • Utilizing CNS equipment in different ways • Following standard procedures (ICAO and/or local) for the occurrence
	Contingency planning	• Timely requesting the use of a restricted airspace for military traffic • Writing a list with all the sectors/units that must be informed and kept updated during an emergency • Employing a "sterilized sector" policy (i.e., not accepting additional traffic from other sectors/units)

(*Continued*)

Table 4.2 (*Continued*)

COGNITIVE FUNCTIONS	COGNITIVE STRATEGIES	BEHAVIORAL MARKERS
Workload adaptation	Prioritizing tasks	• Prioritizing a number of simultaneous incoming and outbound calls (e.g., what units/sectors need to be informed first about the situation) • Shifting attention to emerging tasks • Shifting attention between emergency and normal tasks
	Managing interruptions and distractions	• Avoiding nonessential team communication in high tempo periods • Gauging interruptibility of other team members (e.g., speeding up or postponing communications in relation to the tempo of activities or waiting for a pause in controller–crew communication) • Using hand gestures to acknowledge information and/or reject attempted communication in high tempo periods

situation. Emergencies can occur suddenly when the flight crew formally declares an emergency or evolve slowly over time. In the former case, recognition implies some sort of classification of the emergency and its association to a well-known recovery response; in a sense, decisions are primed by the recognition of a familiar problem (Klein 1998). In the latter case, where the emergency is evolving over time, a model of the situation must be built on the basis of cue observability (i.e., prominent versus subtle cues) and patterns of evolution. Prominent cues are salient data or events that capture the attention of controllers and become conclusive evidence from the onset of an emergency. For instance, an unexpected vertical deviation of an aircraft is a prominent event that may signify an emergency descent. Moreover, the time evolution of this event is quite rapid since it is associated with one radar scan that takes less than 4 seconds. In contrast, subtle or weak cues may not alert controllers of any critical problems as they seem to be unrelated to the main course of events. For instance, a crew request for meteorological terminal air report (METAR) information (meteorological terminal air report) of a nearby airport may be a sign of an unstated intention of a flight diversion due to a technical malfunction. This is also an example of a slow evolution of events since it may take a few minutes before the flight crew explicitly states their intention to divert to a nearby airport.

It appears then that recognition strategies include both elements of cue observation and mental simulation (i.e., projecting ahead and estimating states):

- *Noticing cues and recognizing states* refers to the ability of controllers to timely and accurately detect early cues of an impending problem. Important cues may refer to unexpected course deviations, equipment failures, and even the absence of certain events. Field study data provide some evidence that vertical track deviations are more difficult to observe than lateral deviations. In some cases, the safety of air traffic may be compromised by CNS system malfunctions (e.g., limited radar coverage may endanger traffic in a stripless environment); hence controllers should be able to recognize any cues that implicate a degraded mode of system operation. In other cases, the absence or termination of certain events may be an indication of abnormal situations. For instance, the disappearance of a number of aircraft tracks from the radar screen may indicate a limited radar coverage of the airspace sector.
- *Projecting and estimating states* refers to the ability of controllers to accurately play out the progression of abnormal events and assess their consequences. People make sense of information by making connections to the larger context of present and future states of the system. This requires a mental projection of the system state into the future and the assessment of adverse events. However, state projection cannot be easily observed in normal operations or even simulated scenarios. In some cases, it may be inferred from a number of preparatory activities taken by controllers that seem unrelated to the present situation. Projecting and estimating states calls for coherent mental models that require extensive experience in ATC.

Recognition strategies are based on a mental model of the airspace (Mogford 1997; Reynolds et al. 2002) that classifies aircraft into categories (e.g., aircraft heading to the same destination), portrays critical points of converging traffic, and identifies nonstandard flows (e.g., military traffic, aircraft performing aerial work, rescue and firefighting operations, unmanned aerial systems flights).

4.6.2 Modeling and Critiquing

This cognitive function enables controllers to assemble and critique a model of the situation and the associated safety-related goals. Emergencies and abnormal situations are closely associated with information uncertainty due to their dynamics. Controllers have to assemble a model of situation, formulate goals, and correct any tentative explanations or assumptions by seeking appropriate information.

In general, flight crews may be reluctant to provide conclusive information during emergencies since communication with the ATC units is not their first priority; even if they were willing to communicate their status, it may not be technically feasible. For instance, in the event of an explosive cabin decompression, a flight crew may experience temperature changes, extreme noise levels, and a cluster of objects flying inside the cockpit; in addition, they may be wearing oxygen masks, which hinders communication with controllers. In cases where direct communication with flight crews is temporary terminated or corrupted, controllers will have to rely on traffic information derived from their workstations. Hence, controllers may have to construct a model of the situation on the basis of a minimal set of information on their radar screens.

Uncertainty management can be handled by building and critiquing models of the situation as well as their implicated plans of action:

- *Critiquing models of the situation* requires controllers to build a coherent and complete model of critical events and available resources over a period of interrogations. In particular, emergencies and abnormal situations generate uncertainty and increase complexity. Certain critical cues may be inaccessible and controllers shall have to assemble a coherent model of the unfolding situation without pertinent information. In an emergency descent scenario, for example, controllers may first notice some cues indicative of a vertical deviation but be unable to reach a conclusion without additional information. In trying to make sense of the situation, controllers must tolerate uncertainty and assemble a model of the situation based on assumptions and expectations that can be tested later on. For instance, if an explosive cabin decompression is suspected then they should anticipate new flight crew communications below 10,000 ft.

- *Critiquing goals and responses* requires controllers to build a model of goals at stake and suitable responses for handling the situation. A coherent and flexible set of goals is expected to be formed that will be open to revision as the situation unfolds and new cues are presented. Although the overriding goal will always be the safety of the crew and the passengers on-board, controllers will have to make a number of decisions at other levels (e.g., they may have to decide whether to take immediate action or continue investigation).

Managing uncertainty has been defined according to the R/M model (Cohen et al. 1996) which critiques models and goals in order to make them complete, consistent and reliable.

4.6.3 Anticipation

Controllers should be able to anticipate threats that may come into the fore in the near future and proactively mitigate their consequences. Threats can be defined as events that occur beyond the influence of controllers, increase operational complexity and endanger traffic (ICAO 2005b). Typical threats include: military activity at the borders of a sector, heavy traffic, adverse meteorological conditions (e.g., thunderstorms, low level windshear), airports surrounded by high mountains, congested airspace, aircraft malfunctions, and errors committed by other people (e.g., flight crews, ground or maintenance staff). Anticipation figures as a prominent cognitive function in many performance models in ATC (Reynolds et al. 2002; Oprins et al. 2006).

Anticipating how threats may endanger traffic is a challenging task that requires extensive experience, especially when traffic patterns increase in complexity. For instance, it is essentially unknown when a military aircraft will deviate significantly from a preplanned route and endanger air traffic. In addition, several forms of task difficulty (e.g., sequencing versus crossing traffic patterns) increase uncertainty and require more adaptations. Gronlund et al. (2005) found that approach controllers could anticipate threats in a sequencing problem and could successfully use contingency plans to handle most of the aircraft heading to the same destination. However, when task difficulty was increased in a crossing problem, controllers encountered more uncertainty and required more adaptations of their plans as aircraft were heading to different points and were crossing each other's tracks.

Anticipation involves acknowledging current threats and planning how to handle threats:

- *Threat acknowledgement* requires timely detection and handling of threats. In some cases, it is not obvious whether a traffic activity would give rise to a threat (e.g., route or altitude deviations). Furthermore, as controllers are dealing with several threats on a daily basis, they may become desensitized to certain types and levels of threats. Experienced controllers may use reminders and computerized aids to support their anticipation of evolving threats. For example, they may use velocity leaders to project the lateral path of a flight to the next few minutes and determine the presence of threats. Alternatively, they may use the HALO function that displays a circle around the aircraft of concern of a specified radius (e.g., 2.5 to 5 Nm). An air track with a HALO circle on it can remind controllers that a threat has been detected and its significance acknowledged.
- *Exploiting less busy periods to perform planning* becomes a proactive strategy that allows controllers to utilize low-tempo periods to plan how to handle threats. Planning entails several activities such as envisioning the system state, organizing available resources, and specifying suitable actions. Normally there are quite a few low-tempo periods in order to arrange when to handle threats since many peak traffic periods are known in advance. Responding to a threat is not always a straight forward activity, as it might be in cases of small changes in heading. Sometimes, a more complicated response is required that involves changing the sector exit points for a number of aircraft, which implies more rerouting instructions and coordination efforts for a number of aircraft.

4.6.4 Planning

Controllers have to make plans and in certain cases may replan their actions in order to cope with the demands of an unfolding situation.

Planning may take the form of standard or contingency planning. Depending on the situation, a minimal set of prescribed scripts are normally available to all ATC units. Controllers are trained in certain emergencies and abnormal situations while annual refresher training is part of their competency scheme. Although standard planning is normally available, in many cases the need for contingency planning crops up. It may be a textbook case or an abnormal case but some characteristics of the situation may warrant an additional form of precautionary planning in order to counteract a possible escalation of the situation.

Planning may take different forms depending on the type of situation confronted with:

- *Standard planning* involves the use of operational scripts or actions that are available through training and documentation. Standard planning activities usually follow the identification and classification of the problem into categories for which appropriate plans are available. Even in unfamiliar situations of high uncertainty, standard planning may be still required in the form of providing emergency separation, putting aircraft on holding stacks awaiting the evolution of the emergency, informing the supervisor, and coordinating with adjacent units.
- *Contingency planning* requires controllers to structure the problem and think of new ways how to stabilize or counteract an escalation of the situation. Contingency planning may range from simple and short-term actions to elaborate action sequences. An example of a simple action could be the preparation of a list of the units to be informed in case of emergency. A more complicated action may require controllers to coordinate with adjacent sectors how to transfer traffic in a restricted time window and through nonstandard sector points.

Expert en-route controllers reported that they usually formulate a backup plan in case their initial strategy does not work (D'Arcy and Della Rocco 2001). Although a backup plan may produce a higher workload for controllers, it also provides them with a quick alternative on which to rely. Anticipation and planning are closely related and in some cases a clear distinction of their constituent elements may be difficult to make. In a sense, anticipation and planning involve a stance toward minimizing uncertainty by thinking ahead of possible threats and plans on how to cope with them. This is in contrast to managing uncertainty, where plans are adapted on the fly to manage uncertainty as the situation evolves (Grote 2009).

4.6.5 Workload Management

The main objective of the workload management function is to keep workload below a saturation point. Workload adaptation strategies enable controllers to order tasks, switch attention between tasks, and respond to interruptions. From the onset of an emergency, the workload may increase significantly due to changes in the number of tasks, the available time, and the importance of the tasks to be handled. Workload management regulates how to keep track of tasks, when to interrupt tasks, when to resume tasks, and so on.

Hence, managing workload entails the following:

- *Prioritizing tasks* in order to sequence tasks in a timely and accurate fashion. Abnormal situations change the taskload and create a need for new or modified task sequences. Controllers are expected to devote adequate attention not only to an ongoing emergency but also to the normal traffic in their sector in order to avoid critical conflicts or other safety occurrences elsewhere.
- *Managing interruptions and distractions* in order to maintain a stable flow of information exchanges, especially when some practitioners are involved in an emergency. Controllers forward critical information to their supervisors and all affected units and, in return, they receive instructions for coordination and other pertinent information. In this sense, judging when and how to structure information exchanges becomes a factor in handling a critical situation.

Distractions and interruptions are quite common in real world settings and affect task performance in a negativel way (Loukopoulos et al. 2001). Managing workload is likely to have an impact on several cognitive strategies in ATC. Rantanen and Nunes (2005), for instance, found that certain aspects of workload management, such as mental effort and time cost in performing additional tasks, are likely to affect the scanning patterns of controllers. Managing distractions and interruptions is also likely to allow controllers more time for critiquing their models of the situation and for anticipating future threats.

4.6.6 The Taskwork Model

A proposal of how the cognitive strategies interact with each other is shown in a diagrammatic form in Figure 4.6. Familiar situations allow experienced controllers to retrieve a standard plan from their memory which can be put readily in operation. Some degree of modeling of the situation may be required especially with the increase of uncertainty which can be anticipated to a large extent (e.g., impeding threats may be anticipated followed by standard planning). Unusual situations present controllers with a higher degree of uncertainty, which may require a thorough modeling of the situation and consideration of several contingencies in planning. A critique of understanding and revision of plans may also be necessary as the situation evolves in unexpected ways. The ways that controllers critique their models and revise their plans usually depend on the context of work and remain an important research issue. In many cases, controllers have to proceed with contingency planning for how to react even without a complete understanding of the problem. The interaction of cognitive functions is regulated by the current level of workload; heavy traffic and a large number of tasks, for instance, may reduce opportunities for threat anticipation and standard planning, hence forcing controllers to rely on critiquing and replanning.

The taskwork model (Figure 4.6) can be used to analyze several real-life and simulated occurrences (Malakis 2009). Three case studies are presented below to illustrate the taskwork model.

The case of a CNS system degradation can be used to illustrate the cognitive functions and strategies involved in taskwork as follows:

- *Recognition:* Controllers begin to detect cues of system degradation by observing indications associated with correlation failures in their displays (e.g., loss of flight level information)

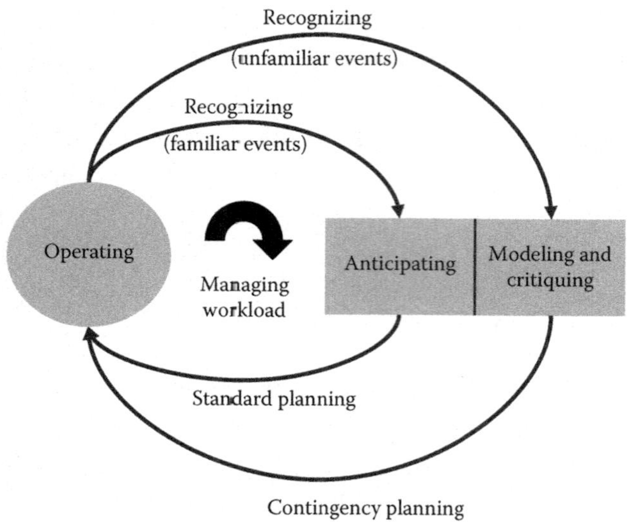

Figure 4.6 A flow of cognitive functions in taskwork (T²EAM).

- *Modeling and critiquing:* Controllers may be uncertain about the system status, duration, and magnitude of degradation. To this end, they may request information about the functional status of the system from the supervisor or make certain assumptions on the basis of their experience of similar events. Consequently, controllers may proceed to the next stage without having completely resolved all levels of uncertainty (i.e., residual uncertainty). Thus, a compelling need to revisit this stage is created.
- *Anticipation:* Acting upon the situation should take into consideration the interplay between current threats (e.g., military activity inside the sector) and emerging threats (e.g., residual uncertainty about the extent and duration of degradation, coordination errors due to aircraft identity confusion in transferring procedures with other sectors).
- *Planning:* Controllers have to plan or replan their actions in order to cope with the unfolding situation. Standard procedures are normally available for passing/receiving handover from adjacent sectors. However, additional factors may be taken into account in order to plan how to cope with uncertainties (e.g., severity and duration of degradation, possible escalation of the event, traffic volume inside the sector). Contingency planning may involve: preparing lists of sectors

to be informed, employing a "sterilized sector" policy, rejecting additional traffic from adjacent sectors, and asking flow controllers to take air traffic flow and capacity management (ATFCM) measures to reduce inbound traffic.

- *Task workload management:* Workload increases as a function of the complexity of traffic due to loss of essential CNS data. Cognitive strategies may include: switching attention between normal tasks and situation related tasks (e.g., controlling the traffic versus receiving radar handover from other sectors) as well as judging interruptibility of other practitioners.

Another case for analysis concerns an airport that is closing for arrivals, requiring all inbound traffic to be placed in a holding pattern. Taskwork functions may involve:

- *Recognition:* Controllers receive notification that the airport is closed and that all inbound traffic already in their sector or about to enter will have to hold in their airspace.
- *Modeling and critiquing:* Uncertainty emerges in relation to the time parameters of the holding pattern due to the airport closure. If the exact time of the opening of the airport as well as the expected approach times (EATs) for all inbound aircraft are promptly provided, uncertainty is minimized. In many cases, the time parameters are not available at the outset of the situation, thus residual uncertainty is created.
- *Anticipation:* Acting upon the situation should take into consideration the interplay between current threats (e.g., military traffic) and emerging threats due to the holding pattern (e.g., stacking aircraft in a confined airspace with minimal buffer for deviations). Experience has shown that many aircraft will raise issues concerning minimum fuel and, thus, request priority for quiting their holding patterns in order to divert to other airports.
- *Planning:* Standard procedures are normally available regarding aircraft entering, maintaining, and exiting a holding pattern. Additional factors may be also considered in planning, such as availability of time parameters, aircraft in holding positions, and traffic density. Contingency planning may also be necessary, requiring controllers to coordinate with other sectors to hold some aircraft outside their area of regard, or coordinate with flow controllers to take measures to restrict inbound traffic.

- *Task workload management:* Issues related to switching attention between normal tasks and situation related tasks as well as judging interruptibility of other controllers.

A third case study regards an aircraft that is compelled to dump fuel due to a technical malfunction. Taskwork functions may include:

- *Recognition:* Controllers are informed by the flight crew (or by an adjacent sector) that the aircraft will have to dump fuel in their airspace.
- *Modeling and critiquing:* Uncertainty regards the potential cues of the problem and the duration of fuel dumping. If the reason and the duration are promptly provided then uncertainty is minimized. However, the duration of the emergency may be decided in the process of fuel dumping and may not be available from the onset of the situation.
- *Anticipation:* Acting upon the situation should take into consideration the interplay between current threats (e.g., high density traffic) and emerging threats due to fuel dumping (e.g., the fuel dumping aircraft needs an extensive area which affects other flights in the sector).
- *Planning:* Standard procedures are available for this event (e.g., ICAO procedures for separation minima between the fuel dumping aircraft and other aircraft). Contingency planning may be required to reexamine the problem, and coordinate with adjacent sectors how to make temporary use of portions of their airspace. The selection of the fuel dumping area is critical since this may increase the complexity of handling the other traffic in the sector.
- *Task and workload management:* Important issues here regard the switching of attention between normal tasks and situation related tasks as well as deciding when to interrupt other sectors and coordinate their efforts.

4.7 Teamwork Functions and Strategies

Teams usually function in environments where task complexity exceeds individual capacity, decisions have dilemmas to be traded off, information uncertainty prevails, errors may have critical consequences, and people's lives depend on collective insights (Salas et al. 2008). Problems in teamwork have been implicated in a number of high profile aviation accidents (e.g., midair collisions at Tenerife and at Ueberlingen). Emergencies present controllers with many challenging issues, such as synchronization of interconnected activities, information exchange within a short time window, balancing of workload, and changing of task priorities. In addition, safety-critical situations are not tolerant of errors, hence controllers should create their own opportunities for error detection and correction.

Traditionally, the focus of controller training has been on fulfilling regulatory requirements. Effective handling of emergencies was considered a natural byproduct of technical skills training. Fundamental elements of formal team training are provided routinely in most air navigation service providers (ANSPs) even though teams are expected to function to a high standard. In the past, research addressing team performance in ATM focused on separate behaviors in isolation from other teamwork processes (Cardosi 1993; Morrow et al. 1993). This approach cannot capture all teamwork functions and interactions, and this requires the development of a comprehensive model of teamwork in ATC.

More recently, a growing body of research in teamwork has emerged in the domain of aviation, military, and acute medicine. The challenges for developing a teamwork model include: (1) how to arrive at precise definitions that are not open to interpretation, (2) how to assist analysts in achieving an acceptable level of reliability in their ratings, and (3) how to tailor team functions to the characteristics of the situation and the culture of practitioners. In our effort to develop a model of teamwork performance for ATM operations, a literature review was undertaken which considered three earlier frameworks of nontechnical skills (NTS):

NOTECHS: nontechnical skills (van Avermaete and Kruijsen 1998; Flin et al. 2003).

ANTS: anesthetists' nontechnical skills (Fletcher et al. 2004; Patey et al. 2005).

The Big Five: Theoretical model applied by Salas et al. (2005).

It should be noted that the dyadic teams of ATCOs differ from flight crews where the captain is vested with the legal responsibilities of the flight leader. The T²EAM model uses the following teamwork functions: team orientation, coordination, information exchange (communication), error management, and change management. In order to help raters recognize the types of behavior associated with a particular scenario, several behavioral markers were specified in the form of "exemplar behaviors" as shown in Table 4.3.

Table 4.3 Teamwork Functions and Strategies (T²EAM)

COGNITIVE FUNCTIONS	COGNITIVE STRATEGIES	BEHAVIORAL MARKERS
Team orientation and shared understanding	Shared situation understanding	• Achieving a shared understanding of the situation with the least effort (e.g., working toward the same planning direction from the onset of the situation)
	Communication of intent	• Providing concise explanations and articulating the intent behind the instructions and/or clearances (e.g., stating the reason behind the selection of a particular diversion route and/or altitude)
Team coordination	Managing dependencies and adopting an assertive stance	• Advocating and defending one's own position as required • Assertiveness in coordination with adjacent sectors (e.g., avoiding the "nice guy" policy and saying "no" to certain requests from other sectors) • Adapting to capabilities of other team members (e.g., defensive vectoring from EC when CC is falling behind the traffic)
	Avoiding information garbling and interruptions	• Using structured formats for recording information • Keeping the size and duration of communication to the required minimum • Selecting low tempo periods to communicate non urgent information (e.g., entry level conflicts in the next 15 minutes, nonurgent altitude, and/or route coordination)
Information exchange (communication)	Unsolicited dissemination of information	• Providing information regarding restrictions or nonstandard patterns (e.g., military activity, route/altitude restrictions, and deviations from letter of agreement (LoA) • Reading the "signs" behind suspicious information that may imply the onset of an emergency (e.g., coordinating with the nearest airport the possibility of a diversion after receiving suspicious cues)
	Updates on situation status	• Providing timely and adequate updates on situation status and actions taken

(Continued)

Table 4.3 (*Continued*)

COGNITIVE FUNCTIONS	COGNITIVE STRATEGIES	BEHAVIORAL MARKERS
Error management	Error detection	• Using CNS resources to enable augmented situation monitoring (e.g., removing altitude filters, zooming out radar screen) • Altering between augmented and normal monitoring in regular intervals.
	Feedback for error correction	• Providing information regarding threats (e.g., military traffic, impending conflicts) that were unnoticed by other team members • Correcting minor coordination communication and human–machine interaction (HMI) errors made by other people
Task distribution (Change management)	Problem detection in task distribution	• Detecting high tempo periods of other team members • Detecting subtle cues about teammates falling behind the traffic (e.g., when unable to locate the aircraft track that is calling or when missing initial radio telephony (RTF) calls)
	Changes in task allocation	• Using nonstandard coordination to relieve the workload of others (e.g., sterilizing the sector to reduce workload) • Changing the sequence of tasks and the allocation of roles (e.g., requesting the watch supervisor to notify other units) • Performing tasks that the accountable controller is unable to perform due to system malfunctions (e.g., the CC may input a level change when the mouse device failed in the EC position)

4.7.1 Team Orientation and Shared Understanding

Team orientation and shared understanding is very important for the development of a common stance toward the direction of problem handling. It involves bottom–up communication and integration of information from divergent sources as well as top–down communication and clarification of intent (that is, concept of operation). This function involves sharing information about the situation and the intention to act:

- *Shared situation understanding* refers to the extent that team members reach a congruent assessment of the situation in a limited time period (Rentsch and Woehr 2004). Any misinterpretations of the situation have the potential to hinder subsequent orientation and planning. Shared understanding is based on shared models that enable team members to make accurate predictions, generate similar explanations, and engage in less overt communication in handling the situation (Entin and Entin 2000).

- *Communication of intent* refers to the extent that controllers communicate regularly and clarify their intentions. In a request for information, there is always the possibility for misinterpretation. Team members trying to follow a request have to figure out what the other team member really wants and to handle several issues that are not explicitly specified or explained (Klein 1998). Clarification of intent increases adaptation of people at the sharp end without any requirements for further authorization from the team leader.

4.7.2 Team Coordination

The nature of team tasks and the allocation of responsibilities can generate many dependencies that require orchestrated action to converge them all toward the master plan (Entin and Serfaty 1999). Keeping the size and duration of communication to the practical minimum is an essential feature of effective teamwork because the structure and length of communication can affect the effectiveness of information exchanges.

The team coordination functions can be achieved using several strategies, such as managing dependencies, adopting an assertive stance, and avoiding interruptions:

- *Managing dependencies and adopting an assertive stance* are very important for team coordination. ATC teams usually comprise two controllers who may function as leaders or followers, depending on the situation requirements. Consequently, each controller could advocate and defend his position in handling the traffic situation. For this reason, assertiveness is an important aspect of coordination especially with adjacent sectors. The role of assertiveness has long been recognized in the aviation sector and several training courses in crew resource management (CRM) have been used in commercial airlines (Helmreich et al. 1999).

- *Avoiding information garbling and interruptions* refers to the extent that controllers manage to exchange information without disruptions or garbling. The structure and length of communications can affect the quality of information exchanges. Keeping the size and duration of communication to the practical minimum is an important feature of effective teams. Garbling information with unnecessary and/or redundant elements may serve only to prolong communications and disrupt teamwork. Mature teams communicate with the least possible overt patterns, using concise operational language at appropriate periods. In mature teams, the CC is able to judge accurately the interruptibility status of the EC and pass him/her with nonurgent information in relatively low workload periods. In addition, this is done with the most concise manner, avoiding ambiguities and follow up explanations. Expert shuttle mission controllers (Patterson et al. 1999) were found to listen in on relevant information exchanges with adjacent sectors and judge the interruptibility of others before communicating with them. By listening to voice loops, mission controllers were able to time their coordination, either by speeding up or postponing their communications.

4.7.3 Information Exchange—Communication

This cognitive function refers to proactive information disseminated between controllers as well as regular updates on situation status. Communication is depended mainly on information exchanges among team members and requires sufficient time and cognitive resources. Critical situations increase the need for dissemination of information about current activities (i.e., what is needed now) and anticipated activities (i.e., what will be needed next). Kanki and Palmer (1993) found that communication patterns may change as a result of changes in the criticality or the workload of the situation.

The most essential strategies of information exchange are providing unsolicited information and regular updates on the situation:

- *Providing unsolicited information* refers to the extent that controllers provide information in advance, without any prior requests. The handling of a critical situation increases the need for information dissemination between team members. Unsolicited information refers to proactive information that facilitates near-future activities and planning. Anticipating information for future activities is an important proactive strategy. The ability to disseminate proactive information before it is requested—and preferable during low tempo periods—reduces the number of communication acts and facilitates planning by broadening the time horizon. Stout et al. (1999) found that providing unsolicited and proactive information increased team performance in a complex military flight task. More importantly, there seems to be a skill associated with shifting between explicit and implicit communication modes which was found to improve team performance in naval surveillance teams (Entin and Serfaty 1999).

- *Providing updates on situation status and its management* is very important in managing abnormal situations. The handling of critical situations can rarely be accomplished in a single stroke of planning and acting, irrespective of the level of maturity of the teams involved. In dynamic work, the requirements of the situation may change as critical information becomes available, missing, or obsolete. In order to achieve a continuum of coordinated action, teams must meet the emerging need for regular updates on situation status and management. Effective teams are able to articulate a concise update on the situation status in certain critical periods and thus establish a continuum of coordinated action.

4.7.4 Error Management

Error management refers to certain self-monitoring functions that enable controllers to detect errors and provide feedback for error correction. Analyses of incidents and simulated emergencies have shown that teamwork can be a major source of detection in erroneous diagnosis and planning (Sarter and Alexander 2000). During the handling of critical situations, some errors may be committed that reduce significantly the safety margins. Errors can be detected and corrected, not only at the individual level but also more effectively through team communications. The error detection process is based on self-monitoring strategies that ran parallel to the normal tasks and this imposes on cognitive resources. In mature teams, people employ efficient monitoring strategies that have been crafted during many years of experience in operating the system. Self-monitoring strategies enable people not only to catch errors but also to correct them and/or provide feedback for error correction without hindering the work of other controllers.

The error management function can be served with several error detection and correction strategies:

- *Error detection* is commonly done by other members who manage to detect errors of others before it is too late. Error detection is facilitated by monitoring information on displays, by listening to voice communication patterns and so on. Normally, the coordinating controller is in a better position to catch errors committed by the executive controller (e.g., inserting a wrong flight level in the computer) who experiences a higher workload in handling traffic.
- *Feedback for error correction* refers to important information that may assist controllers in recovering errors. Correction can be made either by the controller who detected the error or by the same controller who committed the error in the first place.

A thorough discussion of error detection and correction strategies is made in Chapter 6.

4.7.5 Task Distribution or Change Management

Workload is not a constant parameter but follows the changing pattern of work as it escalates; hence the sequence and priorities of tasks can be altered as new tasks are added in the task backlog. Controllers have to manage not only the normal traffic in their sector but also their interactions with other colleagues. Therefore, a critical need arises for developing strategies that keep the workload below the saturation point of controllers. Again, the CC is in a better position to off-load the EC who demonstrates a steeper escalation of workload.

The ability to balance workload problems at the early stages of a critical situation can lead to timely and effective interventions.

Controllers should also demonstrate the ability to sense and anticipate critical fluctuations in the workload of other team members (e.g., falling behind the curve). Several pilots have argued that they can read the body language of their colleagues and understand that others are not in the loop or are not aware that something is coming up (Thomas and Petrilli 2004). Noticing cues about a degrading mental state of others can make controllers more alert to potential errors that may creep in, hence they can increase their scanning patterns of others. A field study of en-route controllers found that "controllers may not ask for help too early because they may be chastised for doing so, but waiting too long makes it harder for them to get help" (D' Arcy and Della Rocco 2001, 50). Reading signs of performance degradation and knowing when to ask for help are critical strategies in detecting workload problems.

The task distribution or change management function can be achieved by strategies in problem detection and task allocation:

- *Problem detection in task distribution* refers to the extent that controllers have detected any taskload problems of other members. Remaining sensitive to workload changes of others is a central feature of mature teams. Expert controllers demonstrate an ability to sense and anticipate critical fluctuations in the workload of others. The ability to detect workload distribution problems, at the early stages of a critical situation, can lead to timely interventions. Detection of workload problems is supported by accrued operational experience and cohesive mental models. In ATM, controllers refer to a workload problem known as "falling behind the traffic" illustrated by several events (e.g., missing the initial calls of a flight crew or being late in issuing instructions to a crew that called earlier).
- *Changes in task allocation* refer to the extent that controllers adjust the delegation of tasks to enable a better balancing of taskload. When workload problems are observed, controllers may employ task balancing strategies in order to counteract any safety repercussions. Examples may include the utilization

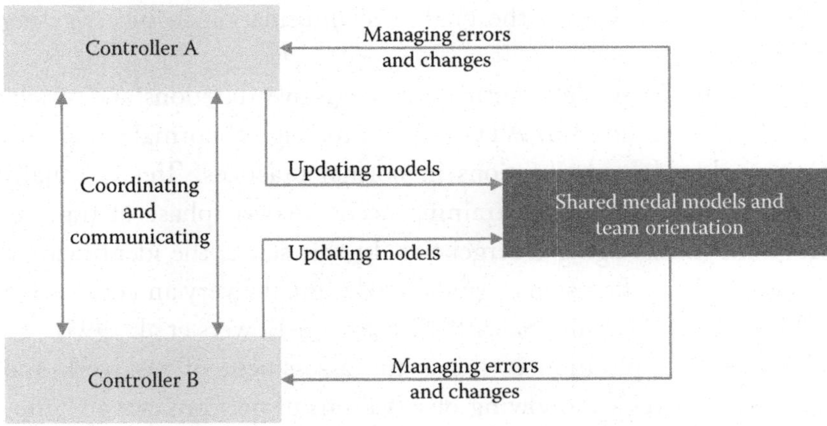

Figure 4.7 Regulation of cognitive functions in teamwork (T^2EAM).

of nonstandard coordination patterns from the CC to off-load the sector or the writing of notes about essential information for an emergency.

4.7.6 The Teamwork Model

In an effort to understand how the five teamwork functions interact, a flow chart has been presented in Figure 4.7. Team orientation and shared understanding can guide coordination in order to plan traffic and resolve conflicts displayed on the radar screens. In turn, feedback of team actions enables controllers to update their models and judge their degree of success, identify errors and change their allocation of tasks to improve team performance. The two teamwork loops are regulated by a shared mental model of the situation and a common team orientation.

4.8 Applications of T^2EAM in Training, Debriefing, and Investigation of Mishaps

The taxonomies of taskwork and teamwork functions can be applied in different ways to improve team performance and safety. Specific applications may include: (1) design of training programs for taskwork and teamwork functions, (2) development of debriefing sessions, after actual practice or simulator training, to identify areas of interventions, and (3) improvement of the quality of information in investigating mishaps.

The T^2EAM model identifies the cognitive functions and strategies that are required by ATC units to manage abnormal situations, high workload, and variations in normal practices. This is usually undertaken as part of the training needs analysis phase of training programs in managing emergencies. In this sense, the identification of core skills is a first step to guide the design, delivery and evaluation of training (Swezey and Salas 1992; Cannon-Bowers et al. 1995).

A related application concerns the assessment of taskwork and teamwork strategies following observations of performance in simulator training or in actual operations. Debriefing sessions after simulator training is a common source of feedback to the operating teams about their strengths and weaknesses that were observed by experienced instructors. Debriefing is also useful in reflecting on actual performance at work on the aftermath of unusual occurrences or even normal operations. Self-reflection alone is unlikely to improve performance without the benefit of a structured list of questions that explore the functions of taskwork and teamwork. The T^2EAM model provides a framework for experienced practitioners to lead debriefing sessions, following observations of task-related behaviors of others.

Finally, the T^2EAM model can be used to improve the quality of information in the investigation of mishaps. The incident reporting systems, in several work domains, produce documents with extensive information; however, the quantity and quality of information concerning the contribution of the human factor are generally poor. The T^2EAM model provides a vocabulary to identify problems in the taskwork and teamwork strategies of controllers and their contribution to the reported incident. To this end, investigators should be trained to use the taxonomies accurately and consistently; this type of training could be conducted jointly with a general training program in human factors.

5

SENSEMAKING

5.1 Introduction

In their everyday operations, controllers are often placed in situations that are either unfamiliar or filled with uncertainty. Without an understanding of the situation, controllers cannot take timely and adequate actions. Making sense of critical situations is difficult, especially when controllers are faced with abundant, conflicted, or limited information. In recent years, the expansion of information technology has increased the amount of information presented to controllers without any assistance on how to make sense of the situation or how to anticipate future trends of the situation. Air traffic control (ATC) is a complex and dynamic environment that requires controllers to attend to multiple events, register fast changing data, diagnose system failures, and resolve conflicts while maintaining resources to handle traffic, and above all to make sense of evolving scenarios. Sensemaking has been viewed as a retrospective activity of individuals and teams bounded by organizational rules and constraints (Weick 1995). Sensemaking has implications for the design of training curricula and decision support systems, especially for major air traffic management (ATM) system-wide interventions (e.g., single European sky ATM research program [SESAR] and next generation air transportation system [NextGen]).

In some respects, sensemaking entails the cognitive functions of recognition, modeling, and critiquing that have been discussed in the taskwork/teamwork for effective and adaptive management (T^{2}EAM) framework. This chapter provides a more elaborate discussion of individual and team sensemaking functions because of the importance of managing modern ATM systems, which present high levels of uncertainty and complexity.

5.2 Frames and Cognitive Functions of Sensemaking

Sensemaking represents one of the key functions of human performance that can be accomplished by individuals, teams, and organizations (Klein et al. 2003). Sensemaking is triggered as a response to situational surprises and failures of expectation. Sensemaking starts when prior understanding is in doubt and further attempts are made to integrate data into a better understanding of the situation. Sensemaking allows practitioners to understand how current accounts of the problem came about and how to anticipate future evolutions through a process of fitting data into an explanatory framework (Crandall et al. 2006). Klein et al. (2005) argue that sensemaking is an essential activity that enables practitioners to reconceptualize the situation and not just fill in gaps to solve the problem at hand. In the context of ATM, the ability to make sense at early stages of the problem may result in timely and effective interventions.

Sensemaking has been based on the concept of a *frame,* which is an explanatory structure that defines entities and relates them to other entities (Klein et al. 2007). Typical frames in the context of ATM include the following:

- *A radar map:* A meaningful pattern that fits in multiple sources of information about aircraft position, flight level, heading, speed, and distances from obstacles or other aircraft
- *An operational plan:* A typical sequence of actions, including how to vector a wave of aircraft into a landing sequence, how to stack aircraft into holding patterns, how to circumnavigate aircraft around areas of convective weather, and so on
- *A script:* A typical pattern of division of work between the executive and coordinating controllers in an air traffic sector
- *A story:* An explanation or a story that a controller develops "why an aircraft executes a rapid descend without prior notice" or "why an aircraft has stopped its taxi out from the runway," and so on

Apart from such typical frames, controllers may construct their own individual frames to make sense of traffic patterns and evolving situations. For instance, controllers can create their own categories of standard and nonstandard flows, group aircraft into units, or imagine possible points of traffic conversion in order to make sense of traffic patterns (Malakis and Kontogiannis 2013, 2014).

According to the data/frame model (Klein et al. 2006) sensemaking is a recursive process that entails six cognitive functions (see Figure 5.1):

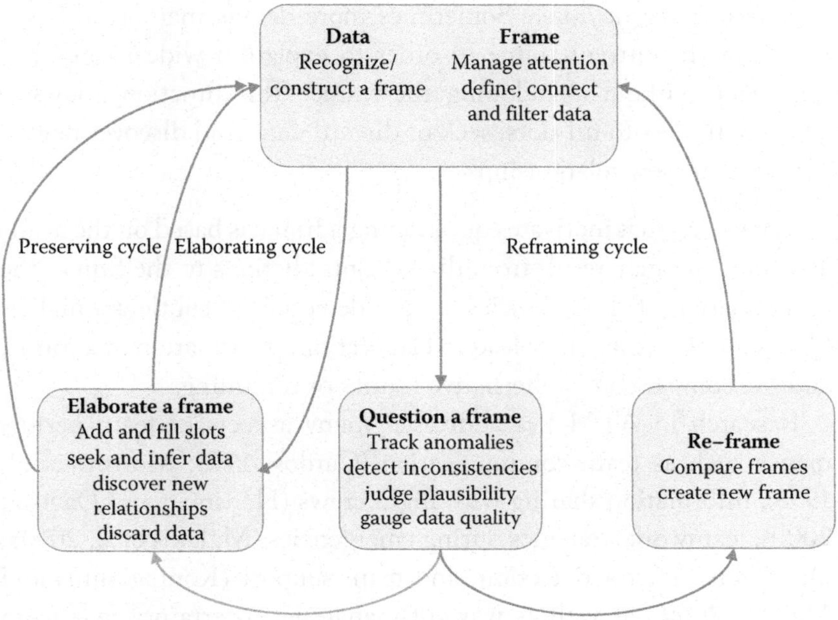

Figure 5.1 Cognitive processes of the data-frame model. (From Klein, G.A. et al., *IEEE Intelligent Systems*, 21, 5, 88–92, 2006.)

1. *Identifying a frame:* This a pattern-matching function where data are fitted into a frame. Expectancies come true and there is a seemingly uninterrupted flow of data that fit into the explanatory frame, with minimal effort.
2. *Questioning a frame:* Inconsistencies between observations and expectations may trigger a process of questioning that entails: gauging the quality of data, tracking anomalies, and judging the plausibility of several scenarios.
3. *Comparing frames:* When more than one frame is plausible, additional data may be sought that allow controllers to choose the most suitable one.
4. *Creating a new frame:* When the current frame may no longer be applicable, the controller may seek to create a new one to accommodate the data.
5. *Preserving a frame:* In many cases, the current frame may be valid and small deviations can be explained to preserve a mindset of operations.

6. *Elaborating a frame:* Sometimes more details may be added into the current frame in order to explain a wide range of data, without abandoning the frame. This function allows controllers to fill slots, seek or discard data, and discover new data or new relationships.

The hunch that motivates questioning a frame is based on the available data and may result from direct contradictions to the frame, the accumulation of discrepancies, or the detection of subtle anomalies. Questioning a frame may lead to elaboration, preservation of a frame, and the comparison of alternative frames or reframing.

Research in ATM has addressed many aspects of team performance such as team communication (Cardosi 1993; Morrow et al. 1993), information sharing with flight crews (Hansman and Davison 2000), teamwork strategies during emergencies (Malakis et al. 2010a, b), aspects of error detection, and team support (Kontogiannis and Malakis 2009), as well as ways of managing uncertainty in a team environment (Corver and Grote 2016). However, these aspects of team performance have been examined in isolation, hence fail to get integrated into the context of team sensemaking.

Team sensemaking refers to the coordination of practitioners as they seek data, synthesize data, and disseminate their inferences in a team environment. According to Klein et al. (2010), the meaning of data becomes the object of negotiation within a team and often triggers a new round of seeking more refined data, hence replacing frames that seemed to be incompatible with data. Team sensemaking is not a stand-alone concept but is related to other team concepts such as team adaptation and shared understanding.

Team sensemaking involves a number of cognitive functions regarding the collection and synthesis of data, the cross-checking of data by other members, the resolution of disagreements, and the dissemination of information among members. Therefore, team sensemaking may include

1. Data synthesis
2. Seeking data
3. Monitoring data quality
4. Resolving disputes
5. Dissemination of information and orders

Individuals and teams may employ similar cognitive functions in sensemaking; however, the particular cognitive strategies for accomplishing these functions may differ (Klein et al. 2010).

This following section presents a case study of tower controllers making sense of low level wind shear phenomena.

5.3 The Challenges of Low Level Wind Shear Phenomena

Wind shear is usually defined as "a change in wind speed and/or direction in space, including updrafts and downdrafts" (ICAO 2005c, 10). It follows that any atmospheric phenomenon, or any physical obstacle to the prevailing wind flow that produces a change in wind speed and/or direction, may cause wind shear. Although wind shear may by encountered at all altitudes, the most hazardous are those that affect the critical phases of takeoff, initial climb, approach, and landing. The spectrum of wind shears and eddy motions associated with different weather phenomena is large and their effects on aircraft performance may be complicated in nature. Two of the most important types of wind shear in aviation are those commonly referred to as low level wind shear (LLWS) and low level turbulence. The term *low level* refers to an altitude of less than 2000 feet from the ground. Windshear is associated with thunderstorms, the passage of a front, a marked temperature inversion, a turbulent boundary layer, and specific topography or buildings around an airport. Normally, LLWS is accompanied with low level turbulence that increases the overall effect on aircraft performance.

The encounter of LLWS by an aircraft is a dynamic situation that depends on many factors. Particular types of aircraft may vary in their reaction to a LLWS event; a light high wing piston-engine aircraft may react in a different way from a heavy four-engine swept wing jet aircraft (NATS 2008). LLWS phenomena challenge flight operations as they can cause drastic changes in aircraft performance that require accurate and timely inputs from the flight crews. First, an aircraft may experience a headwind component that initially enhances its performance. When the pilot reduces thrust to maintain the flight profile, the aircraft may enter into an area of strong downdraft that causes the aircraft to sink; after this stage, an increased tailwind may be experienced that further drains the energy of the aircraft and requires a thrust increase. In short, the aircraft may first encounter a performance-increasing wind shear, then a sinking downdraft, and finally, a performance-decreasing shear (Figure 5.2). These effects may exceed the capability of an aircraft to maintain a safe flight path. The outcome of an encounter with a LLWS event depends on the phase of flight. For instance, in the course of a landing approach, the aircraft may undershoot and collide with the ground just short of the runway. In the process of accelerating for takeoff, the aircraft may not be able to rotate and experience a tail strike and a runway excursion. If the LLWS is encountered during initial climb, the aircraft may stall and collide with terrain. Finally, if an aircraft is approaching and executes a "go-around" it may stall or perform a tail strike.

LLWS events represent a serious aviation hazard and a technical challenge for system manufactures (i.e., on board weather radars and LLWS detectors), airlines (i.e., training in LLWS recovery techniques), and ATC radar system developers (i.e., weather radar and

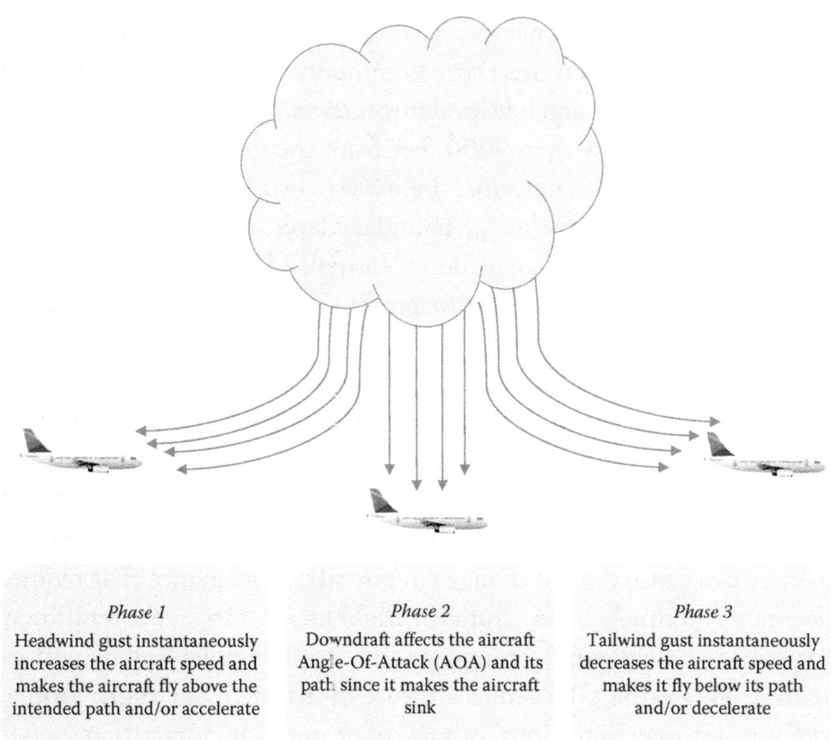

Figure 5.2 Effects of LLWS phenomena on aircraft performance.

LLWS alerting systems). ICAO (2005c) warned that since 1943, low level wind shear has been cited in a number of aircraft accidents that have contributed to over 1400 fatalities worldwide. Airbus (2007) announced that adverse wind conditions have been involved in more than 30% of approach-and-landing accidents and in 15% of events involving controlled flight into terrain (CFIT) events. Windshear has been identified as the primary causal factor in 4% of approach-and-landing accidents and as the ninth cause of fatalities. ICAO (2005c) advised that there is an operational requirement for information on low level wind shear and turbulence to be provided to flight crews to enable them to counteract their effects and maintain safe control of the aircraft. This information is derived from aircraft and/or ground-based meteorological observations or from assessments of weather situations. Airlines and aircraft manufactures focus on avoidance, recognition, and timely application of recovery/escape techniques. For example, Airbus (2007) emphasized crew awareness and alertness as key factors in the successful application of windshear avoidance and recovery techniques. However, recognition of an encounter with a LLWS event may be masked by automation (Dismukes et al. 2007). In the case of the USAir 1016 incident, the airplane's onboard warning systems failed to alert the flight crew of the LLWS event due to the retraction of flaps at that time, which inhibited the generation of the alert. As a result, the flight crew executed a normal missed approach instead of a windshear recovery technique (NTSB 1995).

5.4 Explanatory Frames and Sensemaking Strategies

Practitioners develop their own cognitive strategies throughout their accumulated expertise in their specific domain in order to support the six functions of sensemaking. In questioning a frame, for instance, tower controllers develop rules and set "tripwires" to alert them about the intensity and duration of LLWS phenomena. Practitioners recognize when it is about time to start interrogating a plan by keeping track of events that should not be happening and wait for a predetermined time only. Klein (2004) referred to these alarming events as "tripwires" that indicate when the plan may have some weaknesses that need to be addressed. In our case, the central elements of the organized structure that emerged as a frame in sensemaking regard the intensity and duration of low level wind shear phenomena supported with data from many sources (Figure 5.3).

The intensity and duration of LLWS phenomena can become an explanatory structure or a frame of sensemaking. The frame not only determines the tempo of operations (i.e., arriving and departing flights) but also determines whether controllers should increase the alert of the rescue and firefighting services, whether to coordinate the holding of aircraft that wish to wait for a weather improvement, coordinate and manage the diversion of flights to other airports, and, finally, coordinate with flow controllers to reduce the rate of incoming aircraft.

The data/frame theory claims that the explanatory structure or frame can be used to apply several cognitive strategies and make

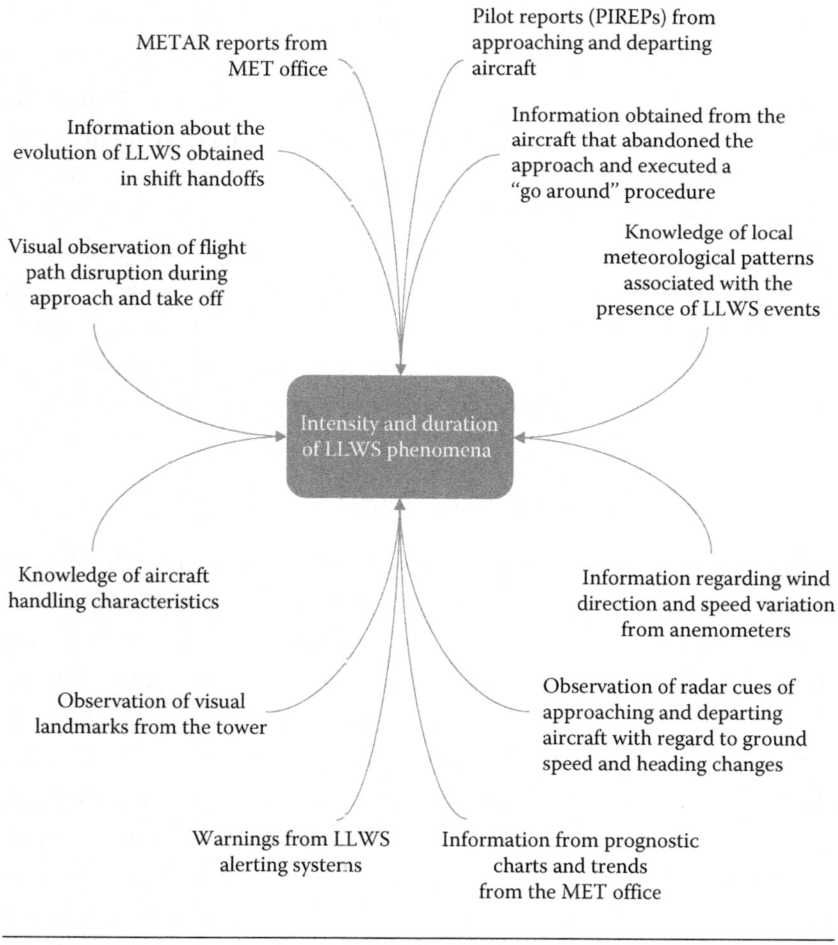

Figure 5.3 Explanatory structure of controller's perception of LLWS phenomena.

sense of the data (Klein et al. 2007). In our case, the explanatory frame about LLWS phenomena was supported by information from meteorological reports, knowledge of local weather patterns, visual landmarks, flight crew reports, handoffs from previous shifts, etc. An analysis of the six functions of sensemaking of LLWS phenomena is provided in the following sections.

5.4.1 Identifying a Frame

In most cases, identifying the presence and intensity of LLWS phenomena is an effortless process and no deliberate sensemaking is required. When encountering familiar LLWS scenarios, controllers may resort to pattern matching. Controllers can estimate LLWS phenomena by synthesizing information from landing aircraft, departing aircraft, and weather reports. This is usually the case when wind conditions are changing slowly and changes in weather patterns are gradual and predictable. Apart from meteorological reports, controllers have a rough idea of the intensity and duration of LLWS from their expertise on local weather phenomena and from visual observations from the tower unit. Controllers observe prominent visual cues (e.g., movement of nearby trees) and are setting visual tripwires to make predictions of the intensity of LLWS event (e.g., the height of waves in the sea close to the threshold of the runway).

Information about the first aircraft to approach can also be used to make estimates about the intensity of LLWS events by observing whether aircraft are able to continue their approach for landing or abandon their approach and perform a go-around. Such expectations are tested against previous approaches and personal knowledge (that is, their data-store of rules). During the approach, tower controllers can visually monitor the roll, pitch, and yaw movements of aircraft as well as the smoothness of the approach. With experience, controllers acquire patterns of aircraft approaches and can easily discriminate changes in speed, roll, or any movement that signify the onset of trouble. Tower controllers can also expect that other arriving aircraft may experience similar movements with some variation. Subsequently, this information is passed onto to approach controllers who may alert the rescue and firefighting services of the airport.

5.4.2 Questioning a Frame

When data violate expectations, controllers may initiate a process of challenging the plausibility and the quality of data they receive. Questioning a frame becomes more difficult at a team level, as Weick (2007) pointed out in many high-profile incidents where questioning the frame failed with grave consequences. Typical triggers for questioning the intensity of LLWS phenomena may include aircraft that abandon their approach at much higher altitudes, and further from the runway, than other aircraft. In these cases, controllers can rely on flight crews and manage to refine their estimation of the intensity of LLWS phenomena.

In questioning a frame, controllers may not know whether the frame is incorrect or the situation took a sudden turn. The violation of their expectations could qualify as novel cases to be added to their knowledge base. For example, the first aircraft that abandons its approach for landing, despite the fact that the previous aircraft continued its landing uneventfully, would trigger a questioning of the frame. As a result, the planning horizon of controllers would be reduced (e.g., releasing only one aircraft at a time from the approach unit of inbound aircraft) to cope with an increase of aircraft resorting to a go-around procedure. This strategy would also allow controllers to look more carefully into the intensity of LLWS events. Hence, controllers may become more sensitive in tracking anomalies (e.g., wind direction, speed variability, and smoothness of the approach), in detecting inconsistencies (e.g., meteorological reports), and in gauging data quality (e.g., flight crews versus meteorological officers). These cognitive strategies are supported by personal knowledge of rules and by collaboration with colleagues and supervisors.

5.4.3 Reframing: Comparing Multiple Frames

Practitioners may track up to three frames simultaneously, with the usual case involving two frames (Klein et al. 2007). Having two or even three explanatory frames requires a mechanism for ultimately settling on one only. Comparison of multiple frames can be initiated by the detection of an anomaly that resembles the function of a bifurcation point (i.e., an unstable state that can evolve into one of many other stable states in the near future). A bifurcation point may be a single go-around event in a series of smooth approaches, or a single smooth landing after a series of go-arounds. However, the direction of change is not clear and extensive expertise is required to track down the possible system states. In general, reframing is triggered by deviations from expectations and by crew reports that contradict their knowledge store of if–then rules. In these cases, controllers may choose between two expectations: (1) a temporal change in the intensity of LLWS phenomena that affects only one aircraft, and (2) a more permanent situation that indicates a rapidly deteriorating situation that could lead to the closure of the airport and the need for holding and diverting flights. Additionally, tower controllers may convey their judgment to the approach controllers so that they could plan for later traffic accordingly. Planning the sequence of approaching aircraft is more difficult when having to divert inbound aircraft to alternate airports than when having to stack aircraft into a holding pattern near the airport, awaiting an improvement in wind conditions.

5.4.4 Reframing: Creating a New Frame

Reframing is a difficult task as it implies aborting a current account and constructing a new one that was not an option in the first place. This process is quite similar to replanning where a whole sequence of tasks has to change in a restricted time window, which implies changes in coordination between tower and approach controllers. Kontogiannis (2010a) argued that replanning requires modifying a plan on the fly, which presents many challenges to teams working in situations of high uncertainty. Replanning involves reinterpreting the situation and reassessing the impact of events and actions on established goals and team functions. Similarly, the creation of a new frame imposes strong demands that may render the process difficult. Controllers face a challenge in choosing between two explanations: (1) a temporal visibility disruption to the approaches of one or two aircraft and (2) an extended period of high intensity LLWS events that will last for several hours. In our example, replanning was supported by a loose tactic (e.g., extending the miles in trail between successive arrivals) and by preserving an airspace volume for holding aircraft. In addition, team supervisors resorted to proactive coordination with neighboring airports to decide on the number of aircraft to accommodate in cases that required a diversion.

5.4.5 Preserving the Frame

When controllers preserve a frame, they may explain away inconsistent evidence, which bears a risk of fixation errors (De Keyser and Woods 1990). Misleading cues, absent indicators, and unusual cues may create an environment that impedes error detection (Kontogiannis and Malakis 2009). For instance, small variations in the intensity of LLWS may be attributed to individual aircraft characterizes or flight crew training and procedures rather than to deteriorating wind patterns. Also, inconsistencies about LLWS intensity may be difficult to understand in cases where lighting conditions or rainfalls create visual distortion of information.

5.4.6 Elaborating a Frame

Elaboration involves preserving the current frame by adding more details and by filling in missing slots. The chances of surprises or inconsistencies are minimized as more details are added. Normally, elaboration is one of the final steps of sensemaking and signals the start of a period of frame stability. In essence, controllers make minor calibrations in their account as new data fit the frame conveniently. The drive for new data is smooth and observed patterns become progressively familiar. For example, in one case, LLWS intensity increased, which caused arriving aircraft to start going around. Tower controllers would infer the increase of LLWS by visually tracking the smoothness of the approach of the aircraft and would advise approach controllers to hold the aircraft until wind conditions improved. The tower controllers would carefully observe the signs from their landmarks and, with the help of data from weather reports, they would better understand the intensity of LLWS phenomena. Finally, controllers would be the first to notice the decrease of LLWS intensity from the smoothness of the approach of the inbound aircraft and they would advise approach controllers accordingly.

5.4.7 Behavioral Markers for Team Sensemaking Strategies

The functions of team sensemaking are supported by cognitive strategies through accumulated expertise. To provide a practical tool for identifying the cognitive strategies of controllers in the six sensemaking functions, Table 5.1 presents several behavioral markers that can be observed during actual performance.

Table 5.1 Team Sensemaking Strategies and Behavioral Markers in Coping with LLWS Phenomena

SENSEMAKING FUNCTIONS	STRATEGIES AND ASSOCIATED BEHAVIORAL MARKERS
Identifying a frame	• Tower and Approach controllers receive routine meteorological reports, prognostic charts, LLWS alerts and formulate an initial estimate of the intensity of LLWS that can be tested by observing the first approaching aircraft • Tower controllers utilize their knowledge store of rules regarding observability of visual landmarks to estimate the intensity and duration of LLWS phenomena • Tower controllers can also observe the smoothness of take-off and initial climb of the first aircraft to verify the intensity of LLWS phenomena • To verify the intensity of LLWS, tower controllers check the smoothness of approach of the aircraft against their visual tripwires when the first aircraft approaches to land
Questioning a frame	• Tower controllers develop rules for visual tripwires to alert them when their current estimate of the intensity of LLWS is no longer valid • Controllers voice any discrepancies derived visually by the smoothness of the approach, which may indicate a need for change • Controllers discuss information derived from the flight crew of an approaching aircraft that may trigger a revision of their understanding
Reframing: comparing frames	• Controllers suggest, negotiate and compare estimates of the intensity of LLWS and select the most plausible frame • Team supervisor listens actively and synthesizes contrasting viewpoints • Team supervisor decides whether to modify, or await for new information, before ending the comparison phase
Reframing: creating a new frame	• Controllers voice modifications in their estimates of LLWS intensity and duration • Team supervisor synthesizes competing estimates of LLWS phenomena • Controllers set revised "tripwires" for new estimates of LLWS phenomena
Preserving the frame	• Small variations in LLWS estimates may be attributed to transient weather phenomena, aircraft performance characteristics, flight crew handling characteristics, and airline SOPs • A single go-around of an aircraft may be attributed to other factors, not related to the understanding of LLWS events • Routine meteorological information, in combination with visual observations, are used to preserve the current frame of LLWS phenomena
Elaborating a frame	• Controllers discuss new information to fine tune the existing frame of the intensity and duration of LLWS phenomena • Meteorological officers are actively engaged in discussions for elaborating existing LLWS frames made by controllers • Controllers collaborate to discover relationships between information derived from all sources that preserve and extend their current frames

5.5 Requirements for Team Sensemaking

Team sensemaking differs from individual sensemaking in how data are collected and how they are cross-checked by different members, how disagreements get resolved, and how information is disseminated among the team. The data/frame model of sensemaking relies on certain task and team requirements with regard to the collection, integration, verification, and dissemination of information among team members. These requirements stem from the collective nature of work in the ATC environment and are discussed below (Malakis and Kontogiannis 2014).

5.5.1 Data Synthesis

Synthesizing data from several sources remains the primary responsibility of watch supervisors who have to collect data from physically remote areas such as the tower and approach units. Watch supervisors can monitor the voice loops between tower controllers and flight crews, which supports a better understanding of the situation but radiotelephony congestion often renders the monitoring process rather difficult or even distracting. During approach, under LLWS phenomena, the number of RTF exchanges between flight crews and controllers increases as the crews request more information updates on wind conditions. Tower provides an unobstructed view of the approach and landing flight phases as well as the smoothness of the flight paths. On the contrary, the radar provides only heading and ground-speed indications that signify the encounter of a LLWS event. This creates a dilemma for watch supervisors, that is, whether to stay in the tower area (where a privileged view of the unfolding situation is obtained) or move to the approach area (where better coordination is achieved for holding and rerouting aircraft). Team supervisors have the added task of deciding when to proceed with the data at hand, or wait for new data, but they are more confident in their judgments as they acquire a more nuanced knowledge store of rules with experience.

5.5.2 Seeking Data

Seeking data concerns individual controllers who often have to coordinate and overcome problems of missing data, unreliable data, and unconfirmed data. Hence, controllers build their own explanatory frames so that their search is not too broad or narrow or even vague. Although controllers have clear job responsibilities, their roles in seeking and synthesizing data are often blended. For instance, the watch supervisor may not only synthesize data but also have some good ideas where to look for useful data.

The function of seeking data takes different forms according to the responsibilities of different control positions. For instance, tower controllers would seek data from direct observation and routine meteorological reports; however, they would mostly rely on direct observation and set visual tripwires to alert them about LLWS changes. In contrast, approach controllers would utilize a combination of radar data and voice loops between the tower and the flight crews. Direct telephone communications between tower and approach controllers can be established to bridge the gap of physical distance and the inability of approach controllers to physically observe the unfolding wind conditions.

5.5.3 Monitoring Data Quality

Data synthesizers have to assess the quality of data in terms of their credibility and relevance. Monitoring of data quality has to take into account the experience of the controllers, the reliability of reports and radars, and the delays in getting the necessary information. Meteorology officers become an official system of detecting and measuring the intensity of LLWS founded on wind direction, speed, and other meteorological variables. Tower controllers may employ a similar referential system but their thresholds could be different because of their own training and procedures. For example, watch supervisors would be very careful how to question the data of inexperienced controllers from the tower in terms of LLWS estimates in order to get a more accurate picture and to find any errors made by junior team members. Apart from identifying relevant risk factors in the quality of data, data synthesizers should develop socially acceptable methods for cross-checking their data.

5.5.4 Resolving Disputes

In a team environment, controllers may develop different account of events or favor different frames of explanation. In questioning a frame, for instance, junior controllers may notice weak signals but fail to mention them to the rest of the team. Similarly, in comparing frames, team members may take different perspectives of what accounts for an accurate frame. Disagreements can be resolved through several means such as hierarchical authority and pressure for consensus. Hierarchical authority and flight crew reports may have the final word through a process of elaborate testing and revision. For example, different estimates of LLWS intensity between tower controllers and meteorology officers could be resolved by the reports of the last aircraft to land. Alternatively, the watch supervisors could make a decision based on a synthesis of data as well as a negotiation with controllers and meteorology officers.

5.5.5 Dissemination

Dissemination of information and orders usually follows the selection of an explanatory framework for making sense of the situation. Dissemination involves verbal communications and written reports about the situation or the means how to control a problem. Controllers need to communicate with flight crews directly using voice communications. In our case, dissemination of information between controllers, flight crews, and meteorology officers was quite accurate based on an operational language that was concise, clear, and meaningful. Controllers were able to appreciate major attributes of information and were able to judge the level of workload and interruptibility of other team members. As a result, tower controllers were not distracted by approach controllers with redundant requests for verifying the intensity of LLWS in critical phases of tower operations.

5.6 Concluding Remarks

Sensemaking is likely to affect the adaptability of practitioners required in complex environments (Klein et al. 2003). Controllers have accumulated expertise in several areas that may be put at the risk of being utterly invalidated from the introduction of new ATM technologies. Specifically, controllers usually develop knowledge store of rules regarding LLWS phenomena that are used as a resource in constructing their frames. With the increasing reliance on ground alerting systems and onboard systems, however, many issues of human competence may arise.

Sensemaking becomes an issue when the flight deck and air traffic control both have similar, but not necessarily identical, information available. An example concerns aircraft that carry onboard radar to detect LLWS ahead of time. The information that controllers have on their displays is not as fine-grained as the information that these flight crews have available. When controllers and flight crews communicate about the weather, they do not have a shared awareness of the situation. In coping with LLWS events, controllers may not have the same quality information about wind conditions that flight crews do; hence, they could not anticipate when a flight crew may initiate a go-around procedure.

This brings forward another critical issue with regard to flight crew decision-making for long time periods. Diminishing controller expertise would put at risk any sudden intervention of controllers when requested by crews who have difficulties in controlling a critical situation. Research into sensemaking can provide useful insights on how to embed operational experience into future ATM systems in order to improve collaborative decision-making. There is also a need for developing appropriate forms of decision-support systems that would enhance sensemaking skills, especially in view of the new developments created by the SESAR and NextGen programs.

6

Human Error Detection and Recovery

6.1 The Concept of Human Error

Air traffic management (ATM) has been a highly reliable system for some decades now. Despite its impressive safety record, studies have shown that many incidents still involve human error. As the air traffic system is being stretched to its capacity limits, safety challenges may increase in the near future. Furthermore, the introduction of new computerized and automated tools may affect the operating methods and coordination patterns of controllers, and hence may change the nature of errors and chances of recovery that have been reported in existing systems. Consequently, aviation organizations should learn from incidents in order to maintain high levels of safety, particularly now with the increasing complexity of operations.

On an intuitive level, many people would feel confident in making judgments about the contribution of human error, especially after the recording of an adverse event. In hindsight, it may seem straightforward to attribute the cause of a problem to the active intervention of controllers. However, the study of human error since the early 1980s has shown that the concept of human error is a rather elusive concept that has been associated with many different meanings. In the following, we briefly review some of the ambiguities associated with the label "human error" and consider different views of human error and error management (for an elaborated discussion, refer to Rasmussen 1986; Reason 1997; Woods et al. 2010).

Human error has been seen as a cause of adverse outcomes in many engineering approaches of risk assessment and incident investigation. A judgment that an outcome was mainly due to human error is an attribution that human performance immediately preceding the incident was unambiguously flawed and led to the negative outcome. The old view that failures are introduced into the system through the inherent unreliability of people (e.g., someone did not pay enough attention, or made a shortcut) gave rise to several error analysis methods (Reason

1990). In practice, however, things have proved not to be this simple. Studies in many work domains show that the label "human error" is prejudicial and underspecified (Woods et al. 2010). It retards rather than advances our understanding of how complex systems fail. It is the investigation of the cognition of people, their tools, and the affordances of the system—and not the attribution of the error itself—that helps us avert the potential for disaster. In the new view, human error is not a cause but a symptom of trouble that resides deeper in the system.

As Dekker (2006) argued, underneath every seemingly obvious story of error there is another deeper story about the system in which people work. Safety is not inherent in systems. Organizations have to pursue multiple goals at the same time (e.g., productivity and financial growth) that may be in conflict with safety. So, people do their best to reconcile different goals simultaneously. This is especially true for air traffic controllers whose mandate is the safe, orderly, and expeditious flow of traffic (ICAO 2007a). Systems are also imperfect because new changes impose more demands or are incompatible with existing operational practices. In the new view, human error is a symptom of imperfect systems, or complex systems, where well-intentioned people try to adapt to complex or unforeseen events. New methodologies of studying "error" focus on how people experiment with their practices to adapt to pressures at work, unavailability of tools, novel events, and coordination problems. Hence, errors are seen as maladaptive changes of practices that worked well in many other situations in the past.

In the context of air traffic control (ATC), Isaac et al. (2001) provided a comprehensible definition of *human error* as: "Any action (or inaction) that potentially or actually results in negative system effects, given the situation that other possibilities were available. This includes any deviation from operating procedures, good working practices or intentions."

There are several benefits with this approach. First, the definition of human error is neutral with regard to any question of blame. Second, an error is judged on the basis of underlying processes and not negative consequences. Third, an action or inaction can only be labeled as error if the person could have acted differently, given the constraints of the situation. Finally, the definition accepts several criteria of performance evaluation such as operating procedures and good working practices.

Although a general definition of "human error" provides a useful framework of analysis, it is very difficult for analysts to reach an agreement on what counts as an error "in the wild." In complex systems, errors become almost indistinguishable from the busy background of real work. Hollnagel and Amalberti (2001) report a study where observers (i.e., psychologists and air traffic controllers) were asked to count errors and categorize them, using a particular taxonomy for air traffic operations. Despite the common taxonomy, the two groups of observers differed substantially in the number and sorts of errors recorded. Air traffic controllers who acted as observers relied on working conditions (e.g., interface design, procedures and time resources) to categorize errors whereas psychologists preferred to locate errors in cognitive processes (i.e., working memory, attention, and judgment).

Moreover, controllers who actually carried out the work claimed that many of the recorded errors were not errors at all but rather normal work. It appears then that an explanation of a possible error shall have to consider the context of work surrounding an action. For instance, an early transfer of aircraft is not always an error; it may be a sensible strategy of managing workload given the unfolding context of work. However it may become an error in certain circumstances. The same applies to the timing of conflict resolution tactics. A controller may delay a de-conflicting action (e.g., a level change or a heading change) in order to solve an additional conflict that may materialize later. To an external observer this may count as a late action to an obvious conflict. The controllers who carried out their jobs had a better assessment of the context of work than the external observers.

Minimizing and controlling risks in the ATM system requires the development of error management strategies. Traditionally, error reduction strategies focused mostly on error avoidance and prevention. This is understandable since many studies of risk analysis have taken the old view that incidents arise from a combination of human errors and system designs that are unforgiving to failures. However, safety strategies narrowly based on error prevention may not be successful for several reasons. First, it is very difficult to anticipate all errors that may occur in a specific context of work. Second, a focus on error prevention may impose some restrictions on human performance, which can compromise effective and adaptive behaviors. Amalberti (2001) argued that error prevention strategies have been overused in ultra-safe domains (e.g., aviation and air traffic control) and that this approach may be ineffective from a safety perspective. Third, studies have shown that many errors are detected and recovered before an adverse event occurs (Amalberti and Wioland 1997). Consequently, error management should also focus on error detection and correction. This chapter aims to explore how practitioners manage to detect or recover errors and what strategies can maintain higher levels of safety.

A framework for studying error management in aviation is introduced in this chapter, followed by a short description of classification schemes of errors, to provide a useful basis for understanding strategies in error detection and recovery.

6.2 Error Management Processes

The field of error management has created an impetus for a new perspective on human errors and the role of humans in the control of complex systems. The traditional view maintains that humans are "intelligent but fragile" agents and, as a consequence, defenses in depth are required for protection from machine failures and human errors (Bove 2004). A more positive attitude toward the human controller has gradually emerged as a result of research in error detection and recovery. Amalberti and Wioland (1997), for instance, have shown that practitioners in aviation develop protections and defenses against their own cognitive deficiencies. In this manner, practitioners play a positive role in returning the system to a safe state, after the occurrence of an error.

To understand the causes of errors and successful recovery of errors, we need a framework that integrates both processes of error production and error recovery. In aviation, Helmreich et al. (1999) developed a model of threat and error management (TEM) on the basis of studies of flight crew behaviors and situational factors on normal flights; later developments applied the TEM model to air traffic control (ICAO 2005b). According to the TEM model, risks may arise from external threats and internal sources of error (Figure 6.1). Threats are defined as events that occur beyond the influence of controllers (e.g., adverse

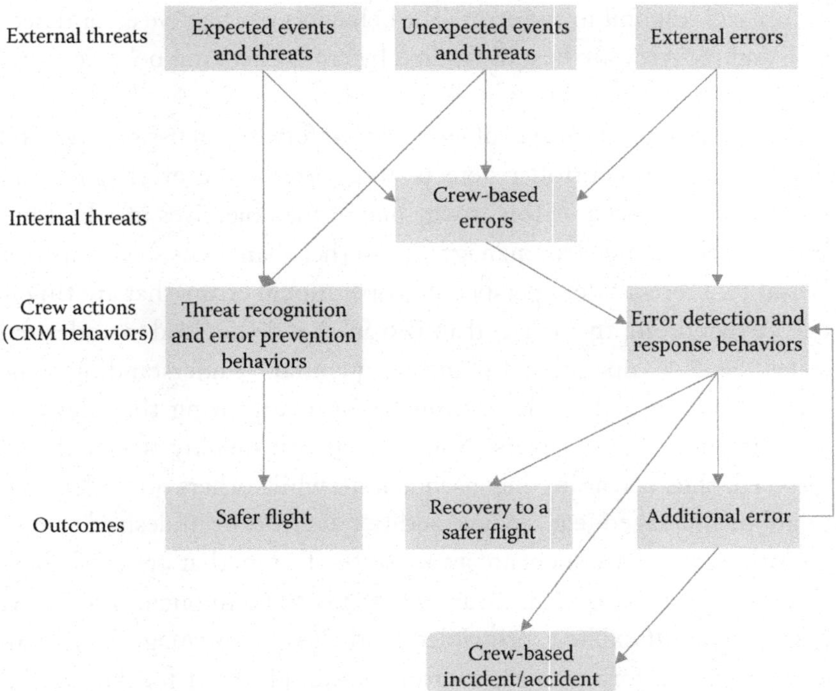

Figure 6.1 The threat and error management (TEM) model. (Adapted from Helmreich, R.L. et al., *Proceedings of the Tenth International Symposium on Aviation Psychology*, The Ohio State University, Columbus, OH, 677–682, 1999.)

meteorological conditions, airports surrounded by high mountains, congested airspace) and/or errors committed by other people (e.g., flight crews, ground staff, or maintenance workers). In contrast, errors are defined as "controller actions or inactions that lead to deviations from organizational procedures or one's own intentions."

When an unexpected threat is recognized, controllers and crews can employ CRM behaviors for error prevention by evaluating the threat's implications and by using decision-making skills to determine a course of action. Threat recognition and error prevention represent a proactive response that can be observed when teams share information, evaluate the situation, and include contextual factors in their planning. For example, a flight crew may recognize the risk associated with adverse weather at their destination airport and practice error prevention by increasing the fuel load or by reconsidering the choice of an alternate airport to conduct a safe landing. In some cases, human error may be inevitable, so when an error occurs, it is the crew's task to detect and respond to the error. The behaviors of effective error detection and recovery are best illustrated by cross-checking behaviors and by evaluating the quality of earlier decisions.

Regardless of the type of error, its safety repercussions depend on whether the controller detects and corrects the error before an undesired state occurs. This is why one of the objectives of TEM has been to understand error management rather than focus solely on error causality. From a safety perspective, operational errors that are timely detected or promptly managed and errors that do not lead to undesired states become operationally inconsequential. Understanding how errors are managed is then as important as capturing the relevance of different types of errors. Some errors are quickly detected and resolved, thus becoming inconsequential, while others go undetected or are mismanaged (e.g., induce additional errors or undesired states).

Although practitioners are aware of the factors that are considered threats in their work, this awareness tends to be implicit. The TEM model makes it explicit, principled, and therefore manageable. In an ideal situation, for instance, controllers should report for duty ahead of the official start of their shift and receive a briefing from the outgoing shift before taking over their position. In practice, however, some controllers report just in time for duty and go straight to their job position, missing an important briefing of any problems encountered in the previous shift. This may have safety repercussions since some issues may continue to exist in the new shift. The real work context is abundant in little threats that are easily forgotten by investigators but that may become precursors of errors and incidents. TEM identifies everyday threats that may lay the ground for new forms of errors or hinder error detection and recovery.

Safety managers of several airlines have embraced a tool called the line operations safety audit (LOSA) that is used to collect information on what types of threats are faced by flight crews, how threats are managed, what errors may result from threats, and how crews manage errors (ICAO 2002). After processing information from LOSA observations, airlines get a clear overview of the strengths and weaknesses of their flight operations with respect to threats, errors, and undesired states encountered by their crews in normal operations. Following the successful implementation of LOSA by a number of airlines, ICAO pursued the development of a similar tool for the monitoring of safety in normal ATC operations. The idea behind the normal operations safety survey (NOSS) was to supply the ATC community with a means of providing robust data on threats, errors, and undesired states to safety managers (ICAO 2008). NOSS provides data from normal operations and is not driven by the analysis of occurrences. Data from NOSS, together with safety data from conventional sources, should make it possible to prioritize recommendations for safety interventions in ATC.

6.3 Classification of Human Error

The identification of human error requires safety analysts to use a systematic method that is based on sound theory about human performance. Several studies have proposed error taxonomies based on situational characteristics and performance failures, many of which are grounded in an information-processing approach (Wickens 1992; Wiegmann and Shappell 2003). In general, the information-processing approach draws on the metaphor of the human as a computer and describes the mental processes that intervene between the registration of cues and the final choice of responses. Each mental process (e.g., attention) receives inputs from earlier ones (e.g., perception) and produces outputs for subsequent processes (e.g., human judgment). At any stage, the transformation process may be in error or information may be lost. In ATC, Shorrock and Kirwan (2002) have developed a classification scheme (technique for the retrospective and predictive analysis of cognitive error, TRACEr) based on Wicken's model. As TRACEr has been applied by many investigators in the analysis of errors in the ATC domain, a short description is provided in Section 6.3.2.

6.3.1 A Model of Unsafe Acts

One of the most prominent models of human performance that has provided a basis for classifying human errors regards the skill-rule-knowledge (SRK) model (Rasmussen 1986). Situations where controllers employ highly practiced routines for everyday tasks call for skill-based behavior (e.g., adjusting the radar screen to monitor aircraft trajectories, annotating flight progress strips, using a headset to communicate with flight crews). For familiar tasks of higher complexity, controllers can resort to rule-based control by using standard procedures and mental rules stored in memory (e.g., conflict resolution of two aircraft). For novel situations, or less frequently practiced tasks, controllers have to work at the knowledge-based level where traffic plans may be based on an adjustment of procedures or even improvisation to create novel plans (e.g., coordinate regular traffic and firefighting aircraft in a fire area adjacent to the airport).

Reason (1990) identified several categories of error for different levels of performance that are applicable to many safety critical industries (Figure 6.2):

1. *Slips of action* are errors at the skill-based level usually described as "actions – not – as – planned"—e.g., performing an action too soon or leaving it until too late, omitting a step in a procedure, providing the correct clearance to the wrong aircraft, or providing a slight heading change to an aircraft while a larger one would have resolved the conflict.

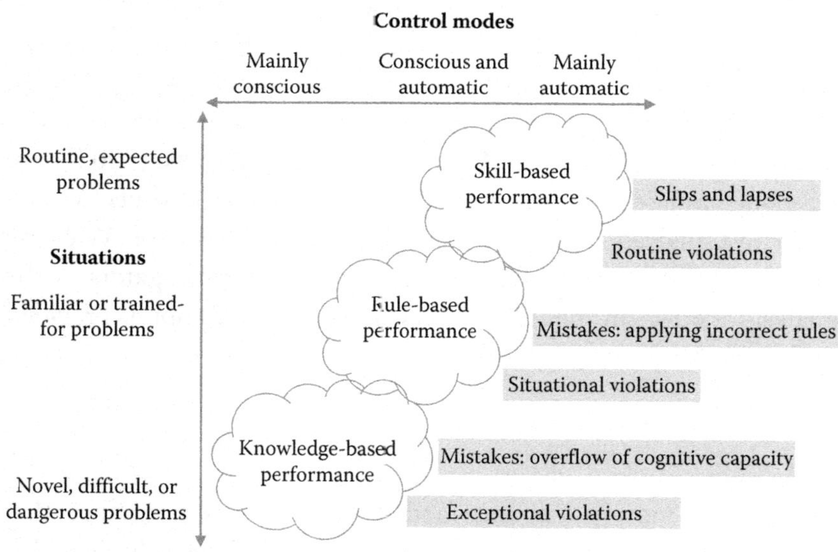

Figure 6.2 Slips, lapses, mistakes, and violations. (Adapted from Reason, J.T., *Human Error*, Cambridge University Press, Cambridge, 1990.)

2. *Lapses* cause people to forget to carry out actions, to lose their place in a task or even to forget what they had intended to do (e.g., forgetting to brief the incoming shift about a military exercise that is still taking place in the sector). Lapses are also errors at the skill-based level that can be reduced by minimizing distractions and interruptions or by proper work design that provides effective job reminders.
3. *Mistakes* concern activities that run according to the plan but where the plan may be inadequate to achieve the desired goal. Mistakes stem from failures of our mental processes to assess information, set intentions, and judge consequences. *Rule-based mistakes* are associated with familiar situations where either a bad rule is applied to a situation or a perfectly adequate rule is applied to a situation that requires a different set of actions. *Knowledge-based mistakes*, on the other hand, can occur in situations where no ready solutions are available and a new plan has to be generated. A typical example regards failures of flow controllers to set appropriate traffic restrictions to a sector due to their limited knowledge of the functioning of the air traffic flow and capacity management (ATFCM) system that does take into account variations or increases in traffic (e.g., favorable winds and route bypasses).

Slips, lapses, and mistakes are essentially defined as failures of information processing. Reason (1990) has also defined another error category referring to "violations" or "deliberate deviations" from the rules, procedures, instructions, and regulations drawn up for the safe operation of the system. Violations occur for many reasons but they are seldom willful acts of sabotage or vandalism. Most stem from a genuine desire to perform work satisfactorily, given existing system constraints and practitioner expectations. Violations are highly susceptible to organizational influences as most causes of violations are either accepted by management or condoned as normal working practice. Very often, pressures to work faster with fewer resources may give rise to procedural violations that are tolerated by senior management. In this sense, deviations from normal methods of work could become the norm rather than the exception.

Violations are also divided into three categories as shown below:

1. *Routine violations* that involve habitual breaking of rules or procedures as this has become a normal way of working within the work group. This can be due to several reasons, such as the desire to cut corners to save time and resources, the perception that rules are too restrictive, or that rules no longer apply. In heavy traffic, for example, a radar controller may routinely vector aircraft to establish the instrument landing system (ILS) with a higher speed than normal and a shorter distance than the International Civil Aviation Organization's (ICAO's) prescribed separation minima.
2. *Situational violations* that occur when particular job pressures make rule compliance difficult (e.g., time pressure, insufficient staff, or high workload). Practitioners are trying to adapt to the pressures at work and may have to deviate from the formal procedures. Furthermore, it may even be the case that working to the rule in such adverse conditions could be unsafe. Situational violations may be reduced by improving job design, supervision, and the working environment. For example, staff shortage may result in pressures to use only one position in the tower unit when a two-person team would have been the most suitable configuration to meet traffic complexity.
3. *Exceptional violations* that occur in circumstances involving familiar features combined in new ways or rare situations not covered in training. Practitioners work at the knowledge-based level as they should recognize new patterns of cues and modify existing rules of operation. For example, a radar controller may issue a clearance to an aircraft to descend lower than the prescribed terrain minima in order to accommodate a nonstandard wave of arrival traffic. This may be safe enough when the terrain is even (e.g., over the sea) and there are no obstacles (e.g., mountains) in the sector area.

The model of unsafe acts provides a useful framework for classifying human failures into a set of broad categories without the need to get into more details about the precise mental processes underlying a particular failure. Although Reason (1990) elaborated this model with a range of causal failures in the processing of information, the previous classification still provides useful insights. Later work by Shorrock and Kirwan (2002) applied this information processing model to the air traffic domain and developed the TRACEr classification scheme.

6.3.2 The TRACEr Classification

A great deal of work has been undertaken in recent years by Eurocontrol to develop tools and methodologies for the analysis of human error in ATM incidents. The technique for the retrospective and predictive analysis of cognitive error (TRACEr) has been developed for the identification of human errors that lead to incidents and the analysis of psychological mechanisms behind human errors (Shorrock and Kirwan 2002). On the surface, human errors may fall into a number of categories associated with the task being performed (e.g., radar monitoring, strip handling). Looking deeper, each error could be related to a number of malfunctions of mental processes (e.g., perception failures, judgment or planning failures, etc.). TRACEr considers human errors in four stages of information processing, as follows:

1. *Perception errors*: Failures in visual detection and visual search (e.g., late identification or no detection) and errors in listening
2. *Memory errors*: Forgetting temporary information, forgetting previous actions, and misremembering planned actions (e.g., forgetting to issue a clearance to an aircraft)
3. *Judgmental or planning errors*: Errors in judging aircraft trajectories, errors in making decisions, and errors in separation planning
4. *Execution errors*: Actions or speech performed not-as-planned or mis-timed

The ultimate analysis of human error involves tracing malfunctions of mental processes that resulted in the practitioner making an error. Examples include cases where controllers focused on a particular situation at the expense of others (e.g., perceptual tunneling) or cases where the human mind was unable to cope with the amount of information. For the analysis of human errors in ATM incidents, Shorrock and Kirwan (2002) developed a taxonomy of psychological error mechanisms (PEMs) to refer to the psychological biases that are known to affect human performance (see Table 6.1). PEMs provide a finer level

Table 6.1 Psychological Error Mechanisms in TRACEr

TYPE OF ERROR	PSYCHOLOGICAL ERROR MECHANISMS
Perception errors	Expectation bias, perceptual confusion, vigilance failure, distraction/preoccupation
Judgmental errors	Incorrect knowledge, misunderstanding, cognitive fixation, false assumption, prioritization failure
Memory errors	Similarity interference, memory capacity overload, negative transfer, frequency bias
Execution errors	Perceptual confusion, habit intrusion, manual variability, misarticulating, distraction/preoccupation

Source: Shorrock, S.T. and Kirwan, B., *Applied Ergonomics*, 33, 319–336, 2002.

of analysis, which is useful for error reduction and mitigation; however, they may require significant understanding of the psychological causes of errors, which may not always be obtainable from incident reports.

This error classification traces the origins of error into certain stages of the information processing system in the human mind. However, explaining errors on the basis of mental process malfunctions would leave other people guessing as to whether the investigator was right or not. Mental processes cannot be observed as they are convenient labels for understanding how the mind works. No one can actually see things such as working memory or perceptual stores and no one can go back into these mental processes of the controllers involved in the work. In this sense, investigators may seem to make justifiable conclusions but these remain largely unverifiable and low on credibility.

To some extent, this problem may be overcome by considering the context of work surrounding the incidence of error. TRACEr proposes another classification of tasks that can help analysts to probe and document the causes of observed behaviors. However, it is still difficult to see how people's mindset is unfolded over time as a function of their interaction with the work environment.

There is usually a dynamic interplay between mental processes and work situations that we need to understand in order to analyze human error. Recent research by Dekker (2006) has shown that people change the situation by acting upon their beliefs but also the evolving situation may change people's behavior. An evolving situation provides new evidence that people use to update their understanding and detect errors. A graphical representation of how people act, monitor the outcome of their actions, detect new problems, and correct their actions, allows verification and debate by other investigators and the practitioners who understand the domain. A large part of understanding human error involves understanding the situation in which practitioners are working, their tasks, and the tools that are used. For this reason, the remainder of this chapter is about how practitioners' assessments and responses evolve in parallel with the changing situation and how they detect and correct their actions.

6.4 A Framework for Understanding Error Detection and Recovery

With the increasing complexity of technical systems, there has been a realization that total elimination of human error may be difficult to achieve. There will always be complex situations in which errors may infiltrate due to high workload, decision-making under stress, and poor team coordination. In these situations, the management of adverse consequences through the detection and correction of errors may be more important than the prevention of errors in the first place. As a result, an increasing emphasis has been placed on how errors are detected, how errors are explained, and how consequences are controlled to maintain safety. Reason (2008) referred to this perspective as the human-as-hero stance, which focuses on the way that people make adjustments, recoveries, and improvisations.

An understanding of the error management process is essential in improving safety and reliability of operations. Since the early 1990s, a growing number of studies have examined error recovery in aviation (Wioland and Amalberti 1996; Sarter and Alexander 2000; Nikolic and Sarter 2007) and ATC (Bove 2004; Kontogiannis and Malakis 2009; Kontogiannis 2011). These studies have shown that a considerable number of errors are never detected or are detected too late for an effective intervention to take place. An observational study of normal airline operations (Thomas 2004) has shown that almost half the errors went undetected by flight crews, although only a small number of errors led to undesired aircraft states. Furthermore, the error detection rate was lower for mistakes but higher for slips and lapses. In addition, many errors were detected incidentally by a routine check rather than a deliberate monitoring strategy on work progress. These findings may indicate that proactive and self-monitoring strategies should become part of error-management training.

Most studies of error management (Rizzo et al. 1994; Kontogiannis 1999b; Kanse and van der Schaaf 2001) have tended to distinguish between three processes in error handling or error management, namely:

1. *Error Detection*—realizing that an error is about to occur or suspecting that an error has occurred
2. *Error Explanation*—identifying the nature of error and explaining why it occurred
3. *Error Correction or Recovery*—modifying an existing plan or developing a new one to compensate

Error handling or *error management* can be used interchangeably to refer to practitioner behaviors that relate to the three cognitive processes.

Error management processes are not discrete processes that are carried out in a fixed order. Kanse and van der Schaaf (2001) identified several patterns of explanation and recovery behaviors following error detection. For instance, recovery actions are often implemented prior to an elaborate explanation of what went wrong in order to keep the system in a stable condition. Observations on pilot recovery from automation problems on modern flight-decks have also illustrated the cyclical nature of error management (Sarter and Woods 2000).

The progression through the error management processes can be represented as a movement between a set of mental states. Figure 6.3 illustrates how error management fits into a state transition diagram of the practitioner's mental states. When an error occurs, the user enters a quasi-normal state where everything seems normal, but where some failure is imminent (Wood and Kieras 2002). The practitioner can continue performing correct actions in a quasi-normal state until something is found to be wrong, prompting him or her to recover. Once in the detection state, the practitioner may try to understand or explain the error (explanation transition) or choose to take immediate corrective action. When error correction is complete (recovery state), the user may return to the normal operation (resumption transition) which concludes the flow of error management.

Consider, for example, a typical aircraft sequencing problem where an approach controller is managing a wave of eight simultaneous aircraft arrivals to an airport. The controller may establish the sequence

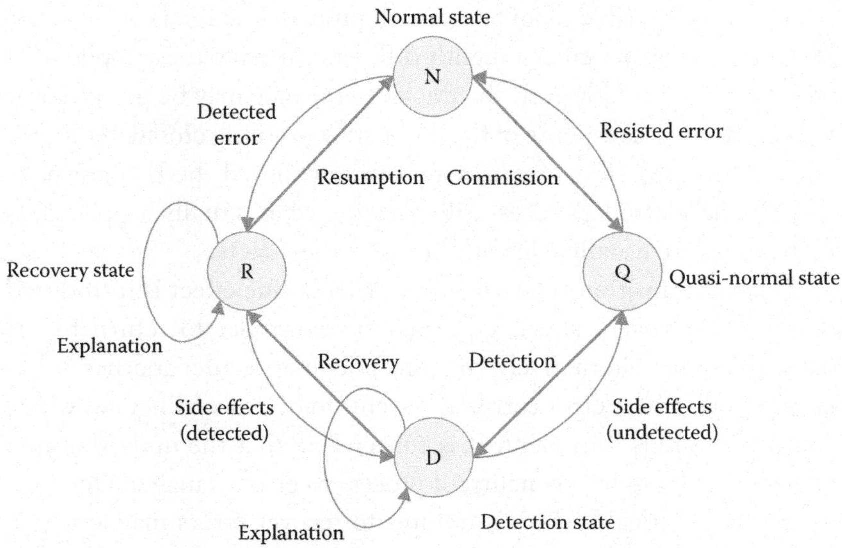

Figure 6.3 A nonlinear process of error detection, explanation, and recovery. (From Kontogiannis, T., *Journal of Safety Research*, 42, 73–85, 2011.)

in which the aircraft approach and their transfer to the tower unit. In this quasi-normal state, certain weaknesses in the approach sequence can be tolerated (e.g., the fourth aircraft is slightly slower than the fifth aircraft). The controller who provides instructions, clearances, and information, may realize at some point that this arrival sequence could lead to separation minima infringement (detection stage). The explanation might be quite straightforward (i.e., the fourth aircraft seemed to move slower than the fifth one, hence their separation distance cannot be maintained). In the recovery state, the controller may have to consider two options: (1) alter the position of the two aircraft or (2) tolerate a delay in the landing of all aircraft following the slow aircraft. After the sequencing problem is solved, the controller may return to the normal operation (resumption transition).

Several other transitions, however, may also occur that reflect the nonlinear nature of error management. For instance, the controller may be able to detect an error as the action is performed and jump directly to the recovery state. In other cases, a function may resist errors and cancel them out, unknown to the controllers. For example, during the last stages of approach, the final vector to establish the ILS proposed by an approach controller may not be at the right angle for interception. Prompted by on-board automation, the flight crew may realize that this vector may not enable them to establish properly the final course of the ILS; hence they may enter a slightly different turn into the autopilot that will do the trick. However, the track discrepancy may be perceived by the controller as a response of the flight crew to strong crosswind conditions at this area. The aircraft may have established the ILS normally without the controller's necessarily knowing what actually happened, as flight crews are usually reluctant to report these facts.

Another transition type can occur when a side effect is introduced during the recovery state, requiring the controller to return to the detection state. Alternatively, the controller may reenter another quasi-normal state where error correction seems to be proceeding, but where another failure is imminent. It is conceivable that the understanding of errors seems to be a concurrent process to error management.

In safety critical systems, attempts to recover errors may lead to a better understanding of the problem rather than to a particular solution. In addition, practitioners may not be able to fully understand the error until system activities are resumed and the system returns to a normal state (not shown in Figure 6.3). Hence, how and why practitioners move from one state to another is critical to understanding error management.

6.5 Cognitive Strategies in Error Detection and Identification

A review of error detection mechanisms (Blavier et al. 2005) has classified them into several forms on the basis of two criteria: (1) whether the agent who detected the error was the same person who committed it, an observer, or a function of the technical system, and (2) whether the error was detected before or after the results of the action appeared on the user interface. Outcome-based detection corresponds to the first form of detection that is triggered by a mismatch between observed and expected outcomes. Difficulties in attending to actual outcomes and in building expectations about effects can be the result of a combination of contextual factors. The action outcomes, for instance, may not be perceptible due to poor interface design, masked by safety logic interventions, or not sufficiently attended due to high workload. On the other hand, expectations about effects may be ill specified because of unfamiliarity with the work domain or may be attributed to other causes.

Errors can also be detected in the execution stage where practitioners notice a mismatch between actions being executed and actions specified in their plans. In ATC, lapses and slips of the tongue can be captured by a comparison between memory of issued instructions and instructions specified in the communication protocol of the operations manual. Sellen (1994) identifies another two forms of error detection based on the availability of "forcing functions" (i.e., design constraints that block deviations from the expected course of action) and cross-checking by another person. While a large number of slips and lapses can be detected by the person who committed the error, the detection of mistakes is more difficult because the same error-producing conditions may hinder error detection. Analyses of simulated scenarios in aviation (Thomas 2004) have shown that team monitoring can be a valuable source in detecting mistakes of other team members.

Error detection and recovery is regarded as a hallmark of expertise as practitioners are able to demonstrate their abilities to catch errors on the fly and come up with resilient methods of plan repairs and modifications. In many cases, people may form the impression that error detection and recovery are spontaneous processes where little preparation has been made by practitioners. This is not actually true, however, since error detection and recovery require that practitioners maintain a state of alertness and mindfulness manifested as: rehearsing tasks for future execution, bringing routine tasks into conscious attention, thinking out possible errors, drawing relationships between data, seeing how trajectories change over time, and cross-checking data for reliability. These cognitive strategies involve deliberate planning for the unexpected and self-introspection that enhances controller resilience. It is important, therefore, that we understand the cognitive strategies of action-based detection and outcome-based detection. Mindfulness has been outlined by G. Kranz, a former mission controller at NASA, as a process of maintaining gimlet-eyed focus on the job while gathering reserves for what lay ahead (Kranz 2001, 308). He coined the term "relaxed alertness" to refer to the state of mission controllers of Apollo 13 prior to the outbreak of events that ended in barely averting a great disaster.

Detection of mistakes can occur while a plan of action is formulated or an assessment is communicated to other team members. Several error detection strategies may be brought into play, such as revising

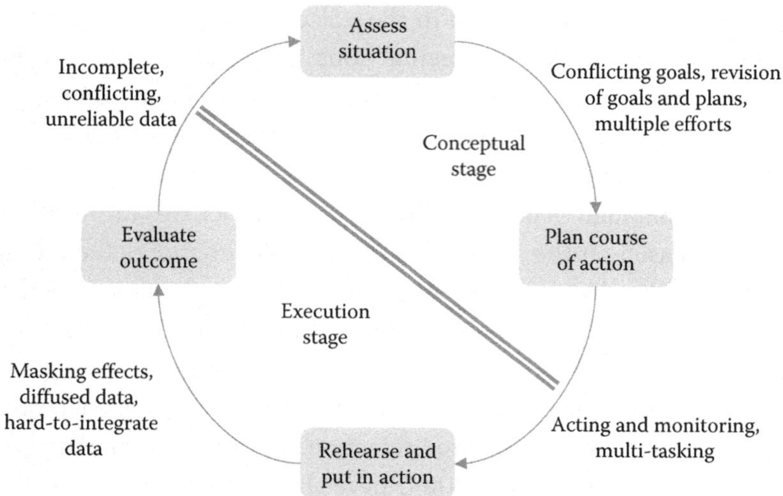

Figure 6.4 A four process model of performance in air traffic control. (From Kontogiannis, T. and Malakis, S., *Safety Science*, 47, 693–706, 2009.)

an assessment that appeared plausible at an earlier stage, finding hidden assumptions, thinking out possible errors, and deciding when and how often to review work progress. These cognitive strategies at the conceptual stage can be termed awareness-based detection and planning-based detection. To put into perspective the various cognitive strategies involved in detecting errors at the conceptual and execution stages, Figure 6.4 presents a simple model of human performance that encompasses four processes: (1) assessment of situation; (2) formulation of plans of action; (3) rehearsal, execution, and adaptation of plans; and (4) evaluation of outcomes of performance. The four processes operate in a circular fashion, so that feedback of performance can alter existing goals or modify earlier assessments of the situation.

This is a nonlinear model of performance since there is no need for practitioners to complete situation assessment before decision-making. Practitioners can live with some uncertainty about the situation and make a decision of how to tackle the problem at an early stage; as more evidence becomes available, the model of the situation can be revised. In the same sense, it is not necessary to specify in advance detailed plans for how to tackle a problem; modifications to plans can be made after some feedback is provided. For example, a flight crew experiencing an emergency may request to remain in a holding pattern in order to do some checks. The plan may be revised later by the crew and priority might be given to a direct approach to an airport since more time would be available now for controllers to respond to the emergency and clear the route between the holding point and a nearby airport.

This chapter proposes four processes in error detection, as follows:

1. *Awareness-based detection* where an assessment of the situation is revised in order to identify hidden and untested assumptions, collect missing data, and formulate a comprehensive account of problem causes.
2. *Planning-based detection* where a plan of action is revised so that new evidence is taken into account, conflicts between goals are balanced, and dependencies between tasks are reduced in order to provide more opportunities for error detection.
3. *Action-based detection* where errors are caught in the act by means of proactive strategies such as rehearsing tasks, thinking out possible errors in advance, and devising barriers at the execution stage.
4. *Outcome-based detection* where mismatches between expected and observed outcomes are identified with cognitive strategies such as seeing how trajectories change over time, spotting rates of change, and cross-checking data.

Figure 6.4 presents an incremental view of performance where an assessment can tolerate certain sources of uncertainty and proceed with a plan of action while remaining vigilant to new evidence. In error detection, controllers have to handle many sources of uncertainty, such as incomplete and conflicting data, goal trade-offs, task dependencies, masking effects, and automation effects. A prominent example of uncertainty sources in ATC regards the intensity and the movement of weather cells within a sector. The ATC radar system may depict some areas of active weather but this is not complete information since turbulence and other disturbing weather phenomena may be encountered well outside the areas depicted in the radar. As a result, flight crews may prefer different routes based on their own radar from the routes offered by ATC. Table 6.2 shows a taxonomy of cognitive strategies in error detection and identification as described in the following sections.

Table 6.2 Cognitive Strategies in Error Detection and Identification

ERROR DETECTION	COGNITIVE STRATEGIES
Awareness-based detection	• Tries to detect missing cues and find hidden assumptions • Does not explain away conflicting evidence • Tests the plausibility of assumptions
Planning-based detection	• Anticipates weaknesses in plans and identifies information needs • Considers a timescale for questioning the plan
Action-based detection	• Carries out preaction and postaction checks on routine tasks • Rehearses tasks that may be carried out later under time pressure • Creates reminders, task triggers, and error barriers
Outcome-based detection	• Examines relational and temporal patterns of changes • Verifies the accuracy and reliability of sensors

Source: Kontogiannis, T. and Malakis, S., Safety Science, 47, 693–706, 2009.

6.5.1 Strategies in Awareness-Based Detection

Detecting errors and misunderstandings in the way that practitioners construe a problem requires a process of introspection or self-monitoring. Studies in making sense of complex problems have proposed that practitioners tend to generate stories and explanations to account for problems that do not seem to fit previous experiences (Cohen et al. 1996; Klein et al. 2003). Story building entails building a mental model of the situation that is subject to critique and correction. Cohen et al. (1996) described story building as a process where experts struggle to construct complete and coherent models of the situation (i.e., the recognition/meta-recognition (R/M) framework).

An assessment is incomplete when key elements of the situation model are missing. Further data can be collected or retrieved from memory, which may result in models of the situation with contradictory elements. Under stress, some individuals may explain away evidence in order to maintain a coherent assessment or model of the situation. In contrast, experts manage to resolve conflicts by testing their explanations or assumptions for reliability. This may entail cross-checking related instruments, waiting for additional data, or inviting colleagues into the assessment; the reliability test usually results in dropping false data and explanations. The R/M framework facilitates critique and correction by reducing considerations into a single common currency: the reliability of assumptions. If data reliability is not acceptable, a new cycle of critiquing may trigger efforts to construct a new model of the situation. Other frameworks for studying the process of critiquing and correcting of mental models (Klein 2004) tend to agree that practitioners should develop skills in coping with uncertainty. Handling uncertainty and revising understanding are important elements of cognitive strategies in detecting mistakes before embarking on a course of action. On the basis of the R/M framework, three types of assessment-based detection strategies have been proposed as follows:

1. *Makes an effort to detect missing cues*: Controllers make sense of a situation by building a coherent story of what has happened; when there is a gap in their story, they look for additional cues. The challenge here is to estimate correctly the urgency of the problem and decide whether it is worth spending more time in collecting data. For example, after departure, a flight may be climbing with a slightly lower rate of climb than expected for this particular scenario. This may not seem initially to be a problem, since it can be attributed to heavy loading, tailwind, or noise restriction procedures; however, it may imply that something abnormal is happening. For this reason, later on, the controller may ask the flight crew directly the reason for the slow climb.
2. *Does not explain away conflicting evidence*: New data may be inconsistent with existing understanding or may reveal contradictions with data that have been trusted in the past. Unfortunately, controllers may explain away inconsistencies particularly under time pressure and this prevents error detection. For example, a controller may attribute small track discrepancies between heading changes and actual tracks to wind conditions when the cause may be a failure of the technicians to update the radar map with the new magnetic variation. Since Navaids are oriented in the magnetic north and not the true north, an uncorrected discrepancy between them may result in tracks displayed with a uniform small error.
3. *Tests the plausibility of assumptions*: Critiquing and correcting models of the situation rests heavily on testing the

trustworthiness or truth of underlying data and assumptions. Due to limitations in time and resources, some assumptions may not be possible to test but this is not a sufficient reason for rejection. In such cases, controllers can acknowledge the risks in their current assessment but take corrective actions so that their plans do not depend upon these assumptions.

6.5.2 Strategies in Planning-Based Detection

Planning is a process that ranges from setting goals and directions to scheduling detailed tasks. This section focuses mainly on detecting errors at the level of setting directions and goals that may be difficult to modify once a decision has been reached. In contrast, task scheduling is more amenable to error detection when practitioners have already chosen a correct goal and direction for action. According to the threat and error management (TEM) model, practitioners first try to avoid threats by anticipating "points of concern" and weaknesses in existing plans. Established plans are questioned and revised to address issues of completeness, consistency, and reliability. Plans can be adapted by regulating their level of specificity (that is, complexity) and modularity (that is, coupling). When changing a plan is not feasible, or not practical, then controllers could try and minimize the threats or error consequences. Two types of cognitive strategies for planning-based detection are presented below:

1. *Anticipates weaknesses in plans*: Practitioners are preoccupied with failures and make continuous efforts to anticipate adverse events that threaten the viability of their plans (Amalberti 1992; Klein 1998). In the ATM context, controllers often employ a threat acknowledgement strategy about certain flights or events and subsequently provide special handling instructions to flight crews. In revising an approach sequence, controllers may set "gates"—that is, certain airspace points where aircraft must be at the correct altitude, heading and speed. Revision of the approach sequence may be very difficult after the aircraft passed these gates. Similarly in aviation, a decision to continue an approach or perform a go-around has to consider the aircraft's position, height, speed,

and configuration. The flight crews assess these flight elements at "gates" which are established between 1500 ft.—500 ft. in order to determine whether to continue or abandon the approach and perform a go-around.

2. *Considers a timescale for evaluating progress*: Complex and dynamic environments make it difficult to think out a detailed plan that would work first time around. In most cases, a plan should be revised to take into account new developments of the situation. Hence, controllers may consider a timescale for revising a plan that prevents them from becoming absorbed by the situation and ensures that other colleagues are available in time to assist. An example from tower control concerns the startup of aircraft under restrictions from flow control (ATFCM). In many cases, tower controllers set their own timescales for meeting the actual flow restrictions—e.g., requiring aircraft to be on the main taxiway at least 5 minutes before the expiration of the flow restriction.

6.5.3 *Strategies in Action-Based Detection*

Self-monitoring can also play an important role in assessing task progress during the implementation of a plan. Inadequate self-monitoring can give rise to omissions, forgetting of steps that have been interrupted or deferred and failures to detect errors of others. Self-monitoring is a proactive strategy that can take three forms. The first is a general work habit where a routine check is made on previous actions, current ones, and actions that have been suspended or deferred. The second form assumes a rehearsal or preview of future actions that may be carried out later on, under time pressure. The third form concerns the creation of reminders, task triggers, and barriers in order to prevent errors or catch errors in the act. The reasons why and when an action sequence is checked may depend on the constraints of the task, the context of work and the practitioner idiosyncratic attitudes. The three types of cognitive strategies for action-based detection are presented below.

1. *Carries out preaction and postaction checks on routine task*: Operating in a familiar environment may result in several lapses and slips, as experts have given many routine tasks to

mental automation. Running a conscious check on highly routine tasks implies engaging in mental processes such as retrieving the intent of tasks, recalling withheld tasks, and rehearsing future steps. Another example involves running of postaction checks in order to review whether tasks have been executed, interrupted, or postponed. For instance, the marking of flight progress strips enables the recording of essential data required for the efficient operation of a particular control position (e.g., estimated time over significant points, vertical and horizontal speed instructions, clearance delivery, etc.).

2. *Rehearses tasks that may be carried out later*: Rehearsing tasks that may be carried out later, under time pressure, is a good strategy for preventing slips. For example, traffic volume follows ebbs and flows and controllers may mentally rehearse the tasks to be accomplished before a wave of arriving aircraft gets inside their sector.
3. *Creates reminders, task triggers, and error barriers*: Experienced controllers acquire useful habits that enable them to perform tasks skillfully but, at times, habits can get in the way of safety. Reminders can help controllers detect omissions, particularly in cases where tasks are independent from each other. Another way to combat this natural tendency is to create a barrier so that errors are stopped from having adverse consequences. For example, controllers may draw lines or polygons in addition to the standard layer on the radar map in order to serve as reminders or barriers. Drawing a line parallel to the airport can serve as a reminder for turning arriving aircraft before reaching the line in order to establish the downwind. Equally well, it may become a barrier not to be crossed by arriving aircraft and prevent a separation minima infringement with aircraft departing an airport.

6.5.4 *Strategies in Outcome-Based Detection*

Outcome-based detection relies on observing mismatches between actual and expected outcomes. Detection of mismatches can be difficult in modern technical systems for reasons related to the overload of information and the capacity of practitioners to formulate accurate and timely expectations. Studies on how practitioners cope with the problem of data overload (Woods 1995) have pointed out the role of the context in which data appear and the role of goals and expectations of the observers. A particular piece of information gains significance or meaning mainly from its relationship to the context in which it occurs (i.e., relationships to other data, the trajectory followed over time, the onset of changes, and any actions taken by other colleagues). Two cognitive strategies for outcome-based detection are as follows:

1. *Examines relational and temporal patterns of change*: Although modern systems overwhelm controllers with data, experts are able to see meaningful relationships that point to the semantic properties of the task. Knowing how long to monitor a trend is difficult and depends on the nature of problem. For example, expert controllers can form a fairly accurate prediction about the evolution of weather phenomena inside a sector and how these may affect operations by examining data and trends in the available meteorological information.
2. *Verifies the accuracy and reliability of information*: The patterns of change shown on the interface are produced through the use of sensors that may vary along several dimensions (e.g., sensitivity and reliability). Controllers should always check the accuracy and reliability of data provided by the sensors. Some information may be inaccurate if the update rate of the sensor is much slower that the speed of change of the problem. Other information may be false if the sensor is not functioning properly. The implication is that controllers should adopt a proactive strategy of cross-checking the reliability of information before drawing any inferences about the nature of the problem. A typical example refers to the cross-checking of radar data in a multi-radar environment, where aircraft tracks are constructed from the information derived from more than one radar head. So, when a radar head fails and the maintenance people do not inform the operations room, the controllers would not be aware that some flights tracks may be lost in the radar display. To overcome this problem, expert controllers always verify the accuracy and reliability of radar data by checking whether aircraft tracks are displayed correctly in suspect areas and altitudes.

6.6 Cognitive Strategies in Error Recovery

Error recovery presents many challenges because the original plan has failed, the situation has gone worse, and the available time for reaction is running out. In addition, work demands increase as controllers are required to come up with a better plan, while being in a situation that is even more prone to errors than when applying the initial plan. Although the processes of error detection have attracted a lot of research in the past, relatively little is known about the processes of replanning and error recovery in safety critical domains. For this reason, a framework is proposed in terms of error recovery strategies that would provide hypotheses for empirical research (Kontogiannis 2011).

The replanning or recovery process usually starts at the monitoring phase, where controllers become aware of any weaknesses in their plans. In many cases, the problem can be resolved with minor changes in the sequence of action; in familiar situations, a slip of action or lapse of attention can be corrected without the need to engage in higher-order processes. Most challenges to replanning regard the recovery of mistakes where the problem symptom may not match the mental model of controllers. A reassessment of current perceptions and plans may result in an elaborated or even a new frame with implications for how to repair plans. The mental rehearsal of options and plans can help controllers think of new solutions and foresee new risks. This section presents a taxonomy of cognitive strategies in error recovery.

6.6.1 Planning and Replanning in Error Recovery

In many ways, error recovery relies on a process of replanning whereby a plan is revised while being in a situation that is even more prone to errors. A common view is that experts are able to replan their actions on the fly and adapt to problems. However, this view fails to recognize all the preparations that have been taken by experts in order to build some flexibility in the initial plan of action. In this sense, a model of error recovery should address the processes of task coupling and coordination so that the initial plan remains adaptable to changes required at later moments.

Effective crew recovery requires regulation of task coupling—i.e., time slack and barriers between task components that can absorb disturbances from the environment. In tightly coupled plans, a disturbance in one task may spread out quickly at other tasks because there is no slack or buffer in the plan. In addition, tight coupling may increase coordination demands as controllers must know more about how activities in their scope of responsibility affect others. In contrast, loosely coupled plans can accommodate shocks, failures, and pressures for change without destabilizing effects. Plans with many loosely coupled tasks are referred to as modular plans (Klein 2009) since each task can be modified without affecting the others.

Error recovery is better served by loosely coupled plans that allow existing tactics to be combined in new ways since it is difficult to invent a plan under time pressure and communicate it to other controllers. Presenting the ingredients of heroic recovery, Prof. James Reason stated that "neither prior experience nor new techniques alone seem to be the defining features of heroic recovery; both can be involved, and the approximate balance depends upon those involved and the nature of the emergency situation" (Reason 2008, 235). This section examines three strategies of task coupling that allow practitioners to strike a balance between old practices and new ways of organizing in the process of error recovery.

1. *Uses redundant resources and barriers to absorb disturbances and minimize dependencies*: In general, barriers and buffers can reduce coupling and make it easier to modify plans in progress. Preserving resources and taking care that barriers and redundancies are available when needed is a good way of creating modular plans. Coupling is also affected by several dependencies that may creep up between barriers and tasks in a plan. Two tasks may share the same equipment, may be performed by the same controller, or may rely on the same resources; hence, a failure of common resources may fail both tasks. A combination of redundant equipment, operators and automated agents can be used to minimize dependency (Clarke 2005).
2. *Incorporates time slack to facilitate revision of task progress and plans*: Another challenge in error recovery is how to evaluate task progress in view of many delays encountered in dynamic work. These delays can create a lot of uncertainty of how

to proceed with further tasks in the sequence. A proactive approach would be to make allowances for possible delays and build some slack into the plan. Hence, loosely coupled plans can be more forgiving of delays and may provide a better basis for replanning.

3. *Makes provisions for other means and methods that may increase plan adaptability*: A challenge in error recovery is that a favorable option may become unworkable in the near future, as the situation takes an unexpected turn. Controllers should make provisions for alternative means to achieve the task goal. To increase plan adaptability, alternative means and methods for achieving tasks should be identified, preferably in advance of task performance. In an emergency situation, for instance, controllers may have to negotiate "action constraints" imposed on their work by the system in order to increase the degrees of freedom. In a sense, controllers are constantly keeping in mind a way out in case the preferred plan runs into difficulties.

The environment of error recovery is characterized by time pressure from the limited time window to respond and the psychological stress from the sense of responsibility that comes with the realization that an error occurred. Stress and time pressure, however, set the conditions for missing some factors in replanning a course of action. The implication for error recovery is that practitioners should retain some residual capacity for managing a number of secondary activities that have to do with correcting side-effects and coping with interruptions (e.g., due to alarms and intense communications).

In this sense, anticipatory planning can manage side effects and interruptions that are likely to occur in a demanding recovery process. Experienced controllers use their experience with different types of threats to anticipate problems when implementing their recovery plans. In other situations, controllers may have to anticipate novel events and see ominous connections between independent events. The other side of anticipatory planning is assessing capabilities to respond and making preparations. In his book *Fundamental Surprise*, Lanir (1983) showed that, in many military cases, the warnings and signals were fairly strong. The reason for surprise was that military officers overestimated their own abilities to react. Hence, anticipatory planning blends our abilities to notice weak signals as well as assess our capabilities to respond.

6.6.2 Coordination Tactics in Error Recovery

Recovering errors in complex systems often requires the cooperation of many practitioners, which raises important issues with regard to the delineation of responsibilities, reconciliation of different views, and communication among team members. In coordinating activities in a recovery plan, practitioners should make an effort to smooth out communications and avoid conflicts or side effects arising from their local perspectives. Coordination breakdowns are most likely in boundary areas where two or more practitioners control systems with common boundaries. This section considers three teamwork strategies in orchestrating actions to converge teams toward the overall recovery plan. The first strategy refers to sharing awareness and communicating intent in a team environment. The second has to do with keeping the size and duration of communications to the practical minimum and avoiding conversations that disrupt the thinking of colleagues. Finally, the third one refers to building plans that reduce the amount of coordination required by the operating teams so that more mental resources are reserved for recovering the problem.

1. *Shares awareness and communicates intent*: An error recovery plan is usually associated with a model of the situation that explains how the plan can recover the situation from the current state (Klein 1998). The recovery plan should carry with it some information about the rationale for moving to another state and explanations about the constraints of action. In distributed decision-making, sharing awareness facilitates adaptation as practitioners would be able to modify local plans without violating the rationale and constraints established in the overall plan.
2. *Minimizes information garbling in bottom-up communication*: Team members exchange information to articulate their planning, which requires sufficient time and cognitive resources to be accomplished. A strategy that minimizes workload in teamwork is knowing when to interrupt colleagues and when

to offer information that has not been requested; this is based on team members building a shared understanding of the situation. Several studies have found that providing unsolicited and proactive information can make team communication more efficient, especially at high tempos of works (Serfaty and Entin 1996). In a field study by Malakis et al. (2010b), expert controllers were able to communicate effectively without unnecessary elements that prolonged and garbled communications. They were able to appreciate major attributes of information (i.e., criticality and timelines) and were able to judge the level of workload and interruptibility of other team members.

3. *Selects actions that require less workload and coordination*: Abnormal situations can increase not only controller workload but also impose a higher communication load. In coping with complexity, controllers often choose options that reduce workload (Malakis et al. 2010b). Abnormal situations usually increase demands on team coordination by imposing a higher tempo of operations (e.g., rearranging aircraft routes, or increasing the number of contacts to be made with a large number of aircraft). In regulating their tempo of work, controllers seem to choose options that required less coordination.

6.7 Concluding Remarks

The proposed framework can be used to support error recovery through training, design, and decision aiding, and finally through incident investigation reporting. Error exposure training involves learning both from personal experience to errors and from vicarious exposure to errors (e.g., watching someone else commit errors). This type of training builds more accurate mental models of trainees, prevents repetition of errors, and increases transfer of skills to novel situations. In general, training methods that help people view their plans from different perspectives and decenter from their current vision of the plan would be very valuable. Premortem training (Klein 1998) is a well-tested exercise that prompts people to examine possible reasons for their plans failing.

Another focus of training would be to enable people to manage task coupling and increase opportunities for considering alternative methods to modify plans. For instance, whole-task training allows practice in fitting together task components in alternative ways and managing goal priorities (Means et al. 1993); this can increase the degrees of freedom for each task and facilitate the selection of alternative methods. Coupling also refers to aspects of time and resource constraints. The higher the time dependency, the greater the need of controllers to master appropriate skills for allocating attention and managing time-sharing. To manage time dependencies, training should enable controllers to cope with task interruptions, shifts in goals, and resumption of tasks. Trainees may be required, for instance, to practice the same scenario under a variety of conditions, such as multi-tasking, high time pressure, and many interruptions in order to learn how to develop skills in controlling attention and task management.

Decision aiding systems and operating procedures could also support controllers in adapting work methods to recover from errors and failures. An important aspect of error recovery is the ability of practitioners to retract and switch to the correct plan or move forward and create a new plan. In many safety-critical systems, switching to the correct plan should be done before proceeding too long with an inappropriate plan otherwise damages will ensue. One suggestion to support the switching of plans would be to design procedures so that the early stages of response are generic and shared with other procedures in the same problem domain. Hence, switching between procedures or methods could be made safely provided that this is done early enough and within the first stage of generic response. This can provide a basis for designing decision-aiding systems that allow controllers and flight crews to rehearse alternative tactics. Decision aids should allow practitioners to flip through the procedures in order to abstract the logic of the procedures and preview risks and constraints. In this way, they would be able to make all necessary preparations in order to switch to another tactic when the preferred one is not working any more.

Finally, the error recovery framework can be used to improve the quality of information in the investigation of mishaps. Incident reporting systems produce documents with extensive information but the quantity and quality of information concerning the contribution of the human factor causes are generally poor. Incident reports and accident investigation techniques should not only focus on erroneous performance outputs but also on the reasons why errors were not detected and corrected in time. In some cases, even a correct plan might be the result of tedious replanning after multiple unsuccessful past attempts; hence, reviews of replanning efforts could also provide useful data about system weakness that might endanger future activities. A systems perspective of accidents should capture information filters, mindsets, constraints, dependencies, and couplings of units that could impede the processes of error detection and recovery.

In error recovery, controllers are required to come up with a better plan of action while being in a situation that is even more prone to errors. This makes it difficult to invent a new plan and get a common stance or communicate the plan within the time pressure of the situation. As a result, controllers are more likely to be reviewing their understanding, reorganizing their existing plans, and managing the coordination of changes. In recovering errors, controllers are likely to adopt several cognitive strategies, such as seeing old things in new ways; making simple plans without simplifying the problem; managing task coupling, coordinating, and anticipating the needs of other people.

In error recovery, practitioners should match the complexity of the situation with their own capabilities and plans. This can be achieved either by reducing complexity or by creating modular plans that are adaptive to change. On the one hand, reducing complexity can bring to the fore critical information for replanning as well as enable people to remain sensitive to subtle events in the environment. Teamwork can also reduce the complexity of monitoring the situation but it comes with a cost of coordination; for this reason, it is important that teams smooth out communications and select actions that require less coordination so that more resources are devoted to recovering the problem. On the other hand, recovery is better served by modular and loose plans that allow more slack for error detection and more decision latitude for reorganizing parts to fit into a more effective plan. In this sense, an essential strategy of replanning under time pressure may concern specifying modular plans that can be adapted in progress. In addition, modular plans allow controllers more degrees of freedom in dealing with situational dilemmas and choosing whether to change to a new goal or when to initiate action.

A related strategy refers to anticipatory planning for managing a number of secondary activities that have to do with correcting side-effects and coping with interruptions. Operating teams should reserve some residual capacity to make preparations so that the recovery plan becomes more robust to threats and side-effects in the execution phase. Both anticipatory planning and sharing awareness can support the process of replanning. By sharing awareness, for instance, controllers would be able to amend plans locally without introducing side-effects because the rationale and constraints of the overall plan have been communicated. Both strategies, therefore, can support adaptation of controllers so that the recovery plans become more robust.

In contrast to errors, whose occurrence and forms are relatively predictable, human recoveries are much more unforeseeable events (Reason 2008). Abnormal situations present new challenges, making it hard to find an optimum balance between maintaining coherence and remaining mindful, or between taking immediate action and retaining thoroughness. Probably the greatest challenge to error recovery research is how to support controllers in achieving a balance between coherence and mindfulness in error detection and a balance between efficiency and thoroughness in error recovery.

7

Adaptive Practices in Air Traffic Control

7.1 Introduction

The interplay between formal work organization (e.g., procedures, safety rules, and structures of authority) and practitioner work practices has attracted a lot of research in human factors (Hale and Swuste 1998; McDonald 2006; Nathanael and Marmaras 2008). Formal descriptions of work usually produce general work patterns that allow a great body of operational knowledge to be expressed in a coherent and orderly fashion. However, the actual practices on the job may deviate from formal procedures for several reasons such as unanticipated events, goal conflicts not addressed in formal rules, time pressure, and changes in technology that require modifications to existing procedures (Kontogiannis 1999a, b; Dekker 2006). Very often, the terms "violations" and "workarounds" have been used to refer to work practices that are not in accord with formal procedures. Violations and workarounds occur frequently in all industries, as they can bring some benefits to individuals and organizations. Sometimes, they are tolerated by organizations as their impact on safety can be controlled with other system defenses.

A work practice is a set of recurrent actions over multiple occasions emerged through human–system interaction over time and shared by the organization (Lave and Wenger 1991). With increasing levels of expertise, however, work practices can become habitual behavior with practitioners failing to adapt when circumstances change. In other cases, practitioners may become absorbed in local optimizations that may produce remote side-effects in other parts of the organization (Snook 2000).

Practices may coexist with the formal work organization but are rarely recognized as complementary by management. In general, safety audits performed on industrial systems usually exclude work practices from a thorough consideration or perceive them as work-arounds to be avoided. As a result, work practices cannot be assessed and improved by formal safety audits or training and rely only on occasional opportunities for on-the-job learning. This chapter reviews earlier research in work practices, identifies vulnerability factors in their adaptation, and addresses several challenges in their integration with the formal work organization.

Traditionally, work practices have been associated with work-arounds and violations from procedures and norms of work (Reason 1997); this functionalist approach to work organization assumes that human responses to critical situations should be made explicit in operating procedures. With the increasing complexity of socio-technical systems, however, there has been a wider recognition that many work issues cannot be specified in procedures and remain to be resolved by people at the sharp end (Woods and Cooke 2002). Work practices capture the cognitive flexibility that experts exhibit in work contexts characterized by complexity and uncertainty (Feltovich et al. 1997). Such practices and others that work around a limitation of the system, are also referred to as "workarounds" (Koopman and Hoffman 2002). According to high reliability organization theory (Weick et al. 1999), work practices also include elements of self-introspection and mindfulness, which increase their learning potential across many situations.

A recurrent concern is also whether work practices that lead to danger or harm have different characteristics than other practices that have more successful outcomes. In hindsight, it may appear that work practices with adverse outcomes are, by their nature, different from practices with desirable outcomes. Recent research in safety critical systems (Hollnagel 2009; Hoffman and Woods 2011) converges in their claim that all practices entail an effort to adapt performance to variations of work by making difficult choices between decision trade-offs; however, the complexity of systems and the constraints of organizations do not provide a firm basis to predict that the final outcome of work practices will be delivered without delays and errors.

Rasmussen (1994) and Amalberti (2001) studied how organizational constraints and work complexity are likely to affect the development of work practices in safety critical systems. Their perspective emphasizes that practitioners operate within an envelope of possible actions that is influenced by wider organizational boundaries (see also Chapter 3). Practitioners try to follow the path that seems most useful and productive within this limited space of possibilities. Under time pressure and organizational constraints, practitioners may gradually become accustomed to operating at the margin of safety, which eventually increases the possibility of disaster. Practitioners should be able to compensate for the under-specification of formal procedures and develop work practices that cope with system variations and disturbances. This may be a well-espoused approach, but there are still many challenges about how to integrate formal descriptions of work with actual work practices. This chapter addresses a number of challenges regarding work practices and formal work organization such as the following:

- How are work practices that lead to danger different from other practices that have more successful outcomes?
- What cognitive and social factors make work practices vulnerable to failure?
- Are there any mindful strategies that allow practitioners to replan or modify successfully their work practices on the job?
- How can good practices be recognized and discussed within organizations?
- What are the difficulties in documenting and transmitting good practices across organizations?

7.2 Conditions Creating Performance Variability in Work Practices

Work practices are not uniform across practitioners and also not stable over time because practitioners learn to explore conditions of work and elaborate practices as more opportunities arise. Furthermore, work practices develop through experimentation or trial and error, rather than formal knowledge and instruction, hence producing variable outcomes and remaining vulnerable to error (Rasmussen 1994). Examining the variability of systems and the conditions in which work practices develop can enable us to understand the factors that make them vulnerable to failure. This section looks into several conditions that require controllers to develop different work practices to deal with problems that are not covered in the formal work organization (Kontogiannis and Malakis 2013a).

7.2.1 Variability of Task Characteristics

Work in complex systems is characterized by variations in the nature of tasks, conditions of work, and environmental requirements. The nature of tasks, for instance, may change and require modifications in work practices due to variations in the triggering conditions, shortages of resources, and time constraints. For example, the introduction of a runway change in certain hours for environmental reasons (e.g., noise abatement) may impose time constraints in tower and approach operations. This may require adaptation of practices such as changes in coordination (Nathanael and Marmaras 2008), recombination of existing procedures, or creation of new ones (Klein 2009). Variations in the nature of tasks and professional norms, valuing judgment rather than compliance, usually create the conditions for workarounds or deviations from procedures.

In the air traffic control (ATC) domain there are many sources of task variability, including:

- Updates in operational manuals
- Revisions of letters of agreement (LoAs) with adjacent units
- Revisions of unit competency schemes (UCSs)
- Revisions of contingency plans
- Updates on the communication navigation surveillance (CNS) system infrastructure
- Changes in airlines' operational procedures
- Updates in airport equipment

For example, a revised LoA with an adjacent unit may inflict significant task variability due to changes in coordination points, handover levels, and handoff procedures. Although a training program may familiarize controllers with the new LoA, controllers may be left on their own to develop new practices in applying the LoA in the operational context. For instance, the introduction of a new transfer point between two units requires new conflict resolution practices to be developed by the controllers.

Changes in CNS systems is another source of task variability. The introduction of new equipment requires controllers, not only to learn a new interface, but most importantly requires the acquisition of new operational knowledge. Through their experience, controllers gain a deep insight on the functions of CNS systems, develop effective practices, and acquire comprehensive knowledge of malfunctions, recovery times, and system limitations. For example, a controller may develop a fairly accurate insight of the coverage limitations of the radar system, know the weather patterns that are not displayed correctly, and get a grasp of system malfunctions and recovery times. The introduction of a new radar would require controllers to build new knowledge and operating procedures.

Finally, the introduction of new aircraft types may pose some challenges for controllers. They should build an accurate understanding of its climb and descend patterns, the range of speed at various flight levels, its approach patterns, and its performance in certain contingencies. Eventually mental models should be updated on how the new type of aircraft can be fitted into the traffic flow without any operational problems.

7.2.2 Organizational Changes and Transitions

Organizational changes create instability that causes more variability into the performance of practitioners and more changes in their methods of work. Pettersen and Aase (2008) studied the role of change and restructuring in the line maintenance of a regional airport. Initially, all technical functions were under the same department, which provided some organizational slack in the form of knowledge and competence in matching demands and resources. When the department was merged with another airline organization, some structural walls were built that took away slack and resulted in a loss of operational experience in the new way of organizing. In general, organizational changes may give rise to vague responsibilities about safety, loss of competence, and low morale, all of which can impact the way that safety practices are organized (Pettersen and Aase 2008; Reiman 2010).

In an effort to meet demands imposed by the implementation of new regulations, many air navigation service providers (ANSPs) undergo constant changes in their work organization. In many cases, regulatory changes could affect the whole organizational structure and functioning of an ANSP. Suppose that a new regulation introduces a new safety investigation technique with another incident categorization scheme. The ANSP should have to train safety investigators in the new technique, learn how to use the new incident categorization, and establish new investigation procedures and lines of reporting. However, the application of this regulatory scheme may produce alterations in the classification of causes and lessons learned from incidents. Because safety techniques are quite general by their nature, they may fail to capture the operational complexities of particular organizations. For instance, critical incidents with high informative value for controllers may be downgraded while other less significant ones may be classified as important incidents. This could result in controllers over-reporting incidents of low informative value while under-reporting significant ones that need urgent intervention. Hence, compliance with a new regulatory scheme may come at a cost of under-reporting truly significant incidents.

7.2.3 Goal Conflicts That Cannot be Reconciled

Economic and production pressures in complex systems create the conditions for goal conflicts that are experienced as daily trade-offs by practitioners throughout the organization. Rasmussen (1994) suggested that work in complex systems is bounded by economic, workload, and safety constraints, which leave small room for practitioners to maneuver. As a result of increasing economic pressures and resource scarcities, there is a gradual migration of performance toward the boundaries of workload and safety (see also Chapter 3). Therefore, practitioners have to work by making trade-offs between conflicting goals as well as between values or costs placed on different outcomes of work. Hoffman and Woods (2011) claim that practitioners have to balance five fundamental trade-offs in their work with regard to aspects of efficiency, thoroughness, planning horizon, team roles, and work organization. Since goal trade-offs are usually not addressed in operating procedures or training, practitioners may take certain risks to compensate for inadequate planning, time or resources, that is, matters that should have been handled by the organization.

A typical example is when an airport demands more capacity than the one declared by the Tower unit. This is usually accompanied by other demands to change runways at certain hours of the day, to use preferential taxi routes, or to wave out air traffic flow and capacity management (ATFCM) restrictions. The aim of demands and pressures is usually to increase the efficiency of airline and airport operations. Although each individual demand can be successfully met, their combination is likely to produce goal conflicts that cannot be reconciled. Take for example the case of a flight that faces a delay in its departure due to security reasons in the terminal and misses its departure slot. The ATFCM system is automatically fed with the new data and a new departure slot, two hours later, is allocated to this flight due to capacity restrictions. Unfortunately, Tower controllers may get very busy with a wave of departing aircraft and have to work above their capacity limits. This unexpected turn of the situation creates a lot of anxiety to the affected flight crew who need to take off as soon as possible because their destination airport is closing at night and there is a risk of flight cancellation. To make matters worse, the airport authorities may inform the controllers that the parking stand of the delayed flight has been allocated to other flights that are awaiting on the taxiway. All these economic, capacity, and efficiency pressures may leave controllers with a narrow space to maneuver and make decisions. In the end, the Tower controllers may be able to cancel out this restriction on the delayed flight and allow the flight crew to depart earlier in order to find the destination airport open, later at night.

7.2.4 Unruly Technology

Other sources of variability in work practices can be traced into the disorderly or "unruly" aspects of technological systems after they are released in a field of practice. Engineering design provides an image of tidiness and control over technology which is not true when it comes to operate in the messy field of practice (Dekker 2011). In fact, many designers and engineers recognize the need to hold safety audits and revisions to take on board features of technology that reveal themselves only after years of use. In the periods between these audits and revisions, however, the prescribed use of technology may not protect practitioners from unexpected events. Examples in the air traffic management (ATM) domain may include: unfriendly interfaces, flight data processing systems tailored for Area control that are also transferred to Approach units for cost reasons, safety nets producing false alarms, and latent software faults in system flight servers.

An example of a system wide software failure occurred in the UK airspace (12 December 2014) regarding the computer system that provided information to National Aviation Traffic Services (NATS) controllers to manage traffic flying high over England and Wales. The controllers made use of procedures to prevent traffic from entering their area of responsibility and resorted to manual methods to ensure aircraft separation (Walmsley et al. 2015). At 14:55, all departures from London airports were cancelled followed by a cancellation of all departures from European airports that were planned to fly through the UK airspace. The functionality of the computer system was partially restored after an hour while full recovery required another half an hour. The failure occurred because of a latent software fault that was present from the 1990s (Walmsley et al. 2015).

The fault was in the software check of the maximum permitted number of people using the system (known as "atomic roles"). The software should have checked whether the limit of 193 atomic roles had been reached; instead the check was performed against a civil limit of 151 atomic roles (Walmsley et al. 2015). At the time of the incident, the total number of atomic roles in use was 153, a figure that was reached for the first time because a change was introduced on the previous day in order to include further military controller roles. This change was not sufficient, in itself, to cause the failure. The workstations in the operations room were left either in the "Signed on" mode for normal control or in the "Watching" mode for monitoring traffic as an observer. Due to the inclusion of extra military controller roles on the previous day, a few unintentional requests were made to enter the "Watching" mode, which led to the recording of 153 roles in total. This exceeded the limit of 151 roles and raised an "exception" in the running of the software that resulted in the shutdown of the system (see Chapter 12 for a systemic analysis of the incident).

7.2.5 Professional Norms and Social Context

Responding to the variability of the environment requires organizations to develop flexible practices and this may create a tension with the requirement to rely on standard procedures. This tension between compliance and adaptation has been reported by many practitioners in their daily work. For instance, McDonald

(2006) summarized the results from a series of European projects concerning aircraft maintenance and concluded that practitioners would find "better, quicker and even safer ways of doing their tasks" than following procedures routinely. Practitioners usually consider adaptation and interpretation of procedures to be an integral part of their work. Any tendency to proceduralize their tasks may be perceived as a threat to their job motivation, meaningfulness of work, and ability to carry out daily work. Their professional norms tend to value judgment rather than compliance, confidence in their abilities to solve problems and reliance on teamwork to promote safety. Adaptation of rules and procedures is also affected by the social context of work and the professional cultures. In a study of aircraft maintenance practices, for instance, Pettersen and Aase (2008) found that the formal documentation system had "gray areas" where "trial-and-error" strategies were essential for problem solving. Technicians adapted their practices, which were supported by a cultural imperative to promote safety but maintained awareness of their own knowledge imperfection.

Similar examples can be found in the ATM domain where controllers develop practices to compensate for limitations in CNS systems, bumpy transfer of control between positions, inadequacies of the ATFCM system, and ambiguities in ICAO or national legislation. For example, ICAO has explicitly acknowledged the fact that the circumstances surrounding an emergency preclude the establishment of detailed procedures to be followed; finally, ATC units shall maintain complete coordination and personnel shall use their best judgement in handling emergency situations (ICAO 2007a. 15-1). Hence, controllers should be able to perform a resilient adaptation and a flexible interpretation of procedures to respond to difficult situations which becomes the hallmark of professional expertise.

7.3 A System Dynamics View of Work Practices

This section provides a system dynamics view of the way that work practices are developed, revised, and articulated within organizations. This perspective has been based on the Repetition–Distinction–Description (RDD) model that sees work practices as the product of repetition, regular distinction and guidance from the formal work organization (Nathanael and Marmaras 2008). System dynamics is a technique for depicting interactions of circular processes with different time lags and influential factors (Sterman 2000). System dynamics provide a framework for dealing with complexity where causes and effects are not explicitly related (Leveson 2012). The RDD model has been modified in some respects and recast in system dynamics terms to capture influential factors and processes that reinforce or adapt existing practices (see Figure 7.1).

Nathanael and Marmaras (2008) claim that work communities become progressively familiar with their environment through regular repetition of actions in a variety of situations. Repetition is not simply a recurrence of experiences but a reenactment that gives rise to further elaborations in other situations (e.g., creating classification or testing other uses of artifacts). In many cases, reenactment can increase the chances of success and build more confidence in a practice (i.e., the plus signs imply that all variables change in the same direction). This may create a reinforcing loop (see R_repetition loop in Figure 7.1) where a practice of increases its chances that will prevail

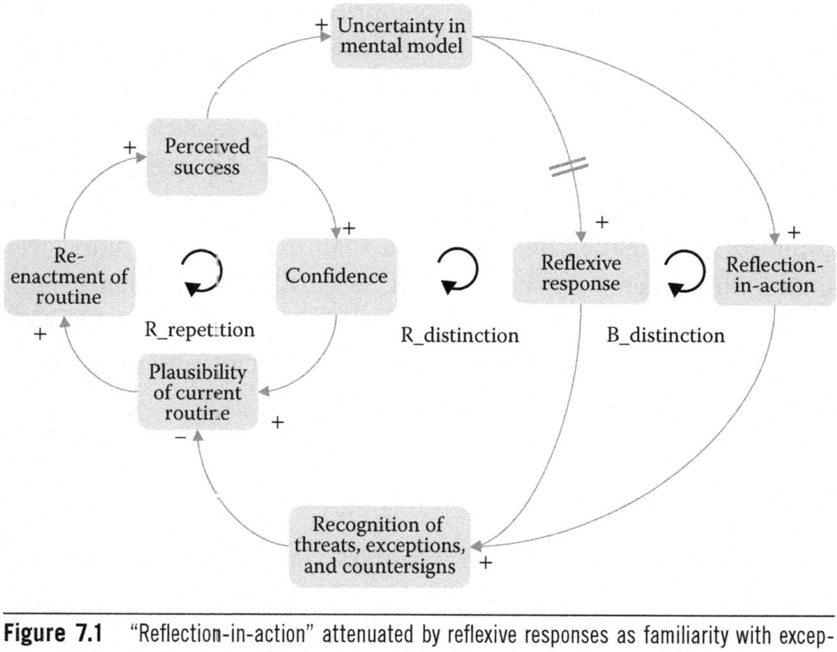

Figure 7.1 "Reflection-in-action" attenuated by reflexive responses as familiarity with exceptions and threats increases with experience. (From Kontogiannis, T. and Malakis, S., *Theoretical Issues in Ergonomics Science,* 14, 6, 565–591, 2013a.)

in similar situations. For example, a Tower controller may develop an elaborate practice of when to issue a take-off clearance with regard to arriving aircraft. The controller learns to create clarifications of distances from the threshold points, distances of the taxiing aircraft from the runway holding position, ground speeds of arriving aircraft and so on. Successful application of a work practice implies that the controller builds confidence that is reinforced by constant operational realization.

When "exceptions to the rules" and "unexpected events" occur, practitioners may change their practices and create new distinctions. Breakdowns in routine practice may increase uncertainty in the mental models of practitioners and foster "reflection-in-action" (Scholn 1983). In this way, practitioners can recognize threats, "exceptions to the rules" or "countersigns" and progressively enhance their practices. When combined with a mindset of alertness, mental models can help practitioners challenge their understanding and remain vigilant to the possibility of failure (Weick and Sutcliffe 2001). For example, the Tower controller may notice that the flight crews of a certain airline may take longer to become airborne from the moment they get the take-off clearance. Similar observations with different wind conditions and other types of aircraft of the same airline may be registered as "exceptions to the rules" how to handle typical patterns of takeoff.

Once a new type of situation or a "way of act" has been identified through this distinction loop (see B_distinction loop in Figure 7.1), it can be gradually exercised in a repetition process, hence enriching or substituting parts of existing practices (Kontogiannis and Malakis 2013a). The delay mark (i.e., the double line) implies that the distinction loop lags behind the repetition loop. The distinction loop (or "refection-in-action" loop) is so highly situated in the context of work that practitioners may fail to recognize and continue to live with some apparent inefficiencies in their work (e.g., redundant actions, overload in communications, poor use of equipment). Nevertheless, as more opportunities are presented, practitioners gradually alter their practices through minute adaptations, without deliberately trying to do so. In the previous example, the Tower controller may explain away an observed delay in becoming airborne due to other operational reasons (e.g., wind conditions or that the cabin crew has not secured cabin on time) This airborne pattern may be classified as an "exception to the rule" that requires a special work practice when it is repeated many times or gets verified with other controllers as well.

An interesting point here is that "reflection-in-action" may be attenuated by several reflexive responses or habitual actions. As variations of the environment are successfully managed through "reflection-in-action," practitioners may develop reflexive responses that bypass the mindful process of reflection. As a result, reflexive responses may reinforce existing practices that are inappropriate in the new context of work (see reinforcement loop R_distinction, in Figure 7.1).

Nathanael and Marmaras (2008) also highlighted another process that affects the adaptation of work practices. At times, practitioners step out of their context of work and start reflecting on it (Scholn 1983). By exchanging stories and reflecting on action, practitioners develop new interpretations of "what they do" and generate new ideas that support anticipation of new ways of acting. These descriptions also lead to the formalization of practice, which includes the expression of "rules of conduct," "tips of the trade," declared responsibilities and also the design of new tools. This "reflection-on-action" process is another balancing loop (not shown in Figure 7.1) that changes existing practices by interacting with the realm of action that is the repetition–distinction double loop. For example, when a new point is established in an Approach sector, controllers get engaged in informal discussions sharing success and failure stories of resolution strategies around this issue. Through elaborate discussions, controllers may generate new ideas on how to resolve conflicts that emerge as a recurring problem. A controller may devise an ingenious practice of reducing complexity by establishing a new traffic flow. The practice may be observed by other controllers who may make important suggestions that lead to further improvements. A final test of this practice on the simulator may lay the ground for turning this practice into a formal procedure.

Overall, the effects of existing practices and past successes can be mediated by the two processes of "reflection-in-action" and "reflection-on-action" The two reflection processes may vary along the same continuum of articulation and testing of practices. "Refection-in-action" refers to adaptation that is situated on daily work experiences whereas "reflection-on-action" is a more general process that requires articulation of daily experience and a more formal critique of practices (e.g., a new practice may be tested through a simulator). The following two sections look deeper into these two adaptive and critiquing processes, propose ways in which they can be put in practical use and raise some challenges for further consideration.

7.4 Reflection-in-Action: Mindful Work Practices and Improvisation

An interesting question is whether the "reflection-in-action" loop should be understood as a deliberate effort by practitioners to alter their practices (Kontogiannis and Malakis 2009) or more like mindful habits that are disengaged from mental models and formal reasoning (Nathanael and Marmaras 2008). The two perspectives have different implications for the training of practitioners. The former relies on comprehensive knowledge to cope with threats whereas the latter relies on mindful habits to avoid threats in old practices. An integration of the two perspectives can be made by using the Threat and Error Management (TEM) model of Helmreich in the context of error recovery (see Chapter 6). The TEM model proposes that practitioners first try to recognize and avoid situations with a high-error potential; second, if this is not feasible or not practical, efforts are made to eliminate some risk factors. Recognizing and avoiding situations with high-error potential requires practitioners to develop a critical mindset that prepares them for the occurrence of errors. Weick and Sutcliffe (2001) referred to this stance of alertness as "preoccupation with failure." Awareness of vulnerability to errors makes practitioners recognize that, although they think they understand the system and the ways in which it can fail, surprises are still possible.

Weick and Sutcliffe (2001) emphasized the traps involved in preoccupation with short-term success and false optimism. Successful application of established workarounds can make practitioners less tolerant of conflicting evidence and may breed overconfidence which reinforces the current plan of action (i.e., the repetition loop R_repetition in Figure 7.2). However, an alertness mindset can create a balancing loop (i.e., the B_repetition in Figure 7.2) that builds a healthy skepticism on the strength and applicability of current practice (Kontogiannis and Malakis 2013a). "Preoccupation with failure" can moderate the effects of the reinforcement loop (R_repetition) as it becomes a useful work habit or "second nature" of practitioners. As the complexity of a system arises out of the difficulty to make predictions about outcomes, work practices that deviate from formal rules must be backed up with comprehensive mental models. Two other mindful strategies seem to rely on good mental models to judge whether threats can be avoided or efforts should be made to minimize their consequences and repair plans on-the-fly. Figure 7.2 shows that forestalling and replanning are two essential skills (see B_distinction) in the second aspect of the TEM model. Mindful strategies can range from rules for judging error potentials and taking precautions to rules for assessing one's own capacity to cope with escalation and error recovery.

Reason (2008) has used the term "error wisdom" to refer to the mental skills required to recognize situations with a high-error potential. Before embarking on a workaround, practitioners should be able to make intuitive assessments of their task, the context of work and the situation. While most practitioners have an understanding of

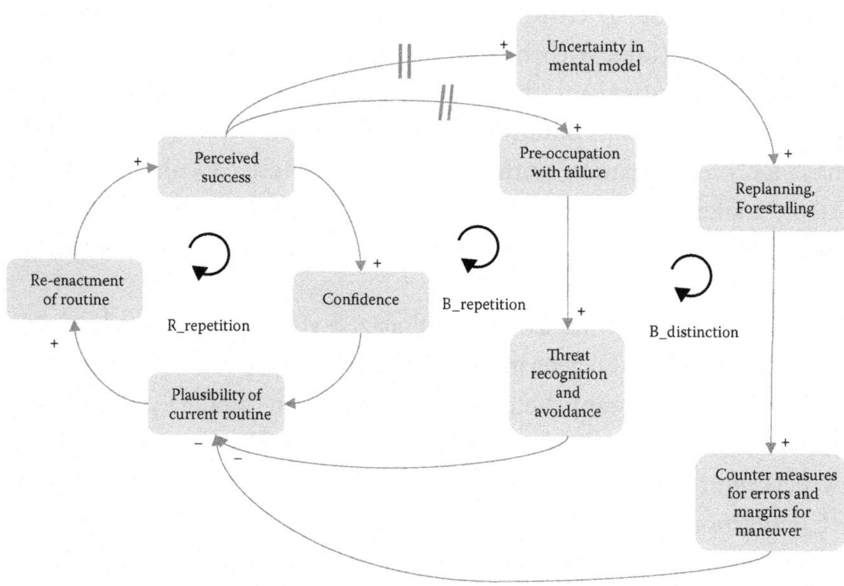

Figure 7.2 Mindful practices that amplify "reflection-in-action." (From Kontogiannis, T. and Malakis, S., *Theoretical Issues in Ergonomics Science*, 14, 6, 565–591, 2013a.)

the situational demands and the context of work, they are less likely to know that workarounds vary in their error potential. Weighing these error factors may be a complex task to proceduralise and other forms of guidance may be used to make decisions about risk potential. "Foresight training" has been a form of guidance that could provide practitioners with mental readiness skills in assessing their capacity to engage in high-risk situations (Reason 2008). Developing an alertness mindset and taking precautions to deal with threats is another useful tactic for mindful workarounds.

When threats or errors cannot be practically eliminated, practitioners should be able to forestall their propagation throughout the system or minimize their impact. The discovery of an error, or the presence of a threat, is characterized by time pressure and psychological stress from the sense of responsibility that comes with the realization that an error occurred. Stress and time pressure set the conditions for missing out some factors in replanning a course of action. This may introduce side-effects and errors in the implementation of the recovery plan. The implication is that practitioners should retain some residual capacity for managing a number of secondary activities that have to do with correcting side-effects and coping with task interruptions (Grote 2009). In this sense, forestalling goes beyond the prediction of impact and includes assessing one's own capabilities and "margins for maneuver." Hence, operating teams should reserve some resources for monitoring and correcting their workarounds. Other sources of support in managing side-effects in the process of error recovery may regard several teamwork functions such as cross-checking team members and shared understanding (Kontogiannis and Malakis 2009).

Expert controllers are not content only with mastering conflict resolution skills for maintaining safety but continue to learn from daily experience. For instance, experts can notice many distinct patterns of traffic that are not strictly related to safety but indicate subtle aspects of aircraft performance; in contrast, novices engage in secondary tasks instead of paying attention to traffic patterns that support further learning. As a result, experts develop better mental models of relationships between aircraft speed, wind or weather conditions, and the resultant rates of descending/climbing and route profiles. This knowledge is not captured in written documents but it is built up from daily experience and feedback from on-the-job learning.

7.5 Reflection-on-Action: Organizational Learning and Practice Communities

In trying to learn from experience, organizations face some challenges on how to select suitable events for learning and how to transform learning opportunities that occur at a local level into knowledge at a global scale. For instance, the choice of learning events may be a difficult decision because critical incidents are not necessarily the ones that produce the most useful lessons; many challenging situations are managed, without leading to severe outcomes, which reveals innovative forms of adaptation. These situations are officially recorded in the ATC domain and they can be shared only through informal discussions between controllers. In general, past situations can provide opportunities to discuss the relevance of procedures and become aware of successful modifications of strategies made on-the-fly. This section addresses three critical learning issues in the collection, documentation, and transfer of work practices, that is,

1. Should all practices be documented and communicated to the workforce, or is it better to rely on other means of communicating knowledge?
2. Can good practices transfer to other organizations with different socio-historical contexts?
3. Should specialized practices communicated to all practitioners or only to a selective pool of experts?

Many organizations tend to document their good practices as a means of encapsulating knowledge at work and disseminate it to a wide range of potential users. However, many practices may contain tacit knowledge that is difficult to extract from the particular situation (Lave and Wenger 1991). Many patterns of coordination and mutual adjustment are not easily reproduced or described in words, which eludes precise codification. An important challenge to organizations is whether to document all good practices or resort to other means of communicating organizational knowledge.

Baumard (1999) argued that tacit knowledge does not need to be verbalized to be acted upon; sometimes tacit knowledge is most influential when left aside from the explication process of organizations. Although the codification of knowledge may enhance the capacity of practitioners to react rapidly to changing conditions, attempts to codify all knowledge could be counter-productive, since flexibility and creativity would be reduced. Baumard (1999) found it useful for organizations to change between different forms of knowledge transfer, according to the context of work. Instead of documenting all practices, organizations may choose to use other means to transfer practices such as relying on demonstration or show-how, rotating practitioners between units, and embedding knowledge in tools and technology (Argote and Ingram 2000).

Work practices developed through experimentation may lead to performance improvements in some organizations but not others, since socio-historical contexts favor the selection of one practice over others. The challenge here is whether good practices can transfer to other organizations with different socio-historical contexts. Organizations tend to move along certain trajectories in which past experience contributes to particular directions of change and reinforces the existing stock of knowledge and expertise (Garud et al. 2010). In this sense, the transfer of practices may be path-dependent, as organizations build upon previous knowledge to acquire new knowledge (Alange et al. 1998). The path dependency of the transfer process can "lock" organizations into a specific learning path that is unable to integrate practices with different perspectives. This phenomenon is so pervasive that the phrase "not invented here" has been coined to refer to practices from different contexts that are less valued in the organization. This is because practices are embedded in social relationships and existing "ways of doing things" which may constrain the assimilation of new practices. It seems that a crucial factor in the transfer of work practices is the organization's ability to deal with a paradox that is using prior knowledge effectively to assimilate new practices while being ready to discard it in favor of new knowledge (Lewis 2000).

Many practices require a high degree of expertise in dealing with risk which implies that not all practitioners should try out and learn such practices. Instead of making a practice widely available to the workforce, organizations may choose to restrict transfer to selected practitioners for dealing with critical situations.

Experienced controllers are usually engaged in many experimentations on how to invent new traffic patterns. This form of "enlightened experimentation" can be undertaken only by experts who try out alternative traffic flows, under controlled conditions, in order to gain significant performance improvements with the least possible resources. In short, this practice requires that practitioners have a deep knowledge of the operational environment, willingly embrace the risk of failing early under controlled conditions, exploit simulators to test their assumptions, and finally get the right lessons from the right experiences (Thomke 2001). This approach appeals to controllers with high levels of expertise, as they are usually the best people qualified to evaluate their performance and explain the causes of any deviant outcomes. According to Hoffman (1998), one can distinguish between journeyman, experts and masters. Journeymen are those who can perform a day's labor unsupervised. Next in the scale, we find experts who are highly regarded by their colleagues, whose judgments are accurate and reliable and can deal with certain types of "tough" cases. Finally, at the top of the scale, we find masters who are highly regarded by experts and whose judgements set the standards of the profession. Enlightened experimentation seems to be a practice restricted to experts and masters.

The following example illustrates how experts build a controlled environment to undertake experimentation and invent new traffic flows. Figure 7.3 shows a well-established sequence routine for arriving aircraft from entry point ALPHA in an airport. The arriving aircraft are vectored from entry point ALPHA to establish the final approach course of the instrument landing system (ILS) from the north side of the airport. All controllers (i.e., journeymen, experts, and masters) of the approach unit are regarded as skilled practitioners at this sequencing routine.

Following a reorganization of the terminal maneuvering area (TMA), a new entry point BRAVO was established east of the airport, thus creating a new arrival flow that was conflicting with the old one. A safety assessment was performed and the change was proved acceptably safe. The new entry point was declared operational during the low season (winter time) to allow controllers sufficient time to familiarize themselves and develop new operational routines before the traffic peaks in the summer season. Now the controllers had to invent a traffic flow from entry point BRAVO and integrate safely the two flows.

In the next days, many junior controllers established a new lengthy traffic flow from BRAVO point without trying to change the established flow from ALPHA point (Figure 7.4). They preserved the entrenched standard ALPHA flow and created a new one by vectoring arrivals

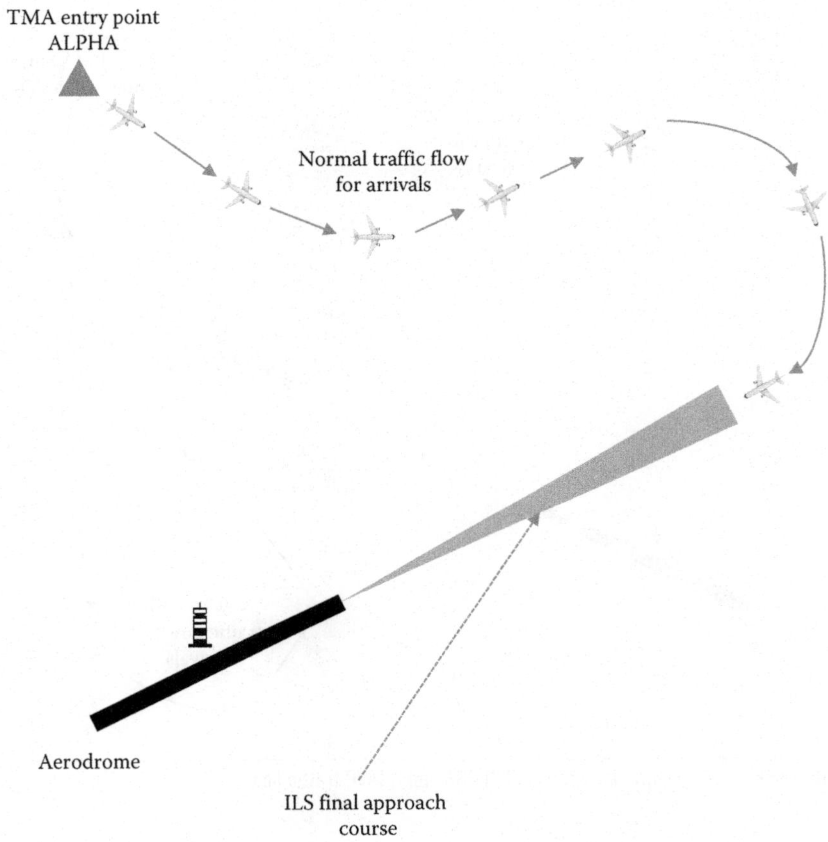

Figure 7.3 Initial traffic flow for the arriving aircraft from ALPHA point.

from BRAVO south of the airport following a lengthy route. The new routine implied that aircraft from BRAVO were initially vectored south and, by the time that aircraft from ALPHA arrival flow were not a conflicting factor, controllers started to vector aircraft north to establish the ILS final approach course. Apparently, this was a bumpy integration that preserved the ALPHA flow and sacrificed operationally the BRAVO flow since aircraft had to follow a lengthy curve, consuming more fuel for their landing sequence. Nevertheless, the new solution was quite safe and was happily accepted by the junior controllers.

However, a small number of controllers (experts and masters) were not happy with it, so they established an informal group and challenged the new practice. They started experimenting with alternative traffic flows in order to integrate the two flows smoothly. They

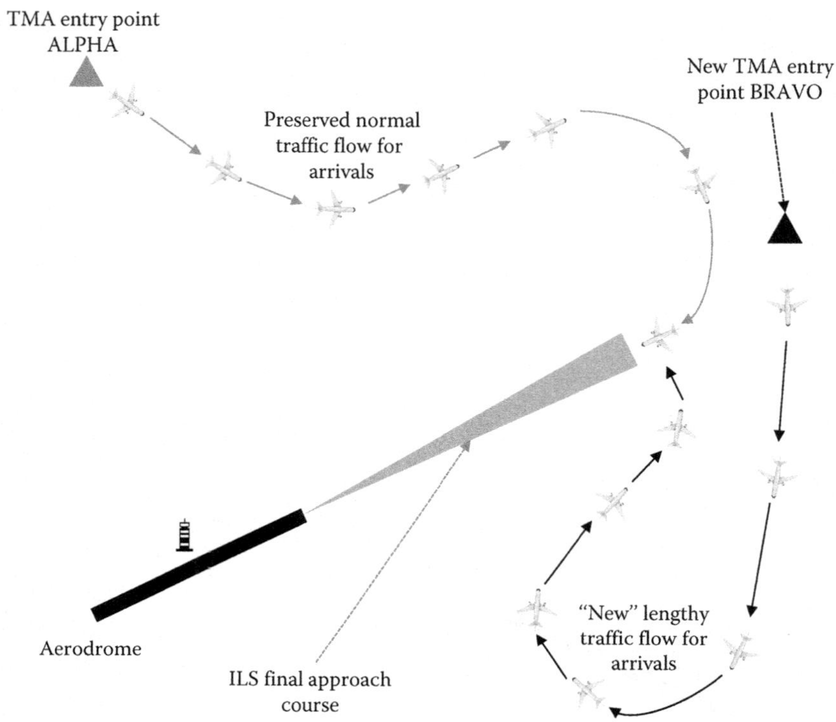

Figure 7.4 A "bumpy" integration of ALPHA and BRAVO traffic flows.

performed a number of trials in the training simulator and carried out "live" experiments under controlled conditions so that things could get back on track if a practice did not work out first time. Learning how to recognize the right conditions to try out new modifications to a practice became an important issue for them too. At first they experimented with only two aircrafts in good weather conditions so that they could fall back on a "go around" recovery plan. In this sense, they started spotting the critical elements that they should exercise control—e.g., relative vertical and lateral positions of the two aircraft, effects of wind and aircraft type, preferred flight profiles of airlines, timing and scale of vectors, and so on. As the live and simulator trials accrued, the experts greatly increased their knowledge of traffic dynamics, which enabled them to design and perform new experiments that were not conceivable in the past.

After some trials, they crafted a solution that integrated smoothly the two traffic flows without sacrificing safety, orderliness, and flight

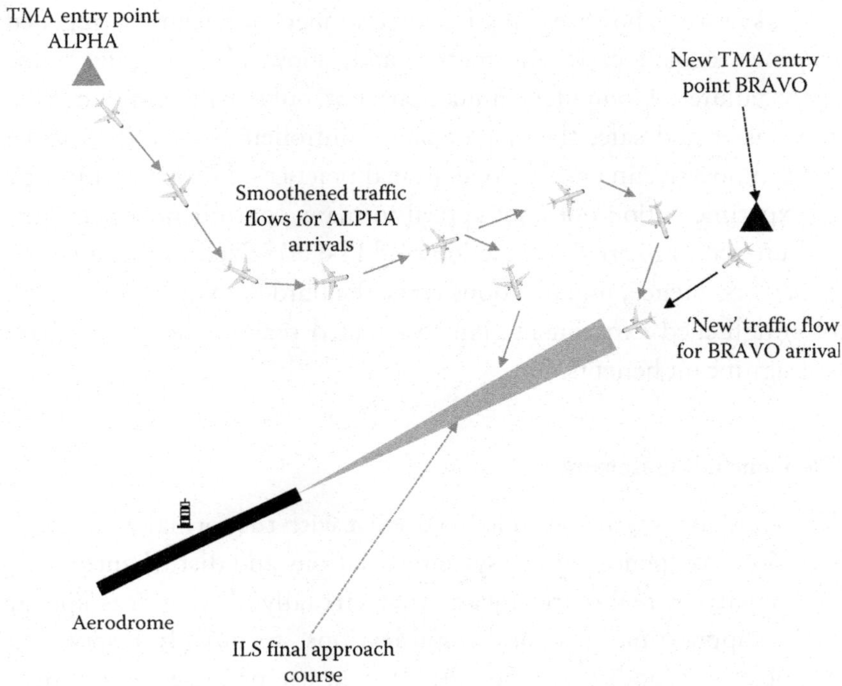

Figure 7.5 A smooth integration of ALPHA and BRAVO traffic flows.

efficiency. Most importantly they came up with a whole network of cues to be spotted and "what–if" actions to be taken in order to perform the smooth integration flawlessly. The solution comprised two variants depending on whether the first aircraft entered from ALPHA or BRAVO point (Figure 7.5). The new solution called for a careful evaluation of various dynamic elements of traffic in order to optimize flight efficiency for both traffic flows and maintain safety at the same time.

It should be noted that experts recognized some early failures (e.g., missed approaches and lengthy vectoring) that exposed important gaps in their expertise that helped them eliminate unfavorable options. They narrowed down the number of traffic flows and then refined the chosen option that appeared to be a viable solution from an operational point of view. Unpromising flows were abandoned early in the experimentation process. Experts benefited greatly from the rapid feedback provided by the "live" and simulated experiments as it took them only two weeks to accept and refine new traffic flow solutions.

This process, however, requires practitioners to revamp entrenched routines, exploit early information and, above all, recognize some early failures. Computer simulations not only make experimentation faster and safe, they also enable controllers and ANSPs to be more innovative in crafting operational practices. However, this type of experimentation does not appeal to many controllers because initial unsuccessful trials on the job could be criticized as incompetent behaviors. Hence, organizations carry the burden to nurture this sort of enlightened experimentation to selected practitioners since there are significant benefits.

7.6 Concluding Remarks

Work practices will continue to co-exist with the formal work organization, compensating for system variations and disturbances. The complexity of real world means that virtually all practices appear to be "approximations" or "simplifications" to quickly respond to a problem (Woods and Cook 2002). Because practices that appear to work well under some conditions may produce errors in others, organizations should set up systems for introspecting work practices, recognizing good practices and revising the formal work organization. The approach taken in this chapter views practices as "approximate solutions" to complex problems that are subject to many vulnerability factors. The problem is not per se that practitioners use workarounds or simplifications, but that they may not recognize situations where their practices are no longer relevant and may not know when to invest more resources to understand the problem and make use of more elaborate mental models that can be built through teamwork.

Practices are important aspects of performance variability and flexibility that can supplement the formal work organization and other rational approaches to safety (e.g., safety audits and risk assessment). In many respects, work practices are similar to recognition-primed decisions (Klein 2009) or "approximate solutions" that are developed through experimentation, or "trial and error," rather than through formal training. As a result, practices remain vulnerable to errors due to many complexity factors (e.g., intractable relations between parts, nonlinear dynamics and unruly technology). Practitioners should be able to recognize subtle differences between situations and transfer or modify their practices accordingly. In this sense, practices should be "mindful approximations" that address aspects of error recovery and replanning. The main focus of this chapter has been on the processes of "reflection-in-action" and "reflection-on-action" that can help practitioners to retain a critiquing stance toward their practices, remain flexible, and develop them further both in their everyday experience and in their interaction with their community of practice.

System dynamics provide a useful basis for examining how reflection-in-action can be attenuated by reflexive responses (Figure 7.1) or amplified by taking precautions and replanning approaches (Figure 7.2). It was proposed that mindful practices should involve two elements: (1) remaining sensitive to the possibility of failure and (2) using mental models to forestall threats or cope with threats by modifying plans on-the-fly. This has implications for the design of operating procedures and refresher training. Procedures should specify more degrees of freedom so that controllers can use them, not only for rote following, but also for general guidance. It is proposed that training should also provide opportunities for testing out work practices that have been learned from everyday experience.

A long-standing issue in learning relates to the specificity of knowledge and skills, which addresses the question: Are practices specific to the personal style of practitioners, the team norms, and the socio-historic context of organizations?. Reflection-on-action is a process helpful for articulating practices and making them widely available to other practitioners in the organization. Practices may be difficult to articulate and document, may be specific to the knowledge path followed by some teams, or may require a high degree of competence in dealing with risk. From this discussion, it appears that a useful proposal for training would be to become a test bed for learning the limitations of existing practices and for making them more adaptive to new situations. In this way, training can bridge the gap between formal procedures based on formal logic and work practices based on daily experience. Finally, there is a risk in putting a lot of emphasis on work practices, since this may obscure the systemic causes of problems and work obstacles. The next chapter provides a framework for addressing the skills and work practices of controllers in systematic training programs for emergencies and abnormal situations.

PART III

Reducing Complexity by Design and Training

8
Training for Abnormal Situations

8.1 Introduction

Emergencies in the air traffic management (ATM) domain range from simple textbook scenarios to system-wide failures that lead to airspace closure. An example of the low end of the range regards aircraft requesting diversion to the nearest airport due to a system fault. This is a textbook scenario where controllers route aircraft to the diversion airport, provide adequate separation with other aircraft, and manage coordination with adjacent units. Of course, several variations can be introduced to make this scenario more challenging (e.g., the flight crew may request to enter a holding stack to complete the relevant checklist). These emergencies happen nearly every day around the world and are successfully controlled. At the more complex end of the range, major system failures may affect ATM functionalities. An example is a software failure that led national air traffic services (NATS, UK) to stop all departures from London airports and alter most flights from European airports that were planned to route through the affected UK airspace (Walmsley et al. 2015). This system failure posed many challenges to a large part of the European ATM network and required large scale changes in coordination and contingency planning in the face of airline and social pressures.

As part of their competency scheme, controllers undertake refresher training courses in managing emergencies and abnormal situations (EAS). Refresher training normally includes classroom courses in standard and contingency procedures, simulator sessions on common technical failures, and courses in team resource management (TRM) that present real incidents for group discussion. Annual refresher courses for operational controllers aim at equipping them with the required skills and knowledge to successfully meet the demands of abnormal situations. In safety-critical organizations, technological innovations are introduced on their anticipated benefits of reducing workload and human error, offering a better representation of the operational environment, and providing assistance with many activities to controllers. Technology-centered approaches offer all-encompassing solutions in a constant struggle to eradicate or contain many sources of vulnerability. Information technology supplies air traffic control (ATC) units with high-fidelity simulators that simulate most technical aspects of controller's working positions (CWPs), including the logic of automation aids and safety nets. In this sense, simulators usually guide regulatory compliance training and drive a need for more "featurism" in synthetic task environments. The choice of scenarios reflects those preferred by regulations. However, studies of abnormal events in envisioned worlds have discovered several hard-to-resolve decisions that call for cooperative human-system architectures (Dekker and Woods 1999). It is very challenging, therefore, to consider training regimes that are both well matched to the cognitive functions of controllers and robust in the face of real-world ambiguities, workload, and time constraints.

8.2 Handling Abnormal Situations in the ATM Domain

A review of emergency and abnormal situations (Burian et al. 2005) has shown that refresher training in aviation provides limited opportunities to flight crews to practice EAS procedures in the context of real-world demands. Time constraints and cost considerations tend to restrict the range and depth of EAS training only to the most common system malfunctions and emergencies.

A review of a sample of EAS reports from the aviation safety reporting system (ASRS) found that the vast majority of aviation incidents were not textbook emergencies but cases of recovering from problems that originated from unsuccessful actions of flight crews, controllers, dispatchers, and maintenance teams (Burian and Barshi 2003). Under the aviation training regulations, flight crews rarely face a situation in simulator training for which a checklist might be difficult to work as expected, although this can be the cause of actual emergencies. In addition, flight crews generally do not get to practice interactions with other participating teams, mainly due to the pressing need to cover a wide range of EAS in a limited amount of time (Burian et al. 2005).

Although modern training approaches offer more opportunities for hands-on learning and practice of realistic scenarios, they still suffer from

1. Inadequate representation of real-world demands and situations (i.e., cognitive fidelity)
2. Lack of integration of cognitive strategies and technical skills.

The issue of cognitive fidelity has emerged as a critical factor in the design of training simulators in aviation. The challenge of cognitive fidelity is treated in this chapter in order to examine how refresher training can represent the work demands of real emergencies so that controllers make decisions and interact as they do in the real world. Meeting the challenge of cognitive fidelity in training requires a clear understanding of the cognitive strategies involved in the tasks that controllers undertake in realistic conditions. Operating factors such as content, timing, and rate of information updates are likely to call for strategies in managing uncertainty rather than simply in pattern recognition. Coping with interruptions, read-back errors, and late transfers while responding to an emergency require not only good communication skills but also error management skills.

With regard to the second issue, a mapping of cognitive strategies to a range of tasks and work conditions can be achieved by using several methods of cognitive tasks analysis (CTA) which have been presented in the literature (Johnston et al. 1997). This chapter proposes a CTA method for studying how cognitive strategies emerge in complex domains and how they can be used to design refresher training that provides practice conditions to integrate technical skills with cognitive strategies.

8.3 Anomaly Response and Cognitive Strategies

An anomaly is an emerging behavior of the system that deviates from prescribed standards and operational expectations. In anomaly response, the triggering event can produce a series of disturbances in the functional and psychical coupling of the system, known as a cascade of disturbances. External or internal disturbances generate demands for practitioners to act and compensate in an adaptive manner (Woods and Hollnagel 2006).

Controllers are expert decision makers whose strategies and tactics have developed over years of operational experience, recurrent training, and accumulation of knowledge. Tactics are used for the normal handling of air traffic, such as aircraft conflict detection and resolution. In contrast, strategies are used for problem solving and coordination that are essential in responding to challenging and novel situations (Woods et al. 2007).

Cognitive strategies go beyond the normative if-then tactics (or well-rehearsed activities) and require skills such as the following:

- Recognizing subtle cues in the environment
- Synthesizing patterns of cues
- Adapting to new constraints of work
- Replanning earlier actions
- Assessing rates of change
- Judging the timelines of interventions
- Communicating intentions behind actions
- Coordinating actions

These cognitive strategies overlap and change over time, creating a cognitive flow that unfolds in an information-intensive and noisy work environment.

Normal daily operations require conflict resolution tactics while abnormal situations call for cognitive strategies presented in the T^2EAM (taskwork/teamwork for effective and adaptive management) model (see Chapter 4). Figure 8.1 shows an integration of taskwork and teamwork strategies to provide a framework for discussing the cognitive strategies that controllers exhibit in the management of emergencies and abnormal situations (EAS).

As can be seen in Figure 8.1, typical situations are handled by a process of recognition followed by standard planning. In contrast, unusual situations present controllers with uncertainty, which can be anticipated to a certain extent, making the situation easier to control from the early stages. Hence, anticipation involves a stance toward minimizing uncertainty by thinking ahead of possible threats and plans to cope with them. However, some uncertainty emerges as the situation unfolds, which requires revision of understanding and

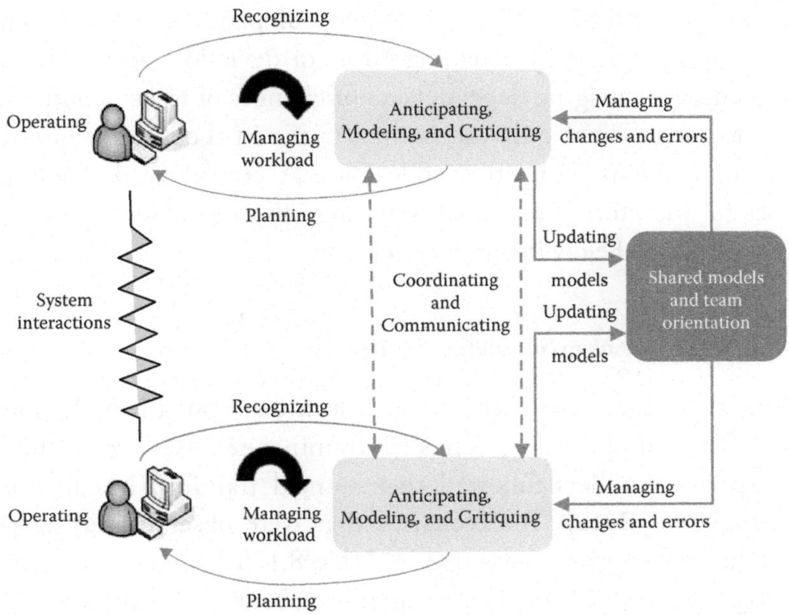

Figure 8.1 Flow control of taskwork and teamwork functions in the T²EAM model. (From Malakis, S. and Kontogiannis, T., *International Journal of Aviation Psychology*, 22, 1, 59–77, 2012.)

replanning of original actions. In managing uncertainty, controllers have to live with uncertainty, critique their initial models of the situation, and adapt their plans on the fly. The choice between minimizing uncertainty and managing uncertainty depends on the context of work (Grote 2008). In particular, temporal aspects of work (e.g., pace of work, number of tasks competing for attention, interruptions, resumption of tasks) are likely to affect all taskwork strategies. In Figure 8.1, managing workload functions as a task regulator that enables controllers to cope with the complexity of the situation.

Team actions are coordinated to plan traffic and resolve conflicts displayed on radar screens. Communication and feedback of team actions enables controllers to judge the degree of success, identify errors, and change the allocation of tasks to improve team performance. Team communication is also functional in building a shared understanding of the problem and a collective stance toward a problem solution. In turn, shared understanding and orientation enables members to predict behaviors and coordinate as well as generate expectations and adjust their communication patterns to avoid delays. This is usually done by detecting errors or deviations from expectations and by managing dependencies or changes of tasks. Controllers have to manage not only the normal traffic in their sector but also their interactions with other colleagues. A critical need, therefore, arises for monitoring and understanding others in order to keep the team workload below its saturation point.

8.4 EAS Scenarios in Simulator Training

The authors conducted field research at a major area control center (ACC) in Europe where refresher training exercises were studied as expert controllers followed their annual training (Malakis and Kontogiannis 2012). The scenarios that were observed during the refresher training are presented in Table 8.1. Two later case studies focus on the cognitive strategies used by controllers in managing the airport diversion and the airspace clearing scenarios.

A common operational requirement in all scenarios is the effective communication with the flight crews and the adjacent ATC units. The coordinating controller (CC) bears the burden of maintaining coordination with the other sectors and informing the shift supervisor. Coordination may range from ordinary and prescribed patterns to time-sensitive coordination. It is the responsibility of the executive controller (EC) to provide the necessary separation minima while CC is responsible for planning how to handle traffic flows in the sector. Escalation patterns of emergencies were generally smooth with varying demands for controllers on how to manage information uncertainty. Earlier familiarization with the emergencies in a classroom course helped controllers to cope with the scenarios in the simulator sessions.

In the refresher course, experienced controllers excelled in many cognitive strategies and especially in recognition, anticipation, and planning. Successful performance could be attributed mainly to the familiarization training and their own expertise. Controllers were briefed about the EAS scenarios and knew what cues to look for, how the situation could evolve over time, and how to respond to the situation. The sheer amount of operational experience, combined with intensive training, resulted in excelling performance.

Table 8.1 EAS Scenarios in Refresher Training

EAS TYPE	DESCRIPTION
Holding scenario	An airport is temporarily closed for arrivals while all inbound aircraft are compelled to enter several holding stacks. The reasons for airport closure may include: bad weather, CNS problems or other system failures. The team has to guide aircraft to published holding patterns, or create new ones by selecting and "freeing" temporarily a part of the sector. The team must coordinate with the adjacent units and provide the holding aircraft with accurate time estimates on when normal operations will be resumed.
Emergency descent scenario	Due to a system failure and a resulting loss of cabin pressure, an aircraft is compelled to initiate an emergency descend. The ATC team has to recognize the emergency descent from its onset, provide traffic information to other affected flight crews and keep clear their aircraft while maintaining coordination with adjacent units. The escalation pattern for this type of emergency is much steeper than all other scenarios.
Radio communication failure (RCF) scenario	An aircraft suffers a radio malfunction that makes the flight crew unable to communicate with ATC. The aircraft is squawking the code 7600 on its mode A of its transponder, which signifies communication failures and problems with its altitude encoding equipment. The controllers have to recognize the RCF situation, discriminate its type, keep the affected traffic clear, and provide greater separation minima to aircraft.
Airspace clearing scenario	Due to a bomb threat received in the ACC, the operations room must be evacuated immediately after the ATC team clears the airspace in a restricted time window (usually 10 minutes). The team has to reroute and transfer all aircraft from their sector to adjacent units while maintaining team coordination.
Fuel dumping scenario	An aircraft is compelled to dump fuel due to a technical malfunction. The ATC team is informed by the flight crew or the previous sector that the aircraft will dump fuel in their airspace. The team has to provide the aircraft with adequate airspace, separate it from other aircraft with greater separation minima and maintain coordination.
Diversion scenario	An aircraft experiences an emergency and requests diversion and navigational assistance to the nearest suitable airport. The escalation pattern is generally smooth provided that a range of informative cues and expectations are available. Upon issuing the request for diversion, the ATC team has to provide direct routings, which usually entail using a restricted airspace between the aircraft and the most suitable airdrome and provide a continuous descent profile to the aircraft.
Hijack scenario	The ATC team has to recognize early cues and synthesize them as a suspected hijack situation, keep other aircraft in the sector clear of the hijacked aircraft, provide increased separation minima, and maintain the required coordination with the other units.

The field study (Malakis and Kontogiannis 2012) also examined a set of real incidents regarding separation minima infringements, since they had very different task characteristics from the simulated events encountered in the refresher course. For instance, the simulator scenarios were very demanding in terms of workload but did not require strong team coordination. All ATM and aircraft systems functioned properly and played a major role in resolving the situation. In most simulations, the short-term conflict alert (STCA) aid alerted controllers and in only one incident did the controllers manage to detect an impending conflict earlier than the STCA system.

In real events, different performance patterns emerged. Controllers made some errors in recognizing problems, anticipating threats, and planning the traffic in the real work environment. Many errors seemed unavoidable, especially when working under a heavy workload and the influence of interruptions and distractions. The result of the incident analysis (Table 8.2) indicated that the most common

Table 8.2 Typical Operational Problems Recorded in Real Incidents

OPERATIONAL PROBLEMS	DESCRIPTION
Readback/hearback errors	Controllers failed to detect readback errors of flight crews. Consequently, the aircraft descended/climbed at the wrong flight but controllers failed to detect quickly the level changes which resulted in a loss of separation. In all cases, controllers failed to detect the readback errors and recognize the unfolding situation before the STCA alerted them to this event.
Early transfers to other sectors	An early transfer is a common practice as controllers may decide to transfer an aircraft to the next sector when they consider it to be no longer a factor in their sector. For instance, an aircraft flying inside a sector was transferred early to the next sector. The color of the radar label of the transferred aircraft was changed and, after a while, the controllers gave an instruction to another aircraft that caused a loss of separation with the "forgotten" aircraft.
Flight plan overlooking	The flight plan of an aircraft was overlooked. Consequently, the aircraft followed a course that was not expected by the controllers, which resulted in a loss of separation with another aircraft in the sector that was considered clear of the expected path of the first aircraft.
Distractions	Controllers were distracted by supervisory duties and failed to attend to their radar screens. The evolving separation loss had been detected by the STCA function two minutes earlier but none of the controllers were attending.
Failure to consider crossing traffic	Controllers gave instructions that did not take into account opposite or crossing traffic.

causes of loss of separation events were problems in error management and task distribution.

The comparison of the simulator training sessions and the incident events have shown that controllers excelled in managing textbook emergencies in simulator training. However, certain variations in the characteristics of textbook scenarios could have been manipulated by instructors to produce a fertile ground for performance problems similar to the ones observed in the recorded incidents.

8.5 Patterns of Resilience, Coordination, and Affordances

This section applies the cognitive systems engineering (CSE) paradigm (Woods and Hollnagel 2006) in order to abstract generic patterns of performance recurring over the simulated scenarios of the field study. A pattern can be described as a relational property that captures problems and opportunities arising from the interaction of practitioners, situational demands and artifacts. To respond to many challenges at work, controllers employ a variety of adaptation strategies that could be assigned into three general patterns:

1. *Resilient taskwork* showing how practitioners manage system demands
2. *Adaptive teamwork* showing how practitioners adapt team coordination
3. *Affordances* provided to practitioners to facilitate their work

A description of the generic patterns offers a good basis for the analysis of EAS demands and the study of cognitive strategies employed by practitioners.

8.5.1 Patterns of Resilient Taskwork

Resilience represents the ability of work systems to adapt to new situations or absorb disturbances, especially those that fall outside the usual operational envelope (Hollnagel et al. 2006). Emergencies and abnormal situations represent critical occurrences close to the margin of safe operation that challenge existing operational practices and supervisory systems. An emergency presents controllers with many challenging issues regarding information uncertainty, safety repercussions, time pressure, and lack of procedures to respond. As soon as a disturbance is detected, a problem-to-be-solved is formulated and the need to replan earlier decisions becomes prominent. To respond to an emergency, controllers should demonstrate problem-detection skills and replanning strategies. As occurrences evolve over time, new threats might appear and current threats might change their demands. The need for gathering new information to fill gaps in understanding, clarify assumptions, and evaluate candidate hypotheses is amplified. This calls for taskwork strategies in recognizing the situation, anticipating how the situation will evolve in the future, and managing uncertainty (Table 8.3).

8.5.2 Patterns of Adaptive Teamwork

ATC requires synchronization of many interdependent activities within a short time window. However, the smooth flow of information between highly experienced practitioners can make external observation of performance very difficult to achieve. Coordination can be exemplified in two dimensions: internal coordination within the team and external coordination between teams or units (Table 8.4).

Table 8.3 Patterns of Resilient Taskwork

RESILIENT TASKWORK	EXAMPLES
Detection of critical cues followed by accurate state projection	Observation of aircraft deviation is followed by accurate projection of its flight path and of other aircraft that could be affected.
Information-based uncertainty is tolerated	EC avoids losing precious time in questioning aircraft deviation, especially when the crew is either not responding or questioning the requests.
Traffic planning is kept simple with open-patterns of traffic	EC uses traffic flows that are easily reconfigurable and avoids extensive and unjustified use of lateral separations.
Threats are detected and resolved at the longest distance	EC avoids the routing of traffic close to areas of observed or expected military activity.
Critical tasks are prioritized without creating task backlogs	CC prioritizes outbound telephone calls (e.g., what sectors need to be informed first about the unfolding critical situation).
In high workload, team members avoid distracting others	CC does not distract EC in critical phases of the situation with minor traffic planning issues.

Table 8.4 Patterns of Adaptive Teamwork

ADAPTIVE TEAMWORK	EXAMPLES
Shared understanding of situation	CC understands and follows how the EC handles traffic by observing the radar screen and by monitoring voice communication with the flight crews.
Communication of intent behind plans	Timely statement of the reason behind the selection of a particular diversion route of an aircraft compelled to divert.
Information exchange within and between teams	Intra- and inter-team information is kept to the minimum by employing a form of operational language that is clear, concise and meaningful.
Error detection and correction management	CC detects and corrects minor errors of EC (e.g., wrong flight level inputs, readback errors) without hindering the flow of tasks.
Adaptive distribution of workload	CC offloads the workload of EC by employing nonstandard coordination with adjacent sectors, without impending air traffic.
Proactive use of restricted resources	In case an aircraft is suspected to divert, the CC requests authorization to use a restricted airspace between the position of the aircraft and the closest diversion airport.

The two coordination types differ in the observability of communication and the exchange of information. For example, coordination between team members is usually implicit. The CC understands and follows the EC simply by observing the evolution of traffic depicted in the radar screen and by monitoring voice communication loops between the EC and the flight crews; this could be achieved without any overt communication and at a minimum cost of information exchange. On the other hand, external communication between units is mainly overt as the CC has to externalize his or her intentions to other units; this explicit form of information exchange comes at an increased cost of communication.

8.5.3 Patterns of Affordances

The ATM environment supports the work of controllers with many artifacts, ranging from simple ones (e.g., the HALO function that displays a circle around the aircraft of concern of a specified radius) to more complex ones (e.g., resolution advisory tools). In the middle

Table 8.5 Patterns of Affordances

AFFORDANCES	EXAMPLES
Available automation functions are used to enable situation monitoring	EC and CC may periodically remove altitude filters and zoom in/out the radar screen.
Automation aids may afford recognition of threats	EC and CC use automation functions to display a HALO circle around an aircraft that needs special handlin.
Simple artifacts can be used for noticing and recognizing critical cues	CC uses simple paper scripts to convey to the EC critical information about the situation status.
Automation aids must be properly used for traffic planning	Controllers utilize velocity leaders frequently to perform tactical panning (i.e., the VERA function)
Sector airspace is an affordance to be used for contingency planning	A certain part of the airspace may be selected and reserved in order to vector aircraft to a holding pattern to dump fuel.
Automation aids can be used to choose options in contingencies	Controllers can use automation aids to estimate distance and bearing to the nearest suitable airport

of the scale, we can find the concept of an airspace sector, which provides important affordances to controllers who can reserve an airspace sector for a specific goal (e.g., for aircraft that needs to dump fuel). According to Hollnagel and Woods (2005), an "affordance" refers to the use of an artifact by a practitioner according to the design properties of the artifact and the preferences of the practitioners. Simple artifact can afford noticing and recognizing critical cues while more advanced automation functions can afford situation monitoring or contingency planning. An affordance is not an attribute of the artifact *per se*, but an *ad hoc* support of a practitioner's goal in the context of an unfolding situation. In this sense, a design attribute of an artifact may be used in different ways by controllers who have different goals and methods of work. Table 8.5 provides examples of several artifacts and affordances operating in the ATM environment.

8.6 Cognitive Tasks Analysis (CTA)

The field study has shown that refresher training was built on a standard set of EAS that are commonly encountered in the ATM domain (Malakis and Kontogiannis 2012). This operational approach to training, however, has not been supplemented with an analysis of cognitive strategies that could be the core of flexible expertise in other realistic situations that were not covered in refresher training. For instance, typical EAS scenarios can become unmanageable when additional factors are introduced (e.g., higher traffic load, unavailability of tools, degradation of equipment, dynamic weather cells, poor communications, etc.). The standard set of EAS scenarios can have many variations in real practice that call for greater flexibility of skills. This implies that a skill-oriented approach with a focus on the training of cognitive strategies should supplement the existing operational approach to training.

It is useful to map out the cognitive strategies that support team performance in abnormal situations. CTA methods can play an important role in identifying performance demands and cognitive strategies that are the building blocks of what has been termed flexible or adaptive expertise (Holyoak 1991). Klein (1998), for instance, proposed the use of decision requirements tables for identifying critical cues to recognize situations, strategies for reducing uncertainty, sources of difficulty in making decisions, and potential errors. EAS scenarios analyzed in this way can provide suitable opportunities for practicing cognitive strategies.

8.6.1 The Airspace Clearing Scenario

This section presents a CTA example in the context of the airspace clearing scenario observed in the field study (Figure 8.2). In this scenario, controllers received a bomb threat alert and had to evacuate the area control center immediately after clearing their allocated airspace in a restricted time window (usually 10 minutes). The controllers had

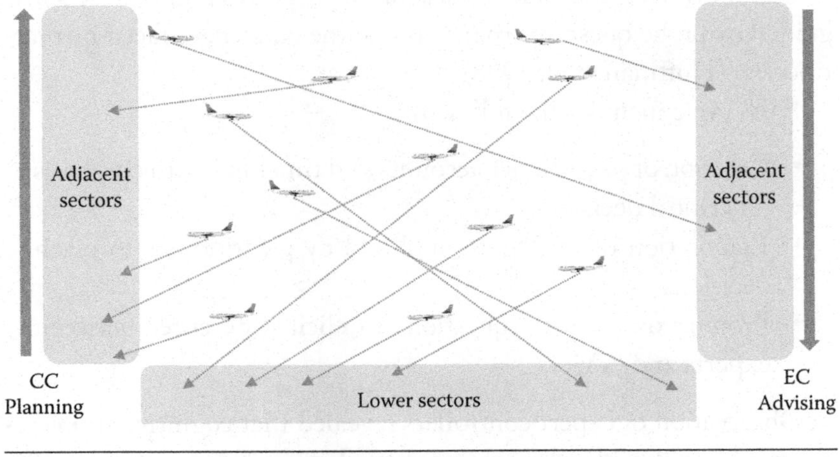

Figure 8.2 A schematic diagram of the airspace clearing scenario.

Table 8.6 Decision Requirements in the Airspace Clearing Scenario

CRITICAL DECISIONS	WHY DIFFICULT?	IMPORTANT CUES	ASPECTS OF EXPERTISE
Decide on which aircraft to reroute first and for which to effect coordination	Letters of agreement and procedures for normal operations do not apply to this situation	Vertical and horizontal positions of the aircraft in terms of adjacent and lower sectors, types of aircraft, wind conditions, and routes to be flown	Start descending the higher aircraft first without making the relevant coordination with other units
Construct most suitable flight path for descending aircraft	Complex unfamiliar conflicting paths	Flight paths of adjacent aircraft	Use velocity leaders to project route
Manage communications	Communication workload	The CC observes the performance of EC in giving instructions to the aircraft The EC monitors the coordination loops of the CC with the adjacent aircraft	Keeping communications to the absolute minimum

to reroute and transfer all aircraft from their sector to adjacent units while maintaining team coordination. The decision requirements of this scenario are shown in Table 8.6, according to the CTA technique proposed by Klein (1998). Challenging decisions included anticipating threats, standard planning, and contingency planning.

CTA involves multiple cycles of retrospection into a scenario guided by probe questions that examine the challenges faced by practitioners (Hoffman et al. 1998).

Each cycle includes the following:

- Creation of practitioner accounts and timelines with emphasis on critical decisions
- Elaboration of strategies employed by practitioners to reach decisions
- Probing on what-if? questions to elicit differences between experts and novices

Observation of expert controllers revealed that cognitive strategies emerged in a flexible manner and shifted in response to the dynamic

Table 8.7 Resilience, Coordination and Affordance Patterns in the Airspace Clearing Scenario

RESILIENCE	COORDINATION	AFFORDANCE
EC initiates descent of higher aircraft first	CC initiates coordination with adjacent and lower sectors first for the lower aircraft	Velocity leaders and VERA functions are used to detect and resolve potential conflicts.
EC resolves unfamiliar types of conflict between aircraft that are not standard for this sector	CC monitors voice loops of EC and follows with coordinating actions	Radar scale and filters are employed for the need of new conflict-resolution strategies
EC initiates changes of level and heading first, leaving fine-tuning, and coordination for later	CC minimizes internal and external communications	

evolution of the scenario. In an effort to map the cognitive strategies over the course of events and actions, Table 8.7 was prepared to represent the cognitive flow of cognitive strategies for this particular scenario. The strategies are divided into three areas, reflecting controller adaptation or resilience, coordination, and use of automated functions (affordances).

As shown in Figure 8.2 and Table 8.7, the EC started descending high flying aircraft to the middle part of the sector first, without waiting for confirmation from the CC. At the same time, the CC would coordinate with other sectors below to expect an early transfer of aircraft that were flying at the bottom of the sector, close to the borders. In a sense, control of the aircraft was initiated from the top of the sector, while coordination was effected from the bottom of the sector. This decoupling of control from coordination was the reason for accomplishing the airspace clearing in a restricted time window. This contrasted with the traditional conflict-resolution method where control of aircraft position and coordination of transfer to other sectors would coincide for the same aircraft. As a result, when the EC and CC were working on the same aircraft a longer period of time was required to evacuate the airspace. Only when the two controllers were managing aircraft at the higher and lower parts of the sector simultaneously were they able to resolve conflicts effectively. A smooth pattern of coordination was also observed where controllers made all the necessary arrangements without garbling the flight crews with unnecessary information.

It is anticipated that the systematic use of CTA methods in the design of refresher training will ultimately lead to a taxonomy of situational demands and provoke cognitive strategies for different scenarios. In summary, the T^{2}EAM model can support CTA methods in extracting situational demands and cognitive strategies that can be practiced in refresher training. This approach is expected to give rise to more resilient and synchronized patterns of response to EAS simulated scenarios.

8.6.2 The Airport Diversion Scenario

The field study has also examined an airport diversion scenario where an aircraft experienced a malfunction and had to be rerouted to another airport different from its original destination. In order to map out the cognitive strategies over the course of events and actions, Table 8.8 was prepared to represent the flow of cognitive strategies for the particular scenario.

In this scenario, an aircraft experienced a malfunction and requested the meteorological conditions (meteorological terminal air report, METAR) of a nearby airport. Although not alerted by the flight crew, the controllers became suspicious of a potential threat and prepared their colleagues in other ATC sectors that supervised possible diversion airports. They also made all necessary coordination for reserving a military airspace that blocked direct access to the diversion airport. Most of the activities are part of the controller strategies for managing uncertainty and planning for contingencies. Controllers avoided asking any questions to the flight crew to let them concentrate on the potential threat and waited until the flight crews declared an emergency. With their planning-ahead strategy, the controllers were able to resolve any potential traffic conflicts and routed the aircraft safely to a diversion airport. This scenario presented controllers with a mix of delayed cues (e.g., delayed declaration of emergency), masked cues (e.g., crew request of the weather of a nearby airport), and attention-diverting cues (e.g., other aircraft on the radar screen). To cope with these demands, controllers recognized the uncertainty of the situation, anticipated possible threats, and planned for a few contingent events (e.g., diversion to an airport and reservation of the military airspace). A smooth pattern

Table 8.8 Cognitive Task Analysis of Controller Strategies across Time in the Airport Diversion Scenario

TIME	EVENTS ON BOARD	RESILIENCE	COORDINATION	AFFORDANCES
T0	The flight crew of an aircraft detects cues of a system malfunction			
T1	Crew considers a diversion and requests the meteorological terminal air report (METAR) of a nearby airport, without alerting controllers	EC provides the METAR to the flight crew	CC records the METAR and hands it over to EC	CC uses automation to obtain METAR
T2	Checklists are applied and company dispatchers are notified	EC and CC suspect a possible aircraft diversion to another airport probably in the next few minutes	CC initiates coordination with the diversion airport and the affected sectors stating that they may face a possible diversion	EC and CC use automation functions to estimate distance and bearing to the diversion airport
T3	Malfunction persists and the crew agrees on a diversion with dispatchers	EC and CC discover that the closest path to the diversion airport is blocked by restricted military airspace	CC initiates coordination with the unit responsible for the restricted airspace and requests permission	EC and CC use velocity leaders to detect potential conflicts with other aircraft in the sector, in the event of a diversion
T4	Crew requests diversion to the nearby airdrome due to technical reasons	EC provides instructions for a direct routing to the diversion airport through the restricted airspace	CC finalizes coordination with military unit, the diversion airport unit and other affected sectors	EC and CC use velocity leaders to perform tactical planning of the traffic for the diversion
T5	The aircraft is routed to the diversion airport	EC handles the diversion and the other traffic in other sectors	CC informs the supervisor and coordinates with the next sectors for the diversion	CC writes down all the necessary information regarding the diversion
T6	The aircraft is transferred to the next sector	EC hands over the diverting aircraft to the next sector	CC informs the supervisor about significant details of the diversion	EC and CC utilizes velocity leaders to perform tactical planning of the traffic

Source: Malakis, S. and Kontogiannis, T., *International Journal of Aviation Psychology*, 22, (1), 59–77, 2012.

of coordination was also observed where controllers made all the necessary arrangements, without garbling the crew with unnecessary information.

8.7 The ABCDE Method of Cognitive Task Analysis

CTA methods are useful for understanding the task challenges and the cognitive strategies employed by practitioners to cope with them. The decision requirements table has been widely used in human factors because it is very practical and provides a basis for looking into how practitioners manage to resolve difficult situations (Table 8.6). In this sense, it is anticipated that further developments can be made to capture more subtle cognitive strategies identified by human performance models. The ABCDE method has been developed by the authors to help practitioners analyze the strategic requirements of ATM situations in a practical manner. The ABCDE method condenses the T^2EAM cognitive strategies into five strategies:

1. *(A)ssessment of situation*—how controllers recognize similar situations experienced in the past and how they manage uncertainty to assess new situations (Table 8.9)
2. *(B)alance of constraints and resources*—how controllers evaluate difficulties, threats, and constraints imposed by the situation as well as how they use resources and affordances provided in the operating environment (Table 8.10)
3. *(C)ommunication*—how controllers communicate information, intentions, and actions to others and how they coordinate with adjacent sectors (Table 8.11)
4. *(D)ecision making and planning*—how controllers make decisions, how they work in smart ways and improvise, how side effects are prevented, and how plans may turn out differently (Table 8.12)
5. *(E)rror detection and recovery*—how controllers make provisions to review their work progress, how they manage to detect errors, and later recover them in a timely fashion (Table 8.13)

Table 8.9 Assessment of Situation in Approach Control

PROBE QUESTIONS	CUES TO RECOGNIZE AND ASSESS THE SITUATION
What features of the situation should be recognized? What was the most important piece of information?	• Top of Descent (ToD) point • The sequence in which the ACC transfers aircraft • Aircraft performance • Track miles remaining to touchdown for each aircraft • Relative position of departures and arrivals • The flight profile followed by the aircraft • Possibility of unstable approaches • Airlines' preferences on descend and climb profiles • Significant variations on wind speed and direction on final approach • Weather information (e.g., cloud base, visibility, precipitation and turbulence) in the TMA
Any other information that might have been used?	• Extra information regarding aircraft flight parameters (e.g., rate of descent, indicated airspeed) • Exact time of transfer by ACC • Exact position of a taxiing aircraft on the ground • Sequence of departing aircraft on the airport taxiways • Information from weather radar
Are cues changing over time or masked?	• All cues are time depended • Aircraft on the ground are not shown on the approach radar screen • Altitude filters and dead radar sectors may cause masking • Wind conditions may mask actual headings of aircraft • Radar may not show areas of adverse weather • Obscured radar track labels (e.g., overlapping of labels)
Were you uncertain, at any stage, about the reliability or relevance of the data?	• Uncertain about the actual point and time of transfer by the ACC • Uncertain about the exact time of a departure • Uncertain about the actual flight profiles (e.g., speed, rate of descent, heading etc.) and the approach patterns • Uncertain about the presence and intensity of weather phenomena • The radar may display ghost aircraft tracks due to clutter (unwanted radar returns) that may "fool" the detection algorithms and display false aircraft tracks • When two aircraft are closely spaced in range and azimuth but at different heights, transponder replies from two aircraft may overlap

The order of presentation has been decided only for the sake of the acronym ABCDE and does not imply the order in which strategies should be applied in practice. For instance, controllers may make a decision first and then communicate it to supervisors for authorization or even to other sectors for a second opinion. In the same way, error detection and recovery may be carried out in parallel to all other strategies. The ABCDE method has been applied in the context of

Table 8.10 Balance of Constraints and Resources in Approach Control

PROBE QUESTIONS	CONSTRAINTS, RESOURCES, AND OTHER INFLUENCES
What makes traffic de-confliction difficult?	• Top of descent miscalculation • Late transfer by ACC • Wrong presequencing by ACC • Calculating track miles to the touchdown • Visual flight rules (VFR) traffic • Constrained airspace • Tight departures of aircraft with dissimilar climb performance • Tight arrivals with dissimilar speed performance • Airlines preferences for Continuous Descent Approach (CDAs) profiles irrespective of the other traffic in the TMA • Critical weather phenomena (e.g., low level wind shear [LLWS] and turbulence) • Unusual traffic scenario
What strategies and time constraints exist?	• Establish an approach sequence when the aircraft are at least 20 NM from the airport • Ask flight crews of the preceding aircraft to accept a visual approach rather than an instrument landing system (ILS) approach • Coordination with the tower to let an arrival cross the extension of the departing runway below 6000 ft. • Request from ACC *ad hoc* level speed and heading arrangements before the transfer points, especially for aircraft that do not fly standard flows, to fit better to the traffic planning
What resources are needed? (e.g., tools, procedures, equipment)	• Aircraft radar labels with their sequence number and the type of intended approach • Backup ADS-B screen
What factors can affect the outcome? (e.g., weather, tools)	• Weather conditions • Aircraft performance • Company procedures (e.g., flight profile, visual pattern) • Proximity of the aircraft • Runway in use • Radar limitations (e.g., dead areas, garbing, inaccuracies in range, and bearing information

approach control units to examine the T²EAM cognitive strategies utilized by the responsible controllers. A complementary study was carried out by Malakis et al. (2014) to provide a list of probes for the CTA of approach controllers that may have wider application for similar domains. Tables 8.9 through 8.13 provide a list of probes for identifying cognitive strategies, which are illustrated in the context of approach control. In each table, the probes (left column) are associated to several behavioral markers (right column) that describe observable

Table 8.11 Communicating Information, Actions, and Intentions in Approach Control

PROBE QUESTIONS	COMMUNICATING INFORMATION, ACTIONS, AND INTENTIONS
When and how much information do you pass to other colleagues?	• Select low tempo periods to communicate non urgent information (e.g., entry level conflicts in the next 15 minutes, nonurgent altitude and/or route coordination). • Keep the size and duration of communication to the required minimum. • Provide concise explanations in terms of the intent behind the instructions and/or clearances (e.g., stating the reason behind the selection of an approach sequence).
What subtle signs in communication may indicate problems faced by others?	• The other controller fails to locate the aircraft track that is calling. • The other controller misses initial radio telephony (RTF) calls. • The flight crew does not follow exactly controller's instruction. • The other controller has problems in selecting an approach sequence.
What errors and dependencies can be made in coordination?	• Accept plan proposed by ACC or Tower controllers without any interrogation ("Mr. Nice Guy policy"). • Let the ACC or Tower transfer an aircraft late. • Accept a tight traffic plan that allows no errors and depends on flight crew following exactly the instructions as told.
What sort of proactive information and action can increase coordination?	• Provides updates on situation status and actions taken. • Provides information regarding threats (e.g., military traffic, impending conflicts) that were unnoticed by others. • Corrects minor communication and interface errors made by others. • Warns adjacent sectors of a suspected emergency.

Table 8.12 Decisions and Plans in Approach Control

PROBE QUESTIONS	SMART STRATEGIES TO MAKE DECISIONS AND PLANS
What strategies exist that allow you to work in smart ways?	• Increase radar range and remove altitude filters to observe incoming traffic from further away • Utilize information from Network Operations Portal (NOP) or Collaboration Human Machine Interface (CHMI) to anticipate incoming traffic • Utilize information from the Automatic Dependent Surveillance – Broadcast (ADS-B) receiver, which shows aircraft taxiing on the ground • Build a "microslack" in the approach sequence (e.g., increase aircraft distance or size of spacing) to accommodate Tower unit difficulties in regulating departures or preparing for a possible go-around
Are there any situations in which the plan of action might have turned out differently?	• When ACC, Tower, or flight crews perform differently than planned, or instructed • When flight crews experience weather conditions that suddenly deviate from earlier reports • When the runway is temporarily closed due to an urgent inspection of the runway for bird remnants and aircraft debris after a bird-strike event • When aircraft performs an unexpected missed approach • When the runway in use has to change rapidly due to a sudden wind change

(*continued*)

Table 8.12 (*Continued*)

PROBE QUESTIONS	SMART STRATEGIES TO MAKE DECISIONS AND PLANS
How can you prevent side-effects for your favored plan?	• Provide slack between successive arrivals • Monitor the RTF exchanges of the Tower unit to detect a problem early (e.g., a missed approach or a bird strike) • Regularly check with the Tower unit for the status of operations
Can you think of examples when you improvised in this task or noticed an opportunity to do something better?	• Use information derived from the ADS-B receiver to look ahead into the traffic situation by 5–10 minutes • Change the approach traffic when runway is temporarily closed for an urgent inspection • Change the approach sequence when preceding aircraft slows down more than anticipated or instructed • Reroute departing aircraft when the previously one climbs much slower than expected by its type and weather conditions

Table 8.13 Error Detection and Recovery in Approach Control

PROBE QUESTIONS	FACTORS IN ERROR DETECTION AND RECOVERY
What errors can be made by novices and experts?	• Underestimate a sudden increase on workload. • Overestimate the ability to handle air traffic. • Underestimate the risk of the agreed plan. • Ignore the effect of wind conditions. • Establish a very tight approach sequence. • Miscommunicate an approach sequence to the Tower unit. • Miscalculate the runway occupancy time of departing and arriving aircraft. • Accept ACC advice without any checks for conflicting traffic and transfer points outside their area of responsibility (AoR). • Underestimate the time needed for the “follow me” car to complete an inspection of the runway. • CC is distracted and does not notice error made by EC. • EC is distracted due to communications with unrelated aircraft. • Accept two aircraft with rapidly eroding vertical or horizontal separation minima from ACC unit. • Request unrealistic rate of climbs or descends to solve conflict geometries between aircraft. • Misinterpretation of clearance by a flight crew. • Rushed action to meet a constrain requested by the Tower units.
How can you detect errors and recover from them?	• Error detection is done by monitoring traffic in the radar screens (wrong speeds, levels) and by using automation tools (e.g., range and bearing, Short Term Conflict Alert). • Error recovery is done by asking the CC to revise the plan and by issuing alternate instructions to aircraft. • Error detection can be made by closely monitoring Tower communications with aircraft on the ground.

aspects of controller performance in collecting data, recognizing familiar situations, assessing constraints and resources, making decisions, communicating actions, and recovering undesirable effects.

Apart from specifying the competencies and strategies required in complex tasks to facilitate training, CTA can provide valuable inputs into the specification of user requirements, prototyping, and evaluation of cognitive aids. Specifically, the analysts can determine situations and tasks for which cognitive aids would be desirable. This investigation can be made by addressing the following questions:

- What other information would be useful to the practitioners?
- Is there a more appropriate form to present the information already used as well as the additional new information?
- Is it possible to increase the reliability of information?
- Could the search for information be facilitated, and how?
- Could the treatment of information be facilitated, and how?
- Could we provide memory supports, and how?
- Could we facilitate the cognitive strategies carried out, and how?
- Could we promote and facilitate the use of the most effective diagnosis and decision-making strategies, and how?
- Could we provide supports that would decrease mental workload and mitigate degraded performance, and how?
- Could we provide supports that would decrease human error occurrence, and how?

Cognitive aids can take several forms, including computational tools for conflict detection, decision aids for conflict resolution, memory aids, and so on.

The ABCDE method can also record what type of information should be collected in the controllers' consoles and how it could serve the particular features of the situation and the controller strategies. A large part of this information is already present at the controllers' workstation and radar screens although it may not always be in an organized form. However, controllers identified additional information that should be provided by special purpose tools that were not readily available at their consoles. Examples include information about anticipating incoming traffic (collaboration human machine interface), information for monitoring aircraft taxiing in the airport (automation-dependent surveillance broadcast), and information from the Tower unit regarding updates of the sequences of departing aircraft.

8.8 Concluding Remarks

Operational reality always contains situations with subtle and infinite variations that can be different from situations replicated in training (Dekker et al. 2008). Consequently, organizations are not expected to train practitioners for all contingencies that may be encountered. However, certain improvements could be made in existing refresher training programs to make the response of controllers more resilient to emergency and abnormal situations (EAS).

From the field study, it appeared that controllers were very competent in handling many abnormal situations that represented the regulatory aspects of training. Although controllers are very good at recognizing or anticipating problems and planning how to counteract them, certain situations of high workload and stress are bound to occur that can lead to errors of diagnosis or planning. Unfortunately, errors could combine with others and escalate into abnormal situations that are difficult to handle. After all, the system comes to rely on the error-management skills and cognitive strategies of controllers in order to contain all adverse consequences.

In the majority of the incidents studied, delays and errors in recognizing the initial cues of critical situations remained unrecovered and led to a rapid transition into abnormal situations (e.g., early transfers that resulted in having to control traffic outside the boundaries of the sector). This finding is not surprising and similar results have been reported elsewhere. For instance, Klein (2006) argued that team members might fail to detect initial problem cues or that, when they do succeed, they might be reluctant to acknowledge them to others, hence losing the opportunity for a critical intervention. In the aviation domain, Dismukes et al. (2007) came up with similar findings from accident studies. In many cases, flight crews were reluctant to voice their concerns when they realized that the situation was rapidly getting out of control. It is essential, therefore, that refresher training identifies a set of EAS scenarios, representative of the situations encountered in the actual work environment.

Several studies have suggested that the instructional facilities embedded in simulators can determine the success of training more effectively than simulation fidelity alone can possibly do (Jentsch and Bowers 1998; Salas et al. 1998). Instructional methods, such as training needs analysis, cognitive task analysis, scenario design, performance monitoring, and feedback or debriefing, are necessary to ensure mastery and evaluation of emergency response skills. Despite the earlier suggestions that aviation training should follow a systematic approach, research presented in this chapter argues that this systematic approach has not been applied effectively to the refresher training in air traffic control. Training needs analysis should be guided by CTA methods to specify the cognitive strategies that will become the focus of the training curriculum.

The ABCDE method of cognitive task analysis can be helpful for specifying the cues, the challenges, the decisions, and the strategies used by experienced controllers in the course of events of a complex scenario. This method can also help instructors to enhance the "cognitive fidelity" of training by identifying events in a scenario that would provide opportunities for controllers to practice specific cognitive strategies. The findings reported here can help air navigation service provider (ANSP) organizations to diagnose weaknesses in their training and seek advice on how to overcome them.

9

Workload and Complexity

9.1 Introduction

To meet increasing traffic demands, air navigation service providers (ANSPs) seek better tools to assess the traffic handling capacity of air traffic management (ATM) systems. This effort requires a better understanding of how complex situations are related to human performance and how controllers intervene to maintain control. Empirical studies and reviews show that controllers cope with complexity by adapting priorities, managing their cognitive resources, and regulating their own performance. This chapter discusses the development of a behavioral marker system to record and evaluate the cognitive strategies that controllers use to cope with complexity.

9.2 Complexity in the ATM System

The continuing growth of air traffic world-wide requires interventions that increase the capacity of the airspace. To assess the appropriateness of design interventions, however, it is important to understand how a particular traffic situation is related to the cognitive difficulty in controlling the situation and the associated workload (Mogford et al. 1995; Pawlak et al. 1996). Studying how controllers adapt their behavior to cope with complexity is very important if we are to understand how modern technology and new demands may affect system performance.

A comprehensive review of complexity in several domains showed that many definitions of complexity rely on the size or number of parts of the system, which is not sufficient to account for the richness of what is meant by the term "complexity" (Edmonds 1999). An important aspect of complexity regards connections between components and their dependencies that make it difficult to predict system behavior in the future. Xing and Manning (2005) proposed that complexity should be understood as a multidimensional construct with attributes ranging from the number and the variety of elements to their underlying relationships. A Eurocontrol report (Hilburn, 2004) identified several metrics of complexity based on regression models that evaluate task demands according to their predictive power.

A well-known set of indicators regards the dynamic density metrics (Laudeman et al. 1998; Masalonis et al. 2003) that attempt to predict changes of mental workload over time. However, a unified dynamic density metric (Kopardekar and Magyarits 2003) was found to account for less than half of the variance in self-ratings of mental workload. Empirically derived metrics, such as the dynamic density metrics, focus on task demands and fail to model the flexibility of controllers to respond; this may explain why a significant portion of variance of performance has remained unexplained in earlier studies (Loft et al. 2007). Recent studies have shown that the relationship between complexity and performance is not linear but it is an emergent property of the complex interaction between controllers and traffic situations (Athenes et al. 2002; Histon and Hansman 2002; Mogford et al. 1995; Loft et al. 2007). This approach reflects earlier views of Sperandio (1978) that the relationship between complexity and performance can be better understood by considering how controllers adapt their cognitive strategies and regulate their workload.

Brooker (2003) postulated an adaptable function of human performance over complexity that has been supported by earlier research. As complexity increases, controllers may adapt their priorities and the quality of service may become less important in favor of maintaining control of the whole stream of aircraft (i.e., fewer variables are taken into account). It is only above an upper traffic density threshold that operational errors would become more frequent and performance would deteriorate. It seems that there is little evidence to suggest that any sudden and uncontrolled fall of performance occurs with the increase of complexity except at very high traffic volumes. A critical review of the literature (Loft et al. 2007) has adopted a systems control model to examine the relationship between complexity and performance (Figure 9.1).

Figure 9.1 shows that, at first, performance can be adjusted by explicit control of the airspace (shown as an outer feedback loop) to

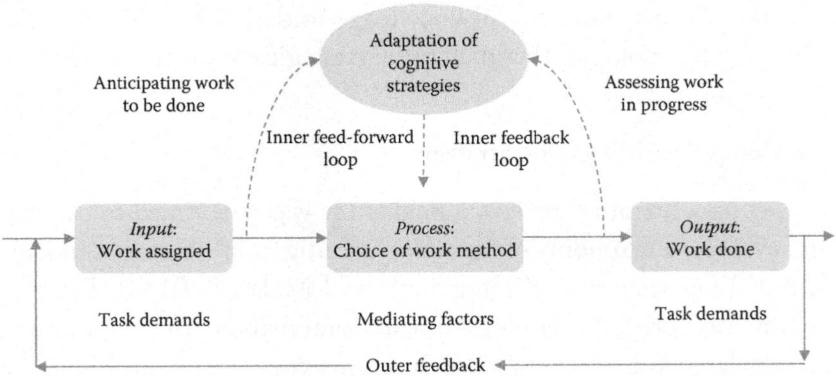

Figure 9.1 A cognitive model of controller activities. (From Kontogiannis, T. and Malakis, S., *Safety Science*, 57, 27–34, 2013.)

reduce complexity. In this sense, controllers can take action to change future task demands fed back through the system. Equally well, complexity can be managed by reorganizing priorities and choosing a different strategy (see dashed lines for feed-forward and inner feedback loops). Looking at the feed-forward loop (at left), controllers may become aware that a large number of aircraft are about to enter the sector and thus adjust their strategy so that the communication load with aircraft on frequency becomes smaller. Looking at the feedback loop (at right), controllers may become aware of potential conflicts and thus adjust their priorities toward achieving safety at the expense of quality of service. The model in Figure 9.1 shows that controllers monitor both the goal-state discrepancies and their own capacity to respond in order to adapt their cognitive strategies.

The present chapter aims to provide a classification of strategies used by controllers to regulate their performance and maintain resilience despite high levels of complexity. The focus has been on the inner feedforward and feedback loops used by controllers to anticipate work to be done and work in progress. Attention should be paid not only to individual strategies, but also to coordination and communication patterns that reduce workload and mitigate complexity.

Complexity mitigation strategies can be illustrated with behavioral markers in order to assess new technologies and foresee weaknesses that may lead to delays and errors in conflict resolution. They can also provide designers with useful knowledge to design flexible tools and emerging technologies that match the strategies of controllers.

9.3 Complexity Mitigation Strategies

Following a literature review, a field study was performed to develop and evaluate a taxonomy of "complexity mitigation" strategies in simulated ATC scenarios (Kontogiannis and Malakis 2013b). The taxonomy was enriched with several behavioral markers that use domain specific language to exemplify performance concepts and facilitate the evaluation of strategies. Behavioral markers are observable behaviors that indicate the quality of performance within a work environment. It is important that behavioral markers describe observable behaviors and have a causal relationship with the performance outcome. Markers should exemplify concepts in a clear manner and relate to each other in a meaningful way (e.g., they may relate to a theoretical model of performance).

A practical scheme has been developed that classifies complexity mitigation strategies according to their role in accomplishing five cognitive functions, as shown in Table 9.1. Adjustments in monitoring involve making subtle adaptations to recognize and anticipate potential conflicts as traffic patterns increase in intensity. Replanning involves changing goal priorities to sustain high levels of complexity without compromising safety, modifying plans on the fly and critiquing mental models to cope with information uncertainty. Managing workload and restructuring refer to changes in task allocation within and between sectors, especially at the highest levels of complexity. Finally, changes in communication and coordination are important teamwork functions that support adaptations at all levels of complexity.

9.3.1 Adjustments in Monitoring and Anticipation

Traffic monitoring strategies enable controllers to detect signs of impending conflicts, build a model of the situation and play out mentally the progression of traffic (Redding et al. 1991; Seamster et al. 1993). The mental model allows controllers to structure information in the airspace into categories (e.g., aircraft heading to same destination, converging points of traffic or hotspots, and nonstandard

Table 9.1 Prototype Taxonomy of Complexity Mitigation Strategies

COGNITIVE FUNCTIONS	COMPLEXITY MITIGATION STRATEGIES	BEHAVIORAL MARKERS
Adjusting monitoring and anticipation	S1. Relies on past experience to identify streams of aircraft and converging traffic	S1. Groups aircraft as "cleared for ILS approach and landing" vs "left on holding pattern;" notices aircraft converging in a small space; classifies aircraft in a trail
	S2. Simplifies the interface and prepares for future demands	S2. Reduces clutter on the radar screen to cope with future demands
	S3. Looks ahead and anticipates times of heavy traffic	S3. Visualizes or writes down possible conflicts from the start before setting a plan; does not leave tasks for long time periods since traffic may soon get heavy
Replanning and managing uncertainty	S4. Underplays criteria of efficiency and adopts more conservative response criteria	S4. Puts some aircraft on a holding pattern; gives priority to aircraft diversion rather than economy of routes; issues large diversions to one or two aircraft to reduce conflicts rather than issuing "small changes to many aircraft"
	S5. Decides how soon to resolve conflicts	S5. Resolves potential conflicts early to turn attention to other things; issues instructions early to put flights on preferred routes and reduce uncertainty
	S6. As uncertainty goes up, precision of instructions is reduced to allow more scope for changes later on	S6. Route instructions focus on direction rather than precise headings; remains careful with commitment to initial instructions as they may have to be changed later on
	S7. Tightens-up task sequences (or aircraft trajectories), takes shortcuts and identifies leverage actions to reduce conflicts	S7. A number of aircraft are urged for landing before handling the emergency; reduces space between aircraft in same trail or group; decides to by-pass aircraft to final route to avoid entering a busy sector
Managing workload and changes	S8. Selects actions that result in lower workload and manages interruptions	S8. ILS procedure for landing is preferred to VOR approaches as crews can rely on autopilot to execute procedure, which relieves controllers; does not interrupt a priority task to provide a service
	S9. Creates groups of aircraft with similar patterns and actions in order to reduce workload and need for continuous changes	S9. A number of aircraft are placed in trail or put on holding or placed close together but separated by altitude
	S10. Weighs the cost of change as efficiency can be traded off for higher workload and risk	S10. Does not agree to last minute suggestions for changes; new plan may increase workload of communications; new plan may place aircraft in a sequence that is difficult to undo; new plan may result in errors difficult to recover

(continued)

Table 9.1 (*Continued*)

COGNITIVE FUNCTIONS	COMPLEXITY MITIGATION STRATEGIES	BEHAVIORAL MARKERS
Restructuring (reconfiguring)	S11. Adapts or expedites handoff procedures with other traffic sectors S12. Asks support from supervisor and others to change allocation of team roles (in some cases, resorts to resectorization)	S11. Expedites handoffs in order to place aircraft in preferred altitude or heading; requires adjoining sector to handoff aircraft in trail S12. An extra controller takes over all aircraft heading for landing; a sector is divided in two parts
Communicating and coordinating	S13. Decides on most efficient ways to communicate and minimizes information garbling S14. Resorts to proactive communication and coordination	S13. Inflicts with voice to indicate "busy;" issues instructions to multiple aircraft; maintains control of frequency S14. Executive controller requires all information together from crews to avoid need for new instructions. Coordinating controller prepares handouts and writes a list with all units to be notified in a suspected emergency

Source: Kontogiannis, T. and Malakis, S., Safety Science, *57, 27–34, 2013b.*

flows) and hence reduce complexity associated with monitoring traffic (Reynolds et al. 2002; Histon and Hansman 2002). Another strategy that reduces monitoring requirements regards the processing of aircraft in groups or streams, which allows controllers to work with more aircraft simultaneously and use fewer control actions (Amaldi and Leroux 1995; Histon and Hansman 2002; Redding et al. 1991). As complexity increases, controllers may become overwhelmed with traffic and their behavior becomes reactive. Coping with complex situations requires being proactive, staying ahead of the traffic. For this reason, expert controllers constantly anticipate threats (e.g., weather cells and frequency congestion problems) and prepare for future traffic demands.

9.3.2 Replanning and Managing Uncertainty

Experts become increasingly adept at handling complex events by making logical leaps beyond the procedures to quickly reach a solution (Wickens et al. 1997). In making decisions, practitioners tend to run a mental test to determine the time available and their priorities. The criteria for choosing a plan are first of all safety, followed by traffic efficiency and workload of sectors. Plans have to meet these criteria, depending on the time that controllers spend on thinking about them. Hence, by changing their priorities, controllers are able to control their workload and maintain a safe traffic environment (Brooker 2003). Several studies have pointed out the uncertainty surrounding the planning of traffic (Amaldi and Leroux 1995; Weitzman 1993). In busy periods, controllers shift their criteria for classifying conflicts, hence becoming more conservative so that they intervene to ensure separation if there is any uncertainty regarding future separation between aircraft (Kallus et al. 1999). Emergencies and abnormal situations generate uncertainty as flight crews may be reluctant to provide conclusive information. In an emergency descent scenario, Malakis et al. (2010a) reported that controllers noticed that an aircraft was initiating a descent without any prior information from the flight crew. In this case, controllers had to assemble a mental model of the situation (e.g., turbulence, cabin decompression, or engine failure scenario) without increasing the workload of the flight crew with extra queries.

9.3.3 Managing Workload and Change

As traffic complexity increases, so does the number of tasks competing for attention, the pace of work, and the number of interruptions resulting from team communication. A usual coping strategy involves changing the distribution of tasks over time and sector. For example, Kirwan and Flynn (2002) identified various heuristics that controllers use to resolve conflicts, such as the following:

- Using as few control actions as possible
- Giving aircraft initial level changes early and fine-tuning later
- Using solutions that require less coordination
- Using vertical separation for complex conflicts

These solutions have to do with how controllers distribute their tasks over time and space to regulate their pace of work and off-load tasks to adjacent sectors. Every new event has to be considered in the context of work and this may change the order of priorities. Workload is not a constant parameter but follows the changing pattern of the situation as it escalates; hence, task sequences and priorities can be altered as complexity increases.

9.3.4 Restructuring Tasks across Sectors

Another aspect of workload regards managing team interaction with other sectors in addition to managing normal duties in one's own sector. In this regard, the CC is in a better position to offload the increased workload of the EC who is exposed to a steeper escalation pattern and reaches higher workload levels. This can be done by restructuring and adapting tasks across sectors either by expediting handoff procedures or by resizing the area of responsibility. In the first case, controllers may hand off aircraft to the next sector early to enable adjacent colleagues to take control of aircraft according to the plans as soon as possible. Another aspect of early handoff procedures may be to ask other sectors to organize certain aircraft in a configuration (e.g., aircraft in trail) before handing aircraft to a busy sector. In the second case, controllers can attempt to change the airspace boundary in order to accommodate more traffic in adjacent sectors. Although this strategy reduces the area of responsibility of controllers, new problems may emerge in the adjacent sector that assumes responsibility for the additional area of control. Such dynamic resectorization may introduce new risks due to unfamiliarity with the obstacles and constraints of the new sector, possible distractions from pending tasks, and higher memory load. For these reasons, resectorization policies should be specified in advance by adjacent sectors and adequate training should be provided to controllers to become aware of obstacles and constraints ahead of time.

9.3.5 Changes in Communication and Coordination

Several studies have found that anticipating information for future activities and predicting the workload of others can make team communication more efficient, especially at high tempos of work (Entin and Serfaty 1999). Expert controllers are able to communicate effectively without unnecessary elements that garble communications. They are able to appreciate major attributes of information (i.e., criticality and timelines) and judge the level of workload and interruptibility of other team members. Other patterns of adaptation to complexity may regard changes in gathering and communicating information—for example, enquiring all necessary information together rather than in a piecemeal fashion (Koros et al. 2006). In an airport diversion scenario, Malakis et al. (2010b) reported that the coordinating controller (CC) first wrote down a list of all relevant issues related to candidate diversion airports before starting communications. This minimized communications by avoiding many calls for individual aspects of the problem; instead, controllers made all enquiries to find a diversion airport at the beginning. Emergency scenarios can increase team coordination by requiring more contacts and route changes with a larger number of aircraft. In coping with complexity, controllers seem to choose options that require less coordination. Emergencies can also increase complexity by restricting the available time window for response. Controllers may adapt their coordination patterns and reduce workload by warning colleagues of imminent problems. In the same scenario, the CC communicates with the nearest airport and alerts colleagues of a potential aircraft diversion without specifying the precise nature of the problem, hence tolerating uncertainty.

9.4 Selection of Strategies for Different Levels of Complexity

This section presents some observations from a field study on the selection of controller strategies (Table 9.1) for scenarios of different levels of complexity (Kontogiannis and Malakis 2013b). Even at low complexity levels, controllers appeared to create comprehensible mental models of the airspace structure by identifying nonstandard flows (e.g., military traffic) and by perceiving aircraft as groups (S1). Controllers were constantly engaged in the anticipation of threats (S2—e.g., weather, frequency congestion problems) and the preparation of activities that simplify problems that may be encountered in the near future (S3—e.g., reduce clutter on the radar screen).

As the complexity increased, controllers tended to reduce the quality of service offered to flight crews by adapting their performance criteria to the situation (S4). Experienced controllers adopted a critical stance and critiqued their understanding of the problem on a regular basis in order to adapt their plans to the changing requirements of the situation; in addition, they tended to adopt more conservative responses by resolving conflicts at early stages (S5) in order to conserve attentional resources. At higher levels of uncertainty, the precision of instructions was reduced to allow more scope for changes later on (S6).

Another aspect of replanning regarded controllers' tendency to resort to sacrificing decisions—for example, attempting to bypass aircraft to final route in order to avoid entering a busy sector (S7).

In order to regulate their workload and pace of work, controllers selected options that would result in lower workload (S8—e.g., choose vertical over horizontal separation, when both options were valid) and created groups of aircraft with similar patterns and action requirements to reduce the need for continuous adjustment (S9). In managing task interruptions, controllers also used reminders for tasks that were interrupted and had to be resumed at specified time periods. Because new plans could increase workload, controllers tended to weigh the cost of change (S10); for instance, new plans may increase communications or may result in errors that are difficult to recover. For the high complexity scenarios, controllers would hand off aircraft to adjacent sectors early and turn down requests for early transfer from other sectors (S11). Resizing the area of regard was also observed for high complexity scenarios (S12).

With regard to team communication and coordination, experienced controllers tried to expedite communications by resorting to trading language and handed notes to colleagues for essential information to enable efficient coordination (S13). As the complexity of the scenario increased, so did the frequency of using proactive strategies (S14) in coordination and communication. Finally, controllers became more vigilant of possible signs of fatigue in the team and were backing up others in heavy traffic situations.

From this discussion, it appears that there was no sudden and uncontrolled fall of performance as the complexity reached higher levels; instead, controllers managed to maintain performance and where there was a decrement it was quite graceful. Figure 9.2 shows how controller performance varies for three levels of complexity. A working hypothesis is made that controllers exhibit three strategies of resilience to resist performance decrements as complexity increases. Initially, controllers may readjust their routine strategies to maintain quality of service at low levels of complexity. In cases where traffic flow becomes moderate to high, controllers resort to some form of replanning that includes changing priorities, critiquing their models regularly, adapting plans to the changing requirements, and regulating workload. At high levels of complexity, a new configuration of teams may be required in terms of task allocation and team organization.

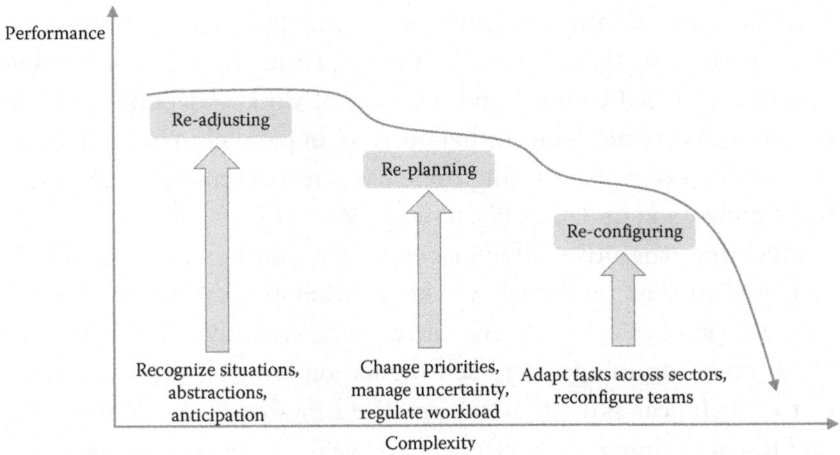

Figure 9.2 Mapping complexity-mitigation strategies to increasing levels of complexity.

9.5 Concluding Remarks

The findings of the field study presented in this chapter indicate that controllers not only react to increasing task demands, but also regulate their strategies proactively so that their performance degrades gracefully with the escalation of complexity. Even at low complexity levels, controllers seem to simplify problems that may be encountered in future (e.g., perceive aircraft as groups, reduce clutter on the radar screen) and anticipate threats (e.g., adverse weather, frequency congestion problems). As the complexity increases, controllers may tend to reduce the quality of service offered to flight crews by adapting their planning to the situation. Experienced controllers adopt a critical stance and critique their understanding on a regular basis in order to adapt their plans to the changing requirements of the situation.

In order to regulate performance and pace of work, controllers may select options that result in lower workload and manage interruptions according to task priorities. Finally, controllers tend to expedite communications by resorting to proactive strategies in communication and coordination. The prototype taxonomy of complexity mitigation strategies (see Table 9.1) provides a structure for evaluating the performance of controllers in complex traffic scenarios.

The benefits should be sought in developing appropriate forms of operational aids, training programs and system design that would support controllers in coping with the complexity of future traffic environments. Controller activities are mainly driven by a process of adaptation so that their recognition, planning, and coordination activities match the tempo and intensity of work. Adaptation can be seen as a control mechanism that operates in parallel to ordinary control activities (e.g., monitoring the radar screen, communicating with flight crews, and updating flight progress strips).

Adapting cognitive strategies requires additional time, which increases further the actual workload from the execution of ordinary activities. A case in point for reducing cognitive burden is error management training where self-regulation is mastered in the context of technical skills that are practiced in simulator training. The traditional training of controllers to become flawless in technical skills can increase preoccupation with success and may deprive them of opportunities to practice skills for regulating workload and hence reduce the likelihood of errors. Error management approaches shift the emphasis from mastering technical skills to building a repertoire of self-regulation strategies.

The taxonomy of complexity mitigation strategies may hold implications for system design and automated presentation tools. In the same sense, an understanding of how controllers mentally structure their airspace, how they simplify clutter on their radar screens to mitigate complexity, and how they manage pending crew requests would enable designers to produce decision-support systems that closely match the information and strategic requirements of controllers.

The complexity mitigation strategies can provide useful input into the design of free flight environments where the separation assurance function is shared between controllers and flight crews. Any changes in the planning priorities and response criteria of controllers introduced by new designs should be carefully studied in advance to avoid contradictions with current controller practices. In the free-flight environment, for instance, controllers may monitor passively the flight crews who would have the primary responsibility for conflict resolution. This change in task allocation can cause many difficulties to controllers who would prefer to resolve conflicts early and move on rather than wait for flight crews to resolve conflicts or wait to be called upon for assistance unexpectedly. The new task allocation between flight crews—controllers and the conditions for task switchover should be considered within the context of controller strategies in coping with complexity.

10

New Challenges in ATM

10.1 Introduction

Expected air traffic increases until 2040 are likely to create a significant capacity problem that airspace systems in Europe and the United States may struggle to accommodate. To advance their aviation systems, the next generation air transportation system (NextGen) has been proposed in the United States together with the single European sky ATM research program (SESAR) in Europe. These next-generation approaches aim at achieving incremental benefits over the next 10 to 20 years, taking advantage of new technologies in the air traffic management (ATM) domain. Advances in route clearance technologies, such as trajectory-based operations and digital communications, have enabled aircraft automation to undertake separation tasks, thus freeing air traffic controllers (ATCOs) to complete other traffic management tasks for a larger number of aircraft.

This vision for future ATM systems is based on a large-scale computer system (Willems and Koros 2007) that would determine and negotiate with flight crews, airline operations managers, and ATCOs, many 4-dimensional (4D) trajectory-based operations (TBOs). A system wide information management (SWIM) system will evolve to become a real-time repository and an archive for all airspace information to promote comprehensive information exchange across all stakeholders. SWIM will support advanced automation, ensure digital data sharing, promote common situation awareness across all users, and enable system-wide collaborative rerouting and other resource allocation functions. New technologies will provide 4D trajectories that would be available to all stakeholders and ensure optimal routings based on nominal schedules, improved methods of weather observation, and cost considerations. Although information sharing will be a positive change, controllers already experience high information load; hence it is crucial to determine what information should be available to them as well as when and how it should be displayed.

Next generation technologies include a range of automated decision support tools and information technology services to flight crews and controllers. In general, two main concepts have dominated new approaches related to the function allocation between flight crews, controllers, and automation: (1) ground-based separation assurance (Erzberger 2006) and (2) airborne separation assurance (Wing 2008). The primary difference between the two concepts lies in the location/distribution of separation functions on airborne separation between aircraft versus ground-based separation in the ATM system.

In ground-based separation, a centralized automated system monitors and manages trajectory-based operations. In exceptional off-nominal operations, controllers shall assume responsibility for conflict detection and resolution. The primary difference in today's systems is that ground-based automation would perform conflict detection and provide conflict resolution trajectories integrated with data link. The modified trajectories can be sent to aircraft either by the controllers or directly by the ground-based automation, whenever certain predefined criteria are met.

In airborne-based separation, flight crews would manage the separation function for their own aircraft supported by an onboard airborne separation assistance system (ASAS). Using airborne surveillance information (i.e., automatic dependent surveillance broadcast, ADS-B), ASAS automation could predict aircraft trajectories, detect conflicts, alert flight crews appropriately, and compute resolution trajectory alternatives. Flight crews will have to select among alternative resolutions displayed in ASAS and execute the new trajectory through the flight management system (FMS) and autoflight system. Both ASAS and flight crews will have to comply with any trajectory constraints set by the air navigation service providers (ANSPs) and the air traffic flow and capacity management (ATFCM) for traffic flow management. Under this user trajectory management, responsibility for aircraft separation and selection of flight paths are either partially or completely transferred from controllers to flight crews (Hollnagel 2007). Dwyer and Landry (2009) discussed a number of plausible changes to separation assurance responsibilities, including: (1) the separation function may be shared between flight crews and controllers, without support from automation, (2) some form of automation that might aid or replace the controller's function, and (3) the possibility that onboard aircraft automation might carry out a separation assurance function or aid flight crews in doing so.

What seems more likely is that centralized and distributed systems will act in concert to provide both separation assurance and collision avoidance services. Certainly some form of collision avoidance will be onboard all commercial aircraft, as it is in today's systems. It is as yet unclear, however, what forms distributed systems might take or what functions they would serve in a mixed concept operation. In general, all future system changes will need to demonstrate that they are at least as safe as today's systems. One way to accomplish this is to ensure that, in the event of failures, the new system will degrade to its present mode of operation. In other words, the system would need to be able to gracefully degrade from an automated mode to a manual mode. Relieved from routine monitoring and control tasks, controllers will be able to devote more time to solving strategic control problems, managing traffic flows during poor weather conditions, and handling other unusual events.

NextGen and SESAR initiatives seem to change both the nature of technological artifacts and the allocation of roles and responsibilities between controllers, flight crews, and automation (Brooker 2007). According to cognitive engineering, a change in the artifacts will alter the cognitive strategies of all human agents in ways that are difficult to predict. In this respect, a theoretical model of taskwork and teamwork strategies can be useful in postulating hypotheses about changes in controller and aircrew strategies to match the demands of the new human–artifact interaction. The taskwork/teamwork for effective and adaptive management (T^2EAM) framework can be used to consider a number of automation challenges in the taskwork and teamwork strategies of controllers as well as organize earlier research findings of the human–artifact interaction presented in the literature.

10.2 Taskwork Performance

10.2.1 Recognition and Monitoring

NextGen technologies will provide better aircraft position information (e.g., automatic dependent surveillance broadcast) that may allow computerized systems not only to track aircraft but also to determine their future position over time. This information can be used by conflict-detection systems to alert controllers without having to scan the radar scope for conflicts. Furthermore, conflict resolution technologies could be used to determine optimal resolutions in terms of safety (e.g., no secondary conflicts, no extreme instructions) and efficiency (e.g., shortest distance around weather), reducing the controller needs to engage in effortful computations. This implies that controllers must continue to monitor traffic although they may not be active in its regulation. Metzger and Parasuraman (2001, 2005) showed that when the role of controllers is shifted to passive monitoring, they take significantly longer to detect conflicts and miss more conflicts under heavy traffic. The challenge for next generation technologies will be to maintain some sort of controller involvement in the monitoring task or, at least, recognize that controllers will be likely to spend time in acquiring a first hand feeling of the way that traffic evolves.

Earlier research has shown that recognition strategies are based on a mental model of the airspace (Reynolds et al. 2002) that classifies aircraft in categories (e.g., aircraft heading to same destination), projects converging points of traffic, and identifies nonstandard flows. By structuring information in the airspace, controllers are able to regulate their workload even in heavy traffic. However, dynamic resectorization policies of new technologies and flexible routings are likely to change the groupings and mental structures used by controllers in the current environment.

Dynamic resectorization involves an adjustment of airspace boundaries to accommodate real-time traffic flow constraints (e.g., weather, equipment outages, or restricted airspace); at present, a limited degree of resectorization occurs in practice (see Figure 10.1). In future ATM systems, airspace changes will occur dynamically in response to weather, demand, and pilot preferences. Today's controllers develop expertise over a period of years and learn to rely on the airspace and route structures to aid their performance. Stein et al. (2006) reported that en route controllers take approximately three years to develop pattern recognition skills that support this sector-specific expertise. Therefore, dynamically adjusting the sector boundaries may potentially negate a lot of this expertise and pose challenges to controller strategies.

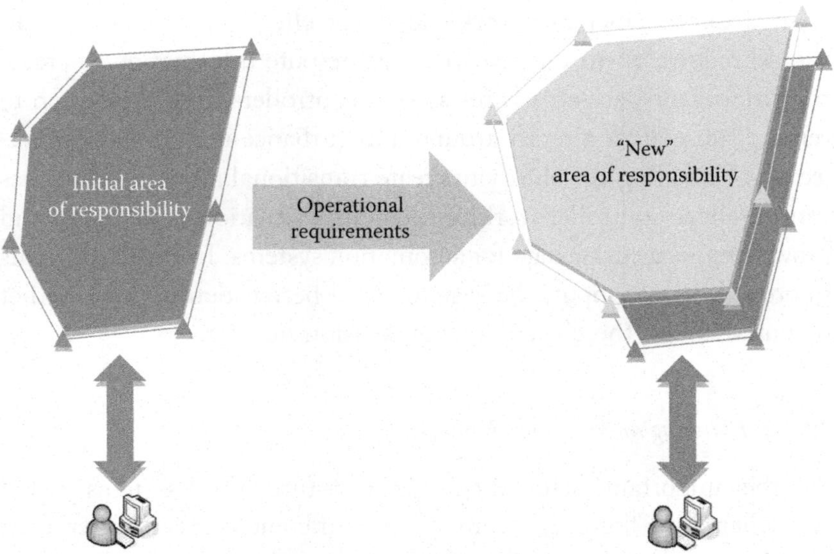

Figure 10.1 The concept of dynamic resectorization.

The introduction of new technologies in surveillance will make it possible for controllers to rely less on ground navigational aids. Although this may benefit flight crews and other users, there is a concern that it will have a negative effect on the mental structures that controllers have used for many years in the past (Brooker 2005). Controllers have spent considerable time in learning their airspace and in acquiring valuable expectations of traffic flows through their sectors. They have learned how to recognize subtle cues in a stable environment, which gives them an almost intuitive feel for managing particular problems. In a flexible routing context, however, such predictable patterns may no longer exist. We are uncertain how the change from scanning hot spots for potential problems to scanning the whole airspace for potential issues may influence human workload. As shared separation is introduced into the airspace, there will be fewer predictable conflict points within a sector, which may increase controllers' efforts in maintaining awareness of traffic conflicts.

The current regulation does not support transitional route structures to move aircraft around weather disturbances. When weather creates a problem, controllers have to vector aircraft around the weather disturbance. However, controllers do not usually enter these amendments into the host computer system because it requires clearances for each change and this increases workload. To be effective, automation tools should require up-to-date weather and reroute information to create transitional airspace structures so that controllers have the option to reroute subsequent aircraft around a disturbance using a similar trajectory. Providing the ability to create transitional airspace structures would relieve controller workload, enhance situation assessment, and provide more accurate data for automation systems. In turn, using this improved data as input, the system could better identify and predict potential issues, thereby enhancing decision support.

10.2.2 Planning and Conflict Resolution

Another important factor to consider in future ATM systems is that controllers and flight crews may have different decision criteria for resolving conflicts (Neal et al. 2011). One reason why flight crews prefer to make lateral resolutions is that they typically fly at optimal cruise altitude, which makes them more resistant to deviate from that altitude when they have a choice. Controllers on the other hand are more willing to use altitude resolutions and speed adjustments when two aircraft are in conflict (Vu and Strybel 2011). These differences in the decision criteria and planning preferences will affect the transfer of separation tasks between flight crews and controllers since the two professional groups have different expectations about the future evolution of traffic.

In today's ATM system, controller planning is based on rules and expectations from aircraft performance that have been developed and worked out for several decades. In mixed aircraft traffic, the introduction of new aircraft types will require controllers to integrate the characteristics of all aircraft into their procedural knowledge. In handling a mix of aircraft that have drastically different characteristics, it is uncertain how long it will take controllers to predict how these aircraft will move into the airspace. The trajectory evaluation process will also be more complex because there might be a greater mix of separation standards.

10.2.3 Anticipating

Anticipation figures as a prominent cognitive strategy in many performance models in ATM (Reynolds et al. 2002; Oprins et al. 2006).

Anticipation enables controllers to foresee threats that may appear in the near future and proactively mitigate their consequences. Typical threats include: military activity at the borders of a sector, heavy traffic, adverse meteorological conditions, airports surrounded by high mountains, congested airspace, aircraft malfunctions, and errors committed by other practitioners.

Currently, controllers take a more tactical role and operate with a short look-ahead window while the focus of tactical control has been on the immediate problem, without much consideration for the downstream consequences. In future ATM systems, controllers will provide a conflict-free aircraft trajectory with the requirement of ensuring that aircraft proceeds on that trajectory without delay. This requires controllers to take a more strategic approach and adopt a longer look-ahead window, since their interventions will have to consider the impact on the entire trajectory of aircraft as well as any other aircraft that may be affected by that decision (Neal et al. 2011). Furthermore, any modification of the original trajectory is expected to be done in collaboration with the airspace users. This shift in the look-ahead window and the longer term impact of decisions may increase the demands of the anticipation task and possibly the amount of time spent on anticipation versus other tasks.

Controllers tend to anticipate threats and plan for contingencies early. For instance, controllers may intervene early, not only to resolve projected conflicts, but also to ensure that there will be sufficient time to respond in cases of abnormal events. This working style is different from automation that merely detects and intervenes. For example, controllers might place a pair of climbing aircraft on parallel courses separated by 1000 feet (rather than on intersecting courses) to ensure that they avoid a loss-of-separation threat. In this supervisory mode, controllers could be kept apprised of near conflicts and even intervene to lessen the pressure on the sector by reducing traffic capacity or by creating contingency plans.

10.2.4 Critiquing and Adapting to Workload

Occasionally, abnormal situations may arise rapidly, leaving controllers unprepared to moderate the intense workload that is provoked. It is important, therefore, that controllers are able to improvise and adapt plans on the fly in order to manage uncertainty as the situation evolves in a dynamic fashion. One way of adaptation involves shifting the decision criteria for taking action or waiting for more information.

Controllers are trained to operate proactively and resolve conflicts soon after their detection. However, the trajectory negotiation concept requires that resolution tasks are placed in a suspended mode until a later time. In this sense, trajectory negotiation breaks the conflict detection and resolution task for controllers into two separate components. Jha et al. (2011) found that when flight crews responsible for conflict detection failed to negotiate, controller performance in taking over conflict resolution degraded in the new ATM context. Many years ago, Endsley et al. (1997) showed that ATC performance and situation awareness degraded as greater responsibility for traffic information was allocated to the flight deck; these effects exacerbated when intent information was not provided. In a series of air-ground concept simulations at the NASA Ames Research Center, air traffic control (ATC) ratings of situation awareness were highest only in cases where controllers maintained responsibility for separating traffic (for a review, see Strybel and Vu 2013). It also appeared that controllers continued to scan self-separating flight regardless of the fact that separation responsibility was passed to flight crews.

The trajectory negotiation concept requires a shift in paradigm that is contradictory to current training practices where controllers are taught to be proactive and always act when in doubt. Bolic and Hansen (2005) reported that controllers found it easier to resolve conflicts and move on rather than having to continuously monitor aircraft and wait for flight crews to resolve conflicts. Controllers also suggested that the trajectory negotiation concept gave them a false sense of security that aircraft would resolve conflicts and this made them very nervous in the real situation.

In managing workload, controllers often choose options that are quicker to implement, have fewer implications, and impose a lower coordination cost. For example, when there is a choice to control with vectors or with altitude changes, controllers manage heavy workload situations by issuing altitude instructions. These adaptations have the benefit of reducing workload locally but may not work well in new ATM systems where the role of controllers becomes more strategic.

10.3 Collaborative Decision-Making

In future ATM systems, decision-making will be a more collaborative process and involve multiple flight crews and controllers sharing separation roles and mutually agreeing on modifications to aircraft trajectories. This imposes a greater requirement for team coordination and collaborative decision-making as these changes might need to be negotiated not only with the airspace user but also with downstream controllers and other airspace users that might be affected.

10.3.1 Sharing Understanding, Orientation, and Trust

Shared understanding and orientation enable team members to develop expectations of how others are doing, how they can get help, and how to change their communication patterns. In getting used to airborne trajectory management, for instance, flight crews may need to gain a shared understanding of the intentions of other aircraft and maintain collective orientation with ground-based controllers. In testing an airborne trajectory management system, Vu and Strybel (2011) found that flight crews indicated that knowing the intent of the surrounding aircraft was an important factor in their decision-making. To some extent, the automated system provided data tags for aircraft destinations and allowed flight crews to infer the general goals of the aircraft (e.g., aircraft near their destinations are likely to descend when in conflict). Despite this, many flight crews indicated that additional information about other flight crews' goals would have been helpful as well. For instance, a cargo aircraft may be willing to go through mild weather because passenger comfort is not a concern, while cargo delivery time remains a central concern. This type of information could help flight crews, especially those who are engaging in self-spacing responsibilities, to anticipate maneuvers and preferences of lead and surrounding aircraft.

Currently, flight crews are not given the responsibility of resolving conflicts, so this additional responsibility may change how they would respond to conflicts when they arise. In the Vu and Strybel (2011) study, many flight crews expressed concerns that the controllers lacked awareness of their aircraft when requesting help. Controllers acknowledged that they did not know what the aircraft were doing, as they were not given responsibility for aircraft separation. Flight crews were not apprehensive about taking responsibility for separation; however, they did want ATC involvement as a back-up for resolutions and as a monitor for overseeing traffic.

The issues of shared understanding and communication of intent become very important in new NextGen policies where ground-based and airborne separation capabilities may coexist in the same system. In other words, flight crews should be able to choose the method of trajectory management (i.e., ground-based or airborne-based separation) that is most cost effective for their business model, aircraft equipment, and flight optimization objectives. In a mixed mode of operation, for instance, flight crews of self-separating aircraft should be capable of resolving conflicts, not only with other self-separating aircraft, but also with aircraft assigned to ground-based separation. Indeed, a study by Wing et al. (2013) showed that when intent information on the ground-managed aircraft was available to flight crews of self-separating aircraft, they had fewer conflict alerts and fewer required deviations from their trajectories. This mixed operation was made easier when ground-managed separation was made less tight by increasing the distance between ground-managed and self-separating aircraft.

The issue of trust in automated and human agents is very important for overall system efficiency. Trust is affected by how well controllers understand the way that other humans or automated agents behave and whether there is hope that a commonly agreed solution will be found in the near future. In conventional ATC domains, controller expectations of the behavior of flight crews will affect how soon controllers will take action (i.e., early action when pilot is predictable) and how often they will take proactive action for possible lack of crew compliance with controller instructions. In the context of human-automation interaction, trust may affect the extent that controllers make use of automated aids. For instance, Jha et al. (2011) showed that an automated ATC system that provided controllers with negotiation information and resolution aids did not reduce cognitive requirements. It appeared that controllers either did not use the resolution aid, or continued to verify the advisory given by the resolution aid before making a decision. This resulted in small differences in workload, compared to another condition of unaided performance.

10.3.2 Managing Task Allocation

In ground-based separation, management by exception is usually applied to free controllers from routine control of aircraft and allow them to focus on traffic monitoring; at their discretion, controllers could step in to control aircraft only in exceptional situations. The allocation of functions between flight crews, controllers, and automation is the most critical issue in this management by exception policy. This raises questions such as: Who should handle off-nominal operations? When should this transfer of control occur? How transparent are the pilot-automation interaction to controllers? and Who should resume ultimate responsibility for decision-making? NASA Ames has performed several human-in-the-loop experiments with ground-based separation assurance (Prevot et al. 2012). Although automation performed well with increased complexity, controllers did not seem to trust automation, especially as traffic increased. In routine traffic situations, controllers were comfortable with automation but wanted decision-making authority and support to maintain situation awareness of traffic (Prevot et al. 2012).

As traffic complexity increases, so does the number of tasks competing for attention, the pace of work, and the number of interruptions resulting from team communication. In managing workload, controllers are changing priorities, doing urgent and important actions first, and delaying other secondary actions. Every new event has to be considered in the context of work and this may change the order of priority. In the new ATM system, controllers will have the additional task of monitoring how flight crews resolve conflicts on their own and judging when to intervene. In many cases, this additional task may create difficulties in having to change priorities or having to resume other ongoing activities (e.g., accepting aircraft in the sector or initiating handoffs). Therefore, function allocation should focus not only on the number of tasks to offload from controllers but also the additional tasks that are required to maintain awareness of the work of others.

Workload is not a constant parameter but follows the changing pattern of a situation as it escalates; hence, the sequence and priority of tasks can be altered as the complexity of traffic increases. Controllers have to manage not only the normal duties in their sector but also their interaction with other sectors. With the new dynamic resectorization policy, the new ATM environment is likely to impact on the controller strategies for handing off aircraft and for reallocating tasks to sectors.

Stein et al. (2006) identified several problems related to controller familiarity with the obstacles and constraints of new sectors and the demand for coding and storing new geometries in human memory. In addition, the transfer of an airspace to another sector may become a source of distraction from pending tasks. For these reasons, a dynamic resectorization policy may be associated with a high risk and a temporal increase of workload during the transition period. As a result, resectorization policies should be specified in advance by adjacent sectors and adequate training should be provided to controllers to become aware of obstacles and constraints ahead of time.

In busy periods, en route controllers tend to hand off aircraft to the next sector as soon as possible in order to minimize their workload. This strategy, however, deals with local problems and does not take into account side effects that may impact other parts of the aircraft trajectory. In the new ATM systems, controllers should take a more strategic role and consider the whole trajectory. The emphasis on TBOs is likely to change this strategy of controllers.

10.3.3 Team Coordination

Coordination among decision makers will become a critical issue in next ATM systems because the actions of any one party can have implications for others. For example, if a stream of aircraft is flying toward a weather cell, the agents responsible for the provision of separation need to collaborate to develop an effective solution that does not unnecessarily disadvantage any party. The coordination requirements are further increased if the proposed trajectory changes affect the point in space and time where aircraft reenter high-density airspace (Neal et al. 2011). It is currently an open question as to how distributed decision-making can be structured so as to achieve an optimal outcome for all airspace users (Dwyer and Landry 2009; Wickens and Colcombe 2007).

Early work on NextGen automation concepts has shown that attempts to reduce workload may come at an increased coordination cost. In a study of pilot acceptance of automated conflict resolutions, for example, Battiste et al. (2008) found that flight crews still wanted to contact controllers to discuss and clarify the auto-resolutions sent to the flight deck. The number of inquiries would greatly exceed the controllers' capability, especially if put in a backup mode. The results suggested that automating conflict resolution would reduce ATC workload but it could increase coordination needs between flight crews and controllers.

Coordination and shared responsibilities become an issue when the flight deck and the ground control units have available different information on a particular situation. An example already exists for aircraft that carry onboard radar that can detect weather cells. The information that controllers have on their screens is not as fine-grained as the information that the flight crews receive from onboard radars. As a result, the different portrayals of weather information may give rise to difficulties in sharing awareness of the situation. In a similar fashion, traffic information service–broadcast systems may provide flight crews with information that does not correspond to the information displayed on controller workstations.

10.3.4 Multi-Modal Information Transfer and Communication

Data communication (Data Comm) technology has become a prominent feature of NextGen that allows controllers to send digital text commands and communications to flight crews, who can then upload commands directly into their FMS. Adopting this concept would alleviate much of the congestion on radio frequencies and may reduce errors resulting from trying to decipher auditory messages in a noisy environment. However, this decrease in voice chatter may also decrease situation awareness of controllers. Voice communications afford controllers easy access to meanings that may be less accessible when using Data Comm. For instance, the seriousness of a situation may be more difficult for controllers to ascertain because text messages cannot capture the emotional state of flight crews. Consistent with this claim, Lancaster and Casali (2008) found that the workload of flight crews was greater when using text-only Data Comm compared to speech communications. Similarly, speech rate and number of clearances issued by controllers can indicate their workload to the flight crews on the radio frequency. In recognition of this, flight crews may refrain from making inessential requests to controllers. Furthermore, Data Comm commands are likely to represent a fixed stock of commands and hence they may not have the same degree of flexibility that voice communication affords. For instance, flight crews may revert back to voice communications with controllers (and vice versa) when immediate or special action is needed.

Although data-link technology offers significant benefits, earlier research emphasized the importance of addressing some potential human factors challenges. For example, users may lose party-line information (Sharples et al. 2007) during the transition from radio communications to electronic messaging (e.g., not being able to overhear other parties on the frequency). Controllers may spend more head-down time in composing, reading, or responding to data link messages. In addition, flight crews and controllers may experience difficulties in reviewing communications that employ a combination of voice and data link clearances. For instance, Dunbar et al. (2001) found that implementing ATC commands using the mixed-modality of voice and data link may increase transaction times.

The introduction of new technologies and procedures may require that controllers are kept informed about what equipment is available on an aircraft and whether that equipment is in use. Simply adding indicators to the aircraft representation may result in a format that contains a dense set of information and results in more clutter. For instance, Jha et al. (2011) found that providing controllers with information about how flight crews managed conflicts had an advantage when there were fewer free-flying aircraft, but resulted in worse performance with more free-flying aircraft. Hence, before attempting to display more information, designers shall have to determine how this information is used by controllers in order to avoid cluttering the workstations with superfluous data.

10.4 Concluding Remarks

Next generation technologies pose many challenges with regard to the interaction between controllers, flight crews, and artifacts in coordinating their functions and in achieving the overall system objectives. Examples include: authority and responsibility at different stages of managing aircraft trajectory, transfer of control between flight crews and controllers, issues of practitioners' trust on automated artifacts, shared understanding and collaborative decision-making and finally, coordination between professional teams with different decision criteria and planning styles.

The cognitive engineering approach provides a good framework for addressing many challenges of next generation technologies. T^2EAM is a theoretical framework that has been validated in the context of en route and approach control and has been enriched with behavioral markers that make its application easier in the ATM domain. In this respect, T^2EAM has been used in this chapter to address challenges in taskwork and teamwork strategies in the extended team of controllers, artifacts, and flight crews required to work together in the new ATM environment.

The T^2EAM framework can become a useful tool for designers and analysts to envision the potential effects that next generation technologies might have at different levels of the human–artifact–organization interaction. This reflects the cognitive engineering view that work systems require analysis across the "cognitive triad," that is, humans, artifacts, and organizations. Theoretical frameworks of the T^2EAM type are needed because there is a long history of system development programs that did not succeed as planned. The next generation ATM systems that are underway in the United States and Europe should be carefully evaluated using a mix of theoretical frameworks and human-in-the-loop experiments.

PART IV

Systems and Organizational Models

11 Organizational Models of Safety

11.1 Introduction

The cognitive functions of controllers constitute a vital capability of organizations in managing critical operations. Their work at the sharp end corresponds to the first-order performance of organizations, which deals with work adaptations to situational demands. The danger of focusing only on first-order performance is that systemic causes at higher organizational levels may be hindered by successful work at the sharp end. For instance, controllers may adapt their practices to cope with degraded system functionalities; however, if the systemic causes are not recovered, there is a risk that controllers may fail to save the day in other circumstances. Ensuring that the right organizational processes, resources, and policies are in place is taken care of by a second-order performance loop that provides support to people at the sharp end.

In the same way, organizational processes are affected by interactions across organizations (e.g., air navigation service providers [ANSPs], airlines, and airports) and by the overall regulatory framework (e.g., ICAO, EASA, and EU). This is a third-order performance loop that has many subtle interactions with the other two loops within the organization. This systems view of human and organizational performance is very important in order to understand how controller functions are affected by organizational resources and constraints, organizational decision making, as well as coordination and information sharing. This chapter presents three different organizational models that affect the performance of practitioners and hence the efficiency and safety of operations.

The first approach to organizational safety focused on how organizational goals, cultures, and processes can set the latent conditions of human work (Reason 1997). For instance, a blame culture can hinder communication of subtle warnings about safety issues while inadequate supervision may deprive controllers from important sources of error recovery. Organizational goals and culture may include commitment to safety, resolving conflicts of protection and production, worker participation in decision-making, and rewards for safety. Organizational processes for safety refer to methods for identifying hazards, systems for collecting risk information, systems for supervising and auditing safety, and provision of resources to people at the sharp end. Overall, organizational goals, cultures, and processes create many defenses in depth that support systems in eliminating threats, controlling adverse events, and recovering from degraded modes of operation. Some defenses are useful for eliminating threats (e.g., mindful planning of traffic), others are useful for controlling events (e.g., competence of controllers, teamwork, and supervision), while still others help in system recovery (e.g., traffic conflict and advisory system).

This defense-in-depth approach has provided useful taxonomies of organizational factors and defenses that should be in place in order to take a proactive stance to safety. For instance, the quality of defenses can be assessed in terms of safety audits (Hudson et al. 1994). Other approaches to risk assessment have incorporated the impact of organizational factors and defenses into a risk model of the system (Davoudian et al. 1994; Embrey et al. 1994; Pate-Cornell and Murphy 1996; Papazoglou et al. 2003). The risk impact is usually established through a process of rating the quality of organizational factors and weighing their relative effects on performance; the overall effect is fed to a risk model of the technical system to estimate risks. This structural approach to safety has focused on the organizational structures and processes that support human performance at the sharp end.

Contemporary approaches to systems thinking have focused more on the performance dynamics of practitioners and organizations. Systems thinking perspectives have examined interactions of organizational processes, changes of priorities over time, delays in effects, reinforcing influences, and long-term organizational changes (Shorrock et al. 2014). This allows analysts to understand how well-intended efforts to improve safety may generate side effects and how local decisions can become uncoordinated in the long term. Systems thinking views organizational failures as a result of control flaws in managing constraints that hold between technical systems, individuals, teams, and organizational processes. Seen this way, rift into failure is a gradual erosion of constraints or a lack of enforcement on the work of practitioners.

Resilience engineering is a third approach to the study of interactions between technology, practitioners, and organizations. Resilience engineering was founded on high reliability theory (Hollnagel et al. 2006; Hollnagel et al. 2011) and took a sense-making approach on how organizations monitor their safety margin, their resources, and work progress in managing risks. Organizational decision-making has addressed many decision trade-offs in all stages of problem solving. At the steering level, for instance, keeping well away from the safety boundary can minimize risks but may deprive organizations of learning opportunities. At the level of coordination, decentralized decision making can provide flexibility but at the risk of failing to comply with the overall plan (e.g., local workarounds can create side effects elsewhere). At the operational level, a trade-off may exist between resolving a conflict early versus waiting until all data are available to make a decision. These aspects of sensemaking and decision-making at different organizational levels interact in complex ways and influence the outcome of performance.

A short presentation of the three approaches to organizational safety is made in the following sections in order to understand the opportunities and constraints presented to the work of practitioners at the sharp end.

11.2 Defenses-in-Depth and Organizational Safety

11.2.1 Concepts and Applications of Defenses in Depth

The organizational analysis of safety grew out of the root cause analysis of major accidents and the realization that technical failures and human errors could be traced into the latent failures of organizations. Reason (1990) has used the term "latent failures" or "resident pathogens" to refer to failures in organizations that produce negative effects but whose consequences are not revealed until some other enabling condition is met. In the Ueberlingen accident, for instance, the air traffic control (ATC) system was operating in a degraded mode (i.e., the communication system was inoperable at night while one controller was absent from the control room), whose potential remained latent, or not realized, until an enabling condition (i.e., an undetected aircraft conflict) created a path to an incident (BFU 2004).

Latent failures refer to flaws in decisions taken at higher organizational levels that are beyond the control of practitioners. Latent failures are associated with the work of managers, designers, maintainers, or regulators—that is, people who are generally far removed in time and space from handling everyday operations. Examples of latent failures may include: inadequate supervision, poor procedural support, poor training, flaws in the design of equipment, inadequate systems of operational feedback, software flaws, and so on. The consequences of latent failures may lie dormant within the system for a long time, only becoming evident when they combine with other active conditions to breach the system's defenses (Reason, 1990). Some of the conditions that serve as triggers are active failures, technical faults, or atypical system states.

Reason (1997) defined organizational accidents as unfortunate situations in which latent conditions (arising from management decision practices or cultural influences) combine adversely with local triggering conditions or with active failures committed by individuals or teams at the sharp end. In this view, accidents are characterized by a concatenation of small failures and contributing events rather than by a single large failure. One cannot study accidents as separate independent elements, but only as part of the human-technology interaction within the constraints of the organization.

According to the latent failure model, we should look deeper into several defense layers that provide protection from risks, such as organizational processes, line management practices, task and environmental conditions, and human actions involved in operations. Indeed, air traffic management (ATM) has been a very safe environment because a weakness in one defense layer may be compensated by the good operation of another defense layer; a late conflict detection by a controller, for instance, may be compensated for by team communication and short-term conflict alert systems.

Reason (1997) proposed this view of organizational safety as multiple defenses carefully integrated to provide a safe work environment. At the organizational layer, for example, safety is supported by goal setting, organizing, communicating, designing, and maintaining. None of these layers are perfect, while their imperfections or flaws can be represented as holes in the system that leave it vulnerable for a long time. Flaws in separate defense layers are not normally a problem, but when combined with other factors, all layers may be penetrated by a triggering event that could lead to an incident. Near misses, in contrast, happen when the chain of events is stopped by a layer of defense somewhere along the way.

Defenses or barriers are system mechanisms that protect against hazards or lessen the consequences of malfunctions. Hollnagel (2004) classified defenses and barriers into four categories as follows:

1. *Physical barriers* that prevent an event from taking place or mitigate the effects of an unexpected event by blocking the transportation of mass and energy from one place to another. Examples of physical barriers are buildings, walls, fences, railings, bars, cages, gates, containers, fire curtains, and so on.
2. *Functional barriers* that create preconditions that have to be met before an action is carried out (e.g., interlocks, entry conditions) or intervene to prevent adverse events (e.g., traffic alert and collision avoidance system [TCAS]).
3. *Symbolic barriers* that work indirectly through their meaning, and hence require an act of interpretation by someone. Examples include: instructions, procedures, warnings, work permits, clearances, and approvals.
4. *Organizational barriers* that are not physically present at work but depend on the knowledge of users to achieve their purpose. Examples include: policies, rules, regulations, restrictions, and social or cultural norms of work.

The implication of the latent failure model is that organizations should pay attention to barriers or defenses that prevent adverse events from occurring or dampen their consequences. In order to ensure safety by preventing something from happening, it is first of all necessary that the risks are known. For this reason, an emphasis has been placed on risk assessment methods that identify all hazards and critical events that may appear in a work situation. However, as it is impossible to predict all hazards, organizations should have to invest on making improvements in many organizational and line management functions so that failures do not combine to create a threatening situation.

This structural approach has been successfully applied in aviation (Kennedy and Kirwan 1998; Ale et al. 2006; Stroeve et al. 2011) in order to identify deficiencies or latent failures that defeat defenses. Safety management can break the accident trajectory by providing defenses in depth, such as training, ergonomic design and procedures, supervision and leadership, communication networks, and a safety culture to govern the interactions of multiple sectors. The integrated risk picture (IRP) promoted by Eurocontrol (2006) has been using the defense in depth approach to identify human errors and influencing factors at the levels of the workplace and the larger organization so that failure probabilities are tailored to the particular work system. The defense in depth approach remains a prominent safety model for many of the safety initiatives taken by Federal Aviation Administration (FAA) and Eurocontrol.

The human factors analysis and classification system (HFACS) for incident investigation has also relied on the latent failures and unsafe acts of Reason's organizational model of safety (Wiegmann and Shappell 2003). HFACS provides analysts with taxonomies of failure modes across the following four levels: unsafe acts, preconditions for unsafe acts, unsafe supervision, and organizational influences. Working from the immediate causal factors, investigators classify human errors and associated causal factors using the HFACS taxonomies. HFACS increases inter-rater reliability since investigators are given guidance, albeit limited, in classifying errors and contributory factors (Lenny et al. 2008; Li and Harris 2006; Li et al. 2008). The ability to link failures across the four levels is also important in accident analysis because it allows associations between failures at the four levels to be assessed statistically.

The defense in depth model of Reason (1997) is a theoretical model that provides valuable insights into organizational accidents. In hindsight, however, accident analysis methods (e.g., HFACS) may associate errors and violations with many latent conditions that are not capable of bringing disaster by themselves. Every flaw in the organization does not necessarily bring disaster since it may be compensated for by good organizational performance at other levels. The analysis of a specific accident provides valuable insights about a specific combination of latent conditions that increased risk vulnerability. But how common is this combination in the ATC domain? Most accidents are the result of unique combinations of latent conditions that are difficult to predict in advance. Overall, HFCAS may identify areas where organizations should improve because organizational flaws may combine in risky patterns in the future; however, this is different from claiming that the specific flaws have led to a particular accident. Probably the statistical analysis of patterns of organizational flaws may have some value here in finding common syndromes or patterns of organizational breakdowns.

11.2.2 Organizational Resistance and Safety Culture

Reason (2004) acknowledged that more attention should be paid to organizational factors that improve system safety. For this reason, Reason (2001) developed the organization safety space model (OSSM) to address the system factors that make organizations more vulnerable or more resistant to failures. In the OSSM, organizational factors are seen as forces that push organizations in two different directions in the safety space: (1) a state of vulnerability due to increasing latent conditions and error triggering events and (2) a state of resistance due to increasing efforts to achieve higher safety standards. If the organization drifts too close to the vulnerability end, it is likely to suffer an adverse event which, in turn, will bring about internal and external pressures to enhance safety (Figure 11.1). These safety enhancing measures increase organizational resistance and

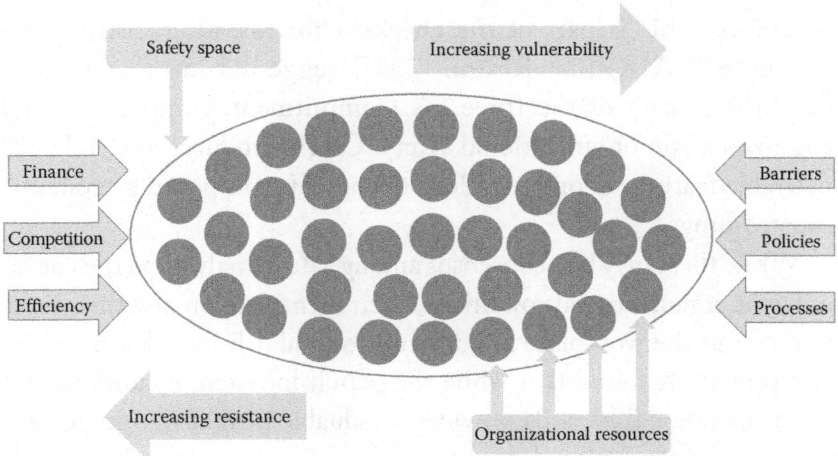

Figure 11.1 Organizational forces of resistance and vulnerability in the safety space. (Adapted from Reason, J., *Managing the Risks of Organizational Accidents,* Ashgate, Aldershot, 1997.)

make the organization a safer place to work. After some time, new economic and efficiency pressures can result in a new drift toward a more vulnerable state. Under the influence of work pressures, organizations may tend to move back and forth between the two ends of the safety space; it will take organizations to commit to a safety policy in order to remain close to the end of accident resistance. That said, however, a tenet of the safety space model has been that organizations should seek to achieve an attainable safety level within their boundaries rather than achieve a zero accident standard.

The safety management system that helps organizations navigate toward accident resistance comprises three qualities: commitment, competence, and cognizance. Commitment comes when the organization strives to be a good model for safety practices by investing financial and human resources in managing risks. The organization must also possess the necessary competence to enhance safety, including hazard identification methods, diverse defenses and redundancies, adaptive organizational structures and systems for disseminating risk information. Cognizance refers to how organizations make sense of their inherent risks and hazards. Cognizant organizations maintain a state of intelligent wariness (Reason 2008) even in the absence of bad outcomes; this collective mindfulness of the ever present risks is one of the defining characteristics of high reliability organizations (HROs). Table 11.1 presents some indicators that help analysts judge the three qualities of organizational resistance. This is part of the checklist for assessing institutional resilience (CAIR) that Reason (2001) suggested in order to consider the impact of the three C's (commitment, competence, and cognizance) on organizational safety. CAIR produces a wish list of desirable features of organizations to combat risks posed by systemic shortcomings.

While there may be a consensus among safety analysts on the nature of latent organizational conditions, their identification and their links to errors at the workplace are difficult to establish in a clear manner. A recent study found that while the search for latent conditions and other operational hazards provided a valuable insight into important aspects of organization, latent conditions were insufficiently clarified regarding the nature of the holes in the accident trajectory (Reason et al. 2006).

Table 11.1 CAIR Indicators for Assessing Organizational Resistance

COMMITMENT	COMPETENCE	COGNIZANCE
Maintain commitment to aviation safety and provide adequate resources	Create competence to achieve enhanced safety (e.g., methods to identify hazards and analyze safety critical activities)	Adopt a proactive stance toward flight safety (i.e., identify recurrent error traps, brainstorm new scenarios of failure, do health checks)
Consider safety-related issues at high-level meetings on a regular basis	Create a system for the collection, analysis, and dissemination of risk information	Accept setbacks as inevitable and train people to detect and recover from errors
Design safety measures to resolve conflicts between protection and production	Improve defenses rather than divert responsibility to particular individuals	Remain mindful of human and organizational factors that can endanger operations
Review past events top-level and make changes a systemic rather than local level	Provide people with cognitive and technical skills necessary to achieve safe performance	Encourage trust of the workforce in regard to reporting systems

The latent failure model and OSSM reserved a special place for violations (that is, deviations from codes of practice and procedures). The term "violations," however, refers to judgements that depend on a performance standard that is externally imposed against which the behavior is compared. Research in cognitive engineering (Dekker 2006; Woods and Hollnagel 2006) has shown that formal procedures are usually underspecified relative to the actual work as well as insensitive to many changes in context, so people always need to bridge the gap by making adaptations. To understand failure and success in safety-critical worlds, it may be more helpful to view deviations from procedures as proximal adaptation to the changing context of work.

Another critique of the latent failures model and OSSM is that they cannot model the dynamics and the buildup of organizational failures into dangerous states that threaten safety (Dekker 2006). Although structural models may be good at identifying the organizational structures and layers of defenses that make the holes in the accident trajectory, they cannot capture the dynamics that lead to the hidden erosion of defenses and the gradual drift of systems out of their safety margins. The need to move into organizational models that are sensitive to the creation process of latent failures has been recognized by modern applications of systems thinking that are described in the next section.

The three qualities of organizational resistance (that is, commitment, competence, and cognizance) reflect a revision of the defense-in-depth approach where Reason (2008) shifted the emphasis from system defenses to aspects of safety culture and sensemaking in organizations. The three qualities of organizational resistance have built on earlier work on safety culture (Reason 1997) that looked into the systems organizations use to collect, process, and learn from risk information. According to Reason (1997), an informed culture collates data from accidents/near misses and combines them with information from proactive measures (e.g., safety audits). When organizations require participation from the workforce to report and become involved in how safety is managed, then a reporting culture is created. However, practitioners may not feel free to contribute unless a just culture is also present, which creates an atmosphere of trust (i.e., rewards exist for reporting, management does not turn a blind eye to negligent acts, and so on). A learning culture is needed to draw appropriate conclusions from the information collected along with the will to implement changes to procedures and equipment as deemed necessary.

Gordon et al. (2007) identified several safety culture dimensions relevant to the ANSPs and related them to safety management systems. A factor analysis on a large sample of responses from many ANSPs has shown that safety culture could be judged according to the following dimensions:

- Commitment to safety at all levels of the organization, especially at the stakeholder level
- Responsibility structures specifying the key people involved in the organization of safety-related activities
- Involvement of people in the development of safety rules and procedures so that they feel they own the safety system
- Teaming and support from others when things are busy, workload is high, or the situation becomes complicated
- Communication of risk information both top-down to manage risks and bottom-up to establish feedback loops to management
- Reporting of risk information and organization learning

Another study by Mearns et al. (2013) provided a multi-factor solution where the first two factors corresponded to prioritization for safety, the third and fourth factors corresponded to involvement in safety, while the remaining two factors corresponded to reporting and learning.

Safety culture focuses on organizational values and safety practices and in many respects supports the implementation of safety management systems. However, the link between a good safety culture and safety performance has been rather difficult to establish, one reason being that the dimensions of safety culture have been identified at an abstract level. Other approaches to safety culture include the work on HROs whose criteria include a culture that encourages interpretation, improvisation, unique action, and a climate of trust and openness (LaPorte and Consolini 1991; Weick et al. 1999). In some respects, the qualities of organizational resistance and HROs take a sensemaking perspective of organizational safety that goes deeper into everyday work practices than other safety culture approaches do. In this sense, a better link between safety records and sensemaking capabilities would be expected.

11.3 Systems Thinking Models

11.3.1 Proponents of Systems Thinking

General system theory (GST) was built on an organism model of system behavior that relied on a regulating mechanism to integrate many components and functions and adapt to the external environment (von Bertalanffy 1950). GST has been based on similar discoveries made in human biology, psychology, economics, and philosophy. GST represents a means of instigating the transfer of systems thinking across scientific disciplines by using unambiguous mathematical laws. After the introduction of GST, other researchers elaborated mathematical systems theories (e.g., Klir 1969; Mesarovic and Takahara 1975) that have been incorporated into numerous disciplines, such as engineering, operations research, economics, and ecology. Other nonmathematical theories have been developed in a variety of fields, such as sociology, political sciences, anthropology, and psychology (Schwaninger 2006).

Systems can be better understood as hierarchies of components and functions. Moving up in the hierarchy provides a deeper understanding of goals, whereas narrowing down on lower levels reveals how systems function to meet their goals. Furthermore, determining the boundary of a system (i.e., what is part of the system and what is part of its environment) is another important aspect of systems thinking. If system goals are to be achieved, components must be controlled via feedback mechanisms when deviations in behavior occur (Skyttner 2005). Systems show dynamic behavior because they have to adapt over time to changing conditions; equally well, systems may migrate towards a state of increased risk and drift into failure (Dekker 2011; Leveson 2012).

When a component is exposed to the environment, it becomes directly or indirectly connected to other components and, therefore, remains influenced by them (Skyttner 2005). The resultant interaction produces emergent behavior that is not predicted by the individual component but by its relationship to others. Therefore, systems may show characteristics and operate in ways not expected or planned by the designers. Consequently, all components, human and technical, must be considered in their potential interactions so that the system is studied in a holistic way.

In systems thinking, organizations maintain equilibrium between interacting components through feedback loops of information and control. A system involves a dynamic process that is continually adapting to achieve its objectives and reacts to internal or external changes. Keeping a dynamic system in equilibrium means that control inputs are continually necessary for the system to stay safe. The system should enforce constraints on its behavior for safe operation and must adapt to changes to maintain safety. In this regard, accidents can be seen as the result of flawed processes involving interactions among people, social and organizational structures, engineering activities, and physical or software components (Leveson 2004).

An early application of systems theory to accident modeling has been Perrow's work on interactive complexity and coupling of systems. Perrow (1984) promoted the idea of system accidents that involve the unanticipated interaction of a multitude of parts in a large system whose complexity can quickly frustrate people's best efforts to predict and mitigate disaster. The thesis of what has become known as normal accident theory (Perrow 1984) is that "accidents are the product of systems that are both interactively complex and tightly coupled." This analysis of complexity has offered new ways of examining how to manage and control complex technologies. Normal accident theory predicts that the more tightly coupled and complex a system is, the more prone it is to suffer a "normal" accident.

Interactive complexity refers to part interactions that are nonlinear, unfamiliar, unexpected, or unplanned, and either not visible or not immediately comprehensible to people running the system. For instance, air traffic can become increasingly complex when regular flights mix with aircraft engaging in firefighting that have less predictable flight paths and different maneuver characteristics. In addition, systems can be loosely or tightly coupled. They are tightly coupled if they have more time-dependent processes, sequences that are invariant (e.g., the order of the process cannot be changed), and little slack (e.g., things cannot be done twice to get it right).

Normal accidents theory (NAT) sees human error as a label for some of the effects of interactive complexity and tight coupling. Practitioners have to live with systems that may conspire against their ability to make sense of what is going on or their ability to recover from failures. However, it must be recognized that complexity is not strictly a property of the domain as proposed in NAT. For example, interactions cannot be considered unfamiliar, unexpected, or unplanned in a system regardless of the people and organizations who need to deal with them. As seen in Chapter 9, controllers may adapt their strategies when complexity and coupling are high yet continue to keep the system in a safe state. In this respect, we need a more comprehensive view of complexity understood in relation to work demands, practitioners, and organizations.

11.3.2 Socio-Technical Approaches

Another group of researchers has advocated an alternative approach (Rasmussen 1997, Woods and Cook 2002, Dekker 2006, Leveson 2004, and Hollnagel 2004). The main features of their systems thinking approach are: (1) a focus on top-down approaches that recognize safety as an emergent system property rather than a bottom-up aggregation of reliable components; (2) a focus on the socio-technical system as a whole and the relationships between the technical, organizational, and social aspects; and (3) a focus on providing ways to model, analyze, and design specific organizational safety structures rather than trying to specify general principles that apply to all organizations. "The goal in organizational safety should be to create system designs requiring the fewest trade-offs between safety and other system goals while considering the unique risk factors involved in the organizational mission and environment" (Leveson et al. 2009).

Socio-technical approaches are concerned with how lack of control allows organizational activities to migrate toward the boundary of unacceptable performance. System control theoretic models (Leveson 2004) examine the adaptive processes that focus on achieving the organization's multiple goals within a set of constraints. The workspace within which practitioners navigate their explorations is bounded by constraints related to administrative, functional, and safety-related requirements. During this exploration, practitioners have many opportunities to identify a performance gradient while management will normally supply a cost gradient. In many cases, migration toward the boundary of unacceptable performance could result in accidents (Rasmussen 1997).

Jens Rasmussen proposed a socio-technical model for modeling the contextual factors involved in organizational, managerial, and operational structures that create the preconditions for accidents (Rasmussen 1997, Rasmussen and Svedung 2000). In addition, Leveson (2004) proposed the systems-theoretic accident model and processes (STAMP), which considers the technical, human, and organizational factors in complex socio-technical systems. The two models have specified certain accident investigation techniques (AcciMap and STAMP) that are further discussed below.

Rasmussen (1997) outlined the AcciMap method, which can be used to graphically represent system failures, decisions, and actions involved in accidents. AcciMap typically focuses on failures across the following six organizational levels:

1. Government policy and budgeting
2. Regulatory bodies and associations
3. Local area government planning and budgeting (including company management)
4. Technical and operational management
5. Physical processes and actor activities
6. Equipment and surroundings

Notably, AcciMap is a generic approach that does not use taxonomies of failures across the organizational levels. AcciMap combines the classic cause-consequence chart and the risk management framework (Rasmussen 1997) that depicts the control of socio-technical systems over six organizational levels (see Figure 11.2).

AcciMap is a good graphic method of depicting a chain of events and the organizational factors that contributed to an accident. However, AcciMap does not provide guidelines for how to conduct an analysis of cognitive and organizational functions for controlling critical situations. For instance, although flawed decisions are presented at the technical and operational levels, their cognitive functions remain hidden. In addition, the dynamics and feedback loops in the organizational processes cannot be identified in relation to their weaknesses. To overcome these problems, a control theoretic approach is presented below that has built on the work of Jens Rasmussen.

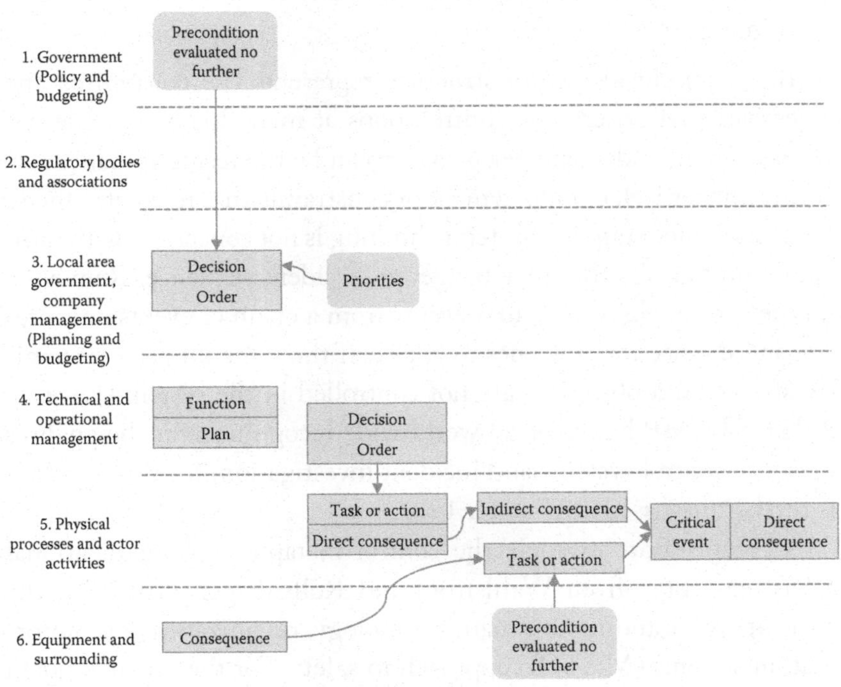

Figure 11.2 Hierarchical model of socio-technical systems on a cause-consequence chart. (Adapted from Rasmussen, J. and Svedung, I., *Proactive Risk Management in a Dynamic Society*, Swedish Rescue Service Agency, Karlstad, 2000.)

11.3.3 Control Theoretic Approaches to System Safety

STAMP (Leveson 2004) focuses on the interactions between control loops that regulate the work system. STAMP views systems as hierarchical levels of control with each level in the hierarchy imposing constraints on the level below. Conversely, information at the lower levels about the appropriateness of controls is communicated upward in the hierarchy to inform high-level decision making. Like Rasmussen's framework, STAMP emphasizes how complex systems evolve over time and migrate toward unsafe boundaries due to physical, social, and economic pressures rather than sudden loss of control capacity.

A STAMP accident analysis requires two activities:

1. Development of the hierarchical control structure that specifies the interactions between control loops and safety constraints
2. Classification and analysis of flawed control actions that identify the causal factors of control problems and dysfunctional interactions.

The hierarchical control structure represents the constraints and objectives that govern the control loops at many hierarchical levels. In this sense, inadequate performance can be the result of inadequate constraints or lack of enforcement of constraints that make it difficult to achieve safety targets. Systems thinking is not concerned with individual unsafe acts that may trigger an accident sequence. Removing unsafe acts, errors, or singular events from a chain of events only creates more space for new ones to appear if the same kinds of systemic constraints and objectives are not controlled in the system (Leveson 2012). STAMP has received world-wide recognition for the analysis of safety control systems and incident investigation; for this reason it is further discussed in Chapter 12.

A recent effort to apply the control metaphor to organizational safety has come from Wahlstrom and Rollenhagen (2012) on the basis of the Scandinavian man, technology, organization, and information systems (MATOI) approach to safety. For the organizational aspects of safety, for instance, Wahlstrom and Rollenhagen (2012) argued that organizations have their own values, goals, and structures that specify the safety model that drives their safety investments. The safety model is also useful for making predictions how to maintain or transfer the system to a safe state, for employing control loops to achieve a safe trajectory, and for avoiding unsafe states. To manage safety, for instance, organizations have to maintain and integrate several control loops at different hierarchical levels. Table 11.2 shows five control loops required to manage safety: risk analysis, feedback from experience, audits, self-assessments, and change management to implement risk counter-measures.

Control loops must achieve certain goals and transfer inputs to outputs with a specific transfer function or algorithm; finally, loops should avoid certain unsafe states. The risk analysis loop, for instance, takes as inputs several event sequences, descriptions of system interactions, and reliability data in order to produce risk estimates, identify risk consequences, and propose risk counter-measures. To this end, risk analysis uses certain algorithms and tools (i.e., bow ties, fault trees, and event trees). In addition, efforts are made to avoid unsafe or unreliable conclusions due to deficient screening or prioritization of risks and inadequate competencies in the use of risk methodologies. Management of safety can also involve other loops that are not described in Table 11.2, such as the assessment of resources, coping with emergency, coordinating with contractors, and safety promotion.

An important aspect of the control metaphor (Table 11.2) is that the design and application of control loops is affected by the model of safety held by the particular organization. The model of safety refers to the awareness of unsafe states identified in risk analysis or audits, the progress made to achieve certain goals, the perception of available resources, and the availability of plans to manage changes. At lower levels in the organization, the team of experts who perform the five control loops (Table 11.2) also have their own models of safety regarding what is safe or not, whether resources and tools are adequate, and how weaknesses and problems can be recovered to achieve a satisfactory product of work. These aspects of the safety model of each work group should also be taken into account when designing the five control loops in Table 11.2. This emphasis on the role of mental models of safety on organizational decision making has been further pursued in resilience engineering and other sensemaking approaches (e.g., high reliability theory).

Table 11.2 Control Loops Required for Managing Safety in Organizations

CONTROL LOOP	GOAL	INPUTS	OUTPUTS	ALGORITHM	UNSAFE STATE CAUSES
Risk analysis	Provide risk estimates, suggest solutions	Event sequences, system inter-actions, reliability data and models	Risk estimates, consequences, alternative counter-measures	Bow ties, fault trees, and event trees	Inadequate competency, deficient screening of risks
Feedback from experience	Collect/analyze experience for recommendations	Reports, comments, discussions	Analysis of human performance	Instructions for event analysis	Too few resources
System audits	Collect/analyze system safety performance	Checklists, observations, interviews, reports	Analysis of system performance with respect to safety	Safety Management System gap analysis	Unreliable data, poor understanding of system
Self-assessment	Assure good performance	Competency problems, staff discussions	Assessment of competences, best work practices	Comparison with guidelines	Inappropriate tools, slow or inadequate self-assessments
Change management	Decide and act on recommendations	Risk estimates, alternative options, recommendations	Modified functions, new man-machine functions	Instructions for managing change, project execution	Side effects from new functions, too few resources

Source: Adapted from Wahlstrom, B. and Rollenhagen, C., *Safety Science*, 69, 3–17, 2012.

11.4 Resilience Engineering

11.4.1 The Proponents of Resilience Engineering

Resilience engineering provides a new perspective to organizational safety by building on earlier approaches to interactive complexity, high reliability theory, systems thinking, and complexity theory. Resilience engineering has built on Perrow's ideas on interactive complexity that consider the difficulties that organizations face in understanding new threats, tracking system interactions, and adapting to complexity. The work of Jens Rasmussen on the safety space created by different work pressures (e.g., efficiency, economy, workload, and safety) has also been very influential since resilience is above all the ability of organizations to recognize how closely they have drifted to their safety boundaries. High reliability theory has been the most influential approach with its emphasis on how organizations make sense of new situations and adapt their resources.

Research on HROs has been based on studies of resilient organizations such as naval aircraft carriers (Rochlin et al. 1987), air traffic control systems (LaPorte 1988), and nuclear power plants (Bourrier 1996). Although there are several variations in the literature regarding the definition of HRO, Weick and Sutcliffe (2001) summarized five common HRO characteristics:

1. Preoccupation with failure and organizational learning (e.g., encourage reporting of errors, remaining wary of complacency and temptations to reduce margins of safety)
2. Reluctance to simplify (e.g., accept that situation is complex, encourage diverse perspectives, remain skeptical of current interpretations)
3. Sensitivity to operations where real work gets done (e.g., less emphasis on strategic goals, more emphasis on situation awareness for operations)
4. Commitment to resilience in order to detect, contain, and bounce back from complex situations
5. Deferring to expertise for complex situations for which procedures are inadequate.

Resilience engineering has been influenced by the HRO approach to organizational sensemaking but looks more carefully into the real constraints in applying this approach (e.g., deference to experis can be constrained by strategic plans to avoid side effects to other units).

These earlier approaches have led resilience engineering to formulate one of the most important principles in controlling complex system, that is, the principle of approximate adjustment or performance variability. As systems grow in complexity, it becomes increasingly difficult for organizations and practitioners to use existing procedures and rules to control situations. In general, procedures and rules are incomplete (or under-specified) as they cannot cover the full spectrum of situations that may be encountered in a complex world. To compensate for this, practitioners should adjust their performance to match current demands, resources, and constraints. The ability to do so is at the heart of successful performance. But since complex systems create situations where information, resources, and time are insufficient, human adjustments will be approximate. Performance variability is thus unavoidable but should also be recognized as a source of success.

Resilience engineering has built on complexity theory and especially the emergence of organizational behavior. For systems that are interactively complex, it is difficult to assign the behavior of the system to individual components, rules, and processes; it is rather the ongoing interaction and feedback loops in the system that create emerging behavior. Complex systems show emergent phenomena that cannot be understood in terms of linear thinking (that is, simple cause-effect chains). The cyclical nature of interactions implies that system behavior emerges in a series of cycles of interaction until a stable state is reached. The principle of emergence also implies that the human-technology interaction is evolved in an incremental way whereby humans continuously adapt to the technological and situational demands.

Finally, resilience engineering has built on systems thinking by looking at the performance of both practitioners and organizations. The important unit of analysis is the system of work that includes human practices, organizational rules, and constraints that interact in complex ways. Systems thinking has provided many useful concepts for understanding how goals and actions interact by means of balancing or reinforcing loops, how system behavior changes over time, how organizations can drift into the safety boundaries, and so forth.

11.4.2 The Four Qualities of Resilience

Although resilience engineering has built on these earlier approaches, its main thrust came from some prominent researchers in the field of human factors and organizational safety (e.g., Dekker 2005; Hollnagel 2004; Cook 2006; Woods 2006). A commonly agreed-on definition of resilience is the "ability of organizations to recognize and respond to regular and irregular threats in a way that is robust and flexible, the ability to anticipate disruptions and work pressures, the ability to monitor their own performance and call into question their models and plans and, finally, the ability to learn from their experience" (Hollnagel 2008). The working definition of resilience can be made more detailed by noticing that it implies four cornerstones of resilience, each representing an essential system capability. The four essential capabilities are: (1) knowing what to do, (2) knowing what to look for, (3) knowing what to expect, and (4) knowing what has happened (Figure 11.3).

1. *Knowing what to do*, that is, how to respond to regular and irregular disruptions by adjusting normal functioning. There are several ways to "respond to the actual," such as: adjust system functions to match new conditions, mitigate the effects of adverse events, prevent further spreading of effects, resume the functioning that existed earlier, change to standby equipment, and so on. Deciding whether to do something and when to do it depends on the competence of practitioners and on the situation in which they find themselves (Dekker

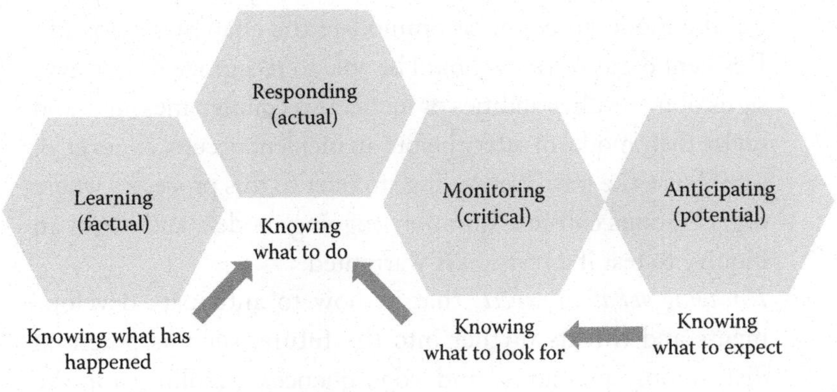

Figure 11.3 Resilience qualities and their interactions. (Adapted from Hollnagel, E., *Remaining Sensitive to the Possibility of Failure,* Ashgate, Aldershot, 2008.)

and Woods 1999). In responding to the actual, organizations should demonstrate some flexibility and manage to restructure their way of functioning. Restructuring may also involve some form of deference to expertise where organizations defer authority to the controllers at the sharp end, when faced with novel and high tempo situations; in contrast, in normal periods of operation, traditional lines of authority may be followed. In complex situations where a recovery plan cannot be implemented, a viable response may be to allow a graceful degradation of system operations and avoid a quick collapse.

2. *Knowing what to look for*: that is, how to monitor events and actions that could become threats in the near term as well as monitor one's own performance. Monitoring should go beyond regular threats and include what may become critical in the near future. So, organizations should continually assess and revise their work in order to remain sensitive to the possibility of failure. HRO researchers refer to this as "preoccupation with failure" or "reluctance to simplify." In order to know what to look for, the most important thing is a set of valid and reliable indicators but this is difficult in complex systems abound with weak signals and delayed indications. In general, unfamiliar situations often call into question the model of safety (e.g., how strategies are matched to work demands) maintained by the organization and demand a shift of strategies, processes and coordination. In this sense, resilience is concerned with monitoring the current model of safety and adjusting or expanding the model to better accommodate the changing demands. Resilient organizations should be able to recognize or interpret signs of new vulnerabilities or ineffective counter-measures and revise their model of safety before an incident occurs. Klein et al. (2007) use the term "reframing" to refer to this processes where organizations call into question ongoing models and begin an inquiry to test if a revision is warranted.
3. *Knowing what to expect*: that is, how to anticipate developments and threats further into the future, such as potential disruptions, pressures, and consequences. Looking for the potential requires the ability to imagine from which direction trouble may arrive and to explore the factors that can affect

outcomes in the near future. Resilient organizations should be able to anticipate when adaptive capacity is falling, when buffers or reserves may become exhausted and when goal priorities should be changed. Many risk analysis methods rely on formal models of past system behavior to make risk predictions; unfortunately, many systems show an emergent behavior coming from complex interactions that is difficult to predict from similar situations in the past. Probably a useful way to know what to expect is to create different viewpoints of the situation so that people develop the big picture of how the situation is evolving and decide whether their plans need revision. HRO researchers refer to this strategy as "collective mindfulness" as they collectively update their assumptions and perspectives on the situation. A final issue is that looking for the potential in itself requires taking a risk, in the sense that it may lead to an investment in something that may happen so far into the future that benefits are rather uncertain. This is a proactive approach to safety where organizations are picking up on evidence of developing problems rather than on reacting after problems are manifested.

4. *Knowing what has happened*: that is, the ability to learn the right lessons from experience. The ability to adjust implies that experiences from past events are used to make decisions about organizational changes so that the system is better prepared for what may happen in the future. An important question in addressing the factual is whether learning occurs whenever something has happened or on a more regular basis. If it only takes place after an adverse event, then nothing is learned from unimportant events that are by far the most frequent. A resilient system should not limit learning to specific failures but should look at the affected system functions, their degree of variability, and the dependencies between system functions.

The question of how well organizations perform on the four qualities of resilience and how well they balance the four qualities may depend on the sort of operations the organizations are engaged. Many organizations usually put some effort into the ability to respond to the actual and learn from the factual. Fewer organizations, however, make a sustained effort to monitor the critical and anticipate the potential, particularly if there has been a long period of stability. For air traffic control, all four qualities seem to be very important not only for periods of heavy traffic and abnormal operations but also for normal operations.

The four qualities of resilience are seen as a model of organizational capabilities rather than a model of performance since there are no regulating mechanisms specified for coordinating the four capabilities. Systems thinking approaches have a relative advantage in this respect since they are more concerned with developing control loops for regulating system performance.

11.4.3 Making Trade-offs in the Four Qualities of Resilience

A common feature of the four cornerstones of resilience is that practitioners and organizations are making several performance trade-offs. Controllers need to make many micro-judgments about thoroughness and efficiency with regard to their tasks. To simply judge events of normal variability as errors, without recognizing the trade-offs that controllers have to do in their job, fails to recognize the complexity of the situation. For the vast majority of cases, controllers perform extremely reliably and keep the system safe. Human error is really just a by-product of normal variability of human performance (Hollnagel 2004). The same variability allows controllers to keep the air traffic moving and recover from adverse events. Organizations should ensure that the system is safe by design and that performance variability is properly handled by making organizational rules more flexible and by enhancing competence and sensemaking at all levels. To cope with an increase in productivity, organizations should strive to maintain a proper margin of safety by raising awareness, by increasing self-monitoring and supervision, by utilizing technological supports, and by rearranging the distribution of work. A more elaborate discussion of managing trade-offs is presented in Chapter 14.

11.5 Complex Adaptive Systems

Another contribution to resilience engineering comes from complex adaptive systems (CAS) that is, a collection of adaptive teams that adjust their behavior to handle demands and disturbances. Complex adaptive systems view operating teams as agents working collectively to achieve their goals with a certain degree of autonomy. Each adaptive team develops a fundamental set of skills and corresponding resources to handle familiar situations, but complex systems often present new situations that challenge this first-order adaptive capacity. When surprising events occur, a second form of adaptation is required that enables adaptive teams to see how closely they operate to the safety boundary and anticipate how a new deployment of resources and competences can handle an unexpected turn of the situation. Although many systems show both forms of adaptive capacity, systems differ widely on the emphasis they put on these capacities. Resilient systems are careful to provide this second-order adaptive capacity.

The concept of margin of maneuver (MoM) has been proposed by Woods and Branlat (2010) to capture how adaptive teams anticipate and respond to new situations and challenging events. Adaptive teams proactively monitor changes in the MoM over time and regulate the MoM in anticipation of potential challenging events to assure that there are available resources. An individual controller, an ATC team, and other adjacent teams can be considered complex adaptive systems at different levels of the organization. The interplay among the CAS agents can be expressed in terms of how the adaptive responses of one agent constrains or releases other agents to adapt locally and how all agents can contribute to the overall system goals. Breakdowns in the ability to regulate the MoM and adapt responses to handle challenges beyond the first-order capacity can fall into three patterns of breakdown (Woods and Branlat 2010).

First, adaptive teams may not be able to extend adaptive capacity because events combine to produce cascades and teams may fail to keep in pace with the tempo of events. This pattern of going solid describes situations where all resources are committed to handle ongoing issues with no reserves remaining to handle other events. To anticipate potential bottlenecks, adaptive teams must pick up early warnings from a stream of cues about how well responses are being matched to challenges. As complex situations present new forms of demands, sometimes effective teams fail to mobilize new resources to enable the deployment of new capabilities. "Being in control" relates to the ability to assess how margins of maneuver are expanding or contracting relative to the uncertainties and the potential for surprise. The literature on CAS has identified three main decision dilemmas or trade-offs that are difficult to manage, that is: (1) optimality-brittleness (i.e., optimal but fragile performance versus good enough but robust performance), (2) efficiency-thoroughness (i.e., narrow but effective search versus in-depth but less effective search), and (3) acute-chronic (i.e., short-term versus long-term goals). In order to manage their adaptive capacity, organizations are faced with difficult strategic decisions: should resources be consumed to address the current situation, or sustained as reserves for new turns of the situation? What critical indicators exist that alert organizations when a switch can be made between these two strategies? and so on.

The second pattern of breakdown concerns the failure of organizations to recognize where they are standing in the space represented by the fundamental trade-offs. An attempt to represent the system's position in a trade-off space is reflected in the safety space model proposed by Cook and Rasmussen (2005). This model represents the boundaries of workload, performance, and productivity in order to describe how work pressures influence the system in one direction or another. The model can track how the system's position evolves in time, in the face of a changing situation.

The third pattern of breakdown concerns maladaptive coordination or vertical interactions between echelons of a system. Multiple adaptive teams exist across organizational levels, each responsible for different sub-goals of the total system. In carrying out their responsibilities, adaptive teams may be absorbed into local workarounds and undermine the system's goals or create side effects to other teams working on separate goals (Woods and Branlat 2010). In a hierarchical organization, where work is done according to predefined plans and procedures, coordination among decentralized teams is especially fallible in situations that evolve dynamically over time (Woods and Hollnagel, 2006). Due to bounded rationality, each adaptive team has finite knowledge that limits its ability to understand the demands of the situation. At the same time, high-tempo situations may hinder coordination efforts to adapt to the situation. Coordination breakdowns among adaptive teams provide the main evidence that this pattern of breakdown has occurred (Patterson et al., 2007; Branlat, et al. 2011).

Adaptive teams need to understand cross-scale interactions in order to avoid undesired side effects in tightly coupled systems. In this sense, amplifying control consists of developing tools that help teams track complex situations and help anticipate how projected actions may propagate (resonate) in a network of goal interdependencies (Woods and Branlat 2010). Hollnagel (2004) has proposed the functional resonance accident method (FRAM) to represent how goals or functions are dependent on each other and how variability in human responses may resonate across the system, giving rise to uncontrolled events and incidents. Studies in ATM have also shown how new representations can support team decision-making in a domain characterized by many function interdependencies (Smith et al. 1997; Smith et al. 2007).

11.6 Functional Resonance as a Model of System Accidents

Further developments in resilience engineering include the modeling of accidents in terms of performance variability and resonance. In a complex system, all its components (i.e., equipment, barriers, and practitioners) may show variable performance for different reasons. For technical components, variability may be due to imperfections in design and operation in the sense that some operating conditions have not been foreseen. For protective barriers, variations may be due to inadequate design, maintenance, or use by people. In the case of humans and organizations, variability may be due to performance adjustments to current conditions, lack of constancy of perceptual and cognitive functions, and so on. The principle of resonance makes it clear that every now and then a number of functions may resonate—that is, reinforce each other's variation—leading to excessive variability in one or more downstream functions; hence, consequences may be spread throughout the system.

The functional resonance accident model (FRAM) examines individual functions and determines their relationships within the total system (Hollnagel 2004). FRAM identifies various functions required for the operation of the system and examines how interdependencies may create problems when functions are combined together in a specific operation. For instance, if the input to a function came too late, or was of the wrong kind, then the source of that input—that is, another function—must be described further. This may in turn require yet other functions to be described, until the total scenario has been considered.

Another critical step in FRAM regards how to describe the functional resonance from the observed dependencies; that is, how variations in the functions may resonate across the system and lead to escalation of problems. Dependencies among functions can be found by matching or linking their aspects (e.g., time, resources, preconditions, controls). For example, the output of one function may be the input to another function, constitute a resource, fulfill a precondition, or enforce a constraint. The result would be a representation of how functions are linked or coupled in an accident scenario and how functional variability propagates through the system. In general, the links indicate how variability in one function may affect another function and how this pattern may propagate throughout the system.

Resilience engineering is a relatively new scientific discipline that has managed to organize several organizational capabilities that contribute to institutional resilience. Yet there have not been many methods for applying the principles of resilience engineering to safety management. FRAM is possibly the most widely used method in resilience engineering both retrospectively for incident analysis and proactively for identifying organizational weaknesses in managing threats. Some of the challenges that FRAM needs to resolve may include the following:

- Offer specific guidance regarding the granularity of functions and the way that functions can be grouped into higher order functions.
- Develop rigorous criteria to distinguish between good and bad couplings between functions. At present, Perrows' criteria of interactive complexity and coupling have been referenced in FRAM but there is no further guidance how to apply them.
- Additional guidance on how to apply the trade-offs between efficiency and thoroughness in specific work contexts (this issue is very important and is further explored in Chapter 14).
- Although FRAM shows the interactions between different functions, it does not allow modeling of reinforcing and balancing loops in the way that STAMP allows.
- There is no systematic representation of the interactions between the individual, team, and organizational levels.

In general, resilience engineering has managed to organize different principles from earlier approaches—NAT, HRO, systems thinking and complexity theory—and presented them in terms of a coherent theoretical framework. It is foreseen that some principles of resilience engineering may also be used to relax the formalism of the systems thinking approach; this effort is undertaken in Chapters 12 and 13.

11.7 Concluding Remarks

Three approaches to organizational safety have evolved since the 1990s, emphasizing the research developments and the practical safety concerns of their time. A final comparison is made here between these organizational safety approaches in terms of their concepts of failure, organizational analysis, safety management implications, and methods of application (see Table 11.3). The three approaches highlight different views of system safety and may be appropriate for systems at different levels of safety maturity and possibly different organizational cultures.

The defense-in-depth approach has been used since the late 1990s and provided useful insights into the weaknesses of organizational processes that create the latent conditions for near misses and accidents; hence, continued efforts of practitioners to avoid a recurrence of similar errors in future would not succeed when latent organizational failures are not recovered. The defense-in-depth approach is still dominant in aviation safety (e.g., the integrated risk picture approach of Eurocontrol) because it provides a useful background of human performance and influencing factors. Human errors have been viewed as limitations of human cognition—that is, poor knowledge, decision biases, limited memory, and outdated attention; for instance, the human error in ATM (HERA) tool (EATMP 2000) undertakes such an analysis of human reliability. In this respect, safety management focuses on how to design and maintain effective safety barriers that prevent failures or at least recover from them before an accident occurs. This approach has been widely applied in aviation because practitioners and managers can audit their safety management so that organizational weaknesses are identified and safety barriers are designed. The increasing safety levels in aviation can be partly attributed to the success of the defense-in-depth approach.

Table 11.3 A Comparison of the Three Organizational Safety Approaches

DEFENSES-IN-DEPTH	SYSTEMS THINKING	RESILIENCE ENGINEERING
Decompose hazardous events in terms of component failures and human errors	Search for relationships between people, technologies and organizations that may give rise to risks	Search for performance trade-offs that may give rise to risks
Emphasis on human errors and organizational holes (weaknesses) that must be identified and prevented with suitable barriers	Emphasis on difficulties in understanding and controlling system dynamics and organizational complexity	Emphasis on the work constraints within which people make trade-offs between different goals and operating modes
Human errors and organizational failures should be identified in advance and prevented early in the design of the system	Errors and failures are unavoidable but the system should create opportunities for error detection and recovery	Errors and failures are due to the same adjustment processes that lead to successes; what went wrong could have previously worked well
Failures are caused by weakness in organizational processes (e.g., poor training and coordination) and human limitations in cognition (e.g., poor knowledge, decision biases, limited memory, constrained attention)	Failures are caused by difficulties in understanding system dynamics (e.g., reinforcing influences, long term, and other delays), complex interactions of organizational processes and changes of priorities over time that can't be followed	Failures are caused by poor management of trade-offs caused by hidden criteria, work constraints, availability of resources, and conflicting requirements that are built in the design and operation of the system
Safety management should focus on how to design and maintain effective safety barriers (e.g., training, operator supports, coordination, control panels and safety interlocks)	Safety management should focus on how to make the system tractable (that is, easy to understand) and ensure more resources and means for practitioners	Safety management should focus on how to make the system tractable and ensure that criteria for trade-offs are clear and widely communicated to workforce
Examples of suitable organizational models: "Swiss model" of safety (Reason 1997) Integrated Risk Picture (Eurocontrol 2006)	Examples of system thinking models: ACCIMAP (Rasmussen 1997), STAMP (Leveson 2004) and VSM (Beer 1985)	Examples of suitable organizational models: Agile organizations and performance trade-offs (Hollnagel 2009; Hoffman and Woods 2011)

However, it appears that safety performance has reached a plateau in the aviation domain and new improvements, particularly in the context of new work systems mandated by NextGen and SESAR, would need to view safety from different angles as well. Systems thinking and resilience engineering provide alternative views of system safety for organizations that have made a good progress in risk management. Applications of system thinking approaches (e.g., AcciMap and STAMP) and resilience engineering (e.g., FRAM) to organizational safety have focused on the complexity and dynamics of systems that make their functioning intractable.

Errors and failures are unavoidable, not only because practitioners and organizations may be limited in their capabilities, but mainly because the work systems are tightly coupled and complex. Failures are caused by difficulties in understanding system dynamics (e.g., reinforcing influences, long term goals, and feedback delays), complex interactions of organizational processes, and changes of priorities over time that are difficult to follow. In complex systems, practitioners and organizations are bound to make errors that are not always possible to prevent; hence, the emphasis has shifted into the processes of error detection and error recovery. The efforts of safety management should be directed not only in designing safety barriers, but also in making the system more tractable as well as ensuring that practitioners have abundant resources to cope with complex situations.

The third approach of resilience engineering has also presented valuable insights into the resilience qualities and adaptive capacities required of organizations to withstand and recover from complex situations. In today's hectic workplace, practitioners often have to produce more, faster, but with fewer resources. The complexity of modern systems creates new work constraints on the available options to do the job. On the other hand, amplifying the variety of organizations by means of multiple perspectives, new automated support systems, and alternative modes of authority and control, does not always make the problem any easier to solve. The alternative modes of operation often require additional knowledge and judgment how to manage trade-offs. Systems thinking proposes ways to amplify variety but does not consider the trade-offs that are created from alternative options and means to do the job. The management of trade-offs shows clearly that errors and failures are due to the same adjustment processes that lead to success; what went wrong may have worked well in past experiences. Therefore, safety management should try and amplify variety but also consider the knowledge requirements for managing performance trade-offs. The other side of the coin is making the work system more tractable by managing its complexity. This may also involve making the system less tightly coupled and maintaining the margin to maneuver (Woods and Brantland 2010).

Overall, the three approaches to organizational safety highlight different views of system safety. The defense-in-depth approach can be used to identify organizational weaknesses (e.g., poor training, procedures, and information systems) and design defenses to improve safety. Other types of problems may be identified by using the systems thinking perspective in order to examine whether system dynamics and complexity may make the work system intractable. This may drive an inquiry into how to relax the constraints of the system (that is, attenuate complexity) and how to equip practitioners with more means to do their jobs (that is, amplify organizational variety). This systems thinking approach can provide additional safety improvements but it may also create more performance trade-offs for the practitioners. The third approach to safety can look into the management of trade-offs and reveal hidden criteria, work constraints, and conflicting requirements that have been built into the operation of the work system. The challenge for safety management may be to look into the trade-offs presented by the three approaches and decide what approach is most suitable for its safety maturity level, or how to create a blend of them for special circumstances.

12

System Modeling and Accident Investigation

12.1 Introduction

It has been widely accepted that the air traffic management (ATM) system has achieved a very high safety record mainly through a structured process of incident analysis and a continuous learning process. The aviation industry's commitment to incident and near miss analysis has been exemplified in a set of detailed International Civil Aviation Organization (ICAO) documents under the collective term of incident investigation and prevention (i.e., Group 176 of ICAO publications). The aim of incident investigation is to map out the accident trajectory and identify root causes in terms of defenses-in-depth that did not work out. The investigation process represents the culmination of an industry-wide belief that a thorough examination and a deeper understanding should be achieved about what really happened and what went wrong. There is still a growing effort to develop incident investigation techniques and accident causation models that look deeper into the systemic factors and their dynamics that make a fertile ground for safety-critical events (Johnson 2003). Examples of systemic factors include the interplay between human and organizational performance, the management of goal trade-offs, and the inherent variability of organizational processes.

Modern accident investigation techniques have shifted their focus from shortfalls in the actions of sharp-end practitioners to the weaknesses in the capabilities of organizations in maintaining a safe system. Systems thinking models (Rasmussen 1997; Leveson 2004) have been particularly useful in helping investigators to probe into the complicated interactions between system elements that lead to performance decrements and unfortunate events. At the same time, another research strand has relied on organizational models to reveal system vulnerabilities and patterns of breakdown that produce flaws in the control of safety processes and in the enforcement of safety constraints.

Perrow's (1984) normal accidents theory, for instance, has been extensively used to look into aspects of interactive complexity and tight coupling in the structure of organizations that make accidents virtually inevitable. Beer's (1985) viable system model has been applied in accident investigation (Santos-Reyes and Beard 2006; 2008) to reveal problems in the way that organizations control their safety processes and manage their requisite variety to respond to adverse events.

The literature that deals with the degradation of organizational capabilities has emphasized that the gradual erosion of capabilities and safety standards may escape the attention of organizations until an adverse event occurs. For instance, in the incubation model, Turner (1978) pointed to the discounting of danger signals and the gradual progression of organizations toward the safety boundary that are not seen in time, until an incident occurs. This degradation has also been linked to the gradual build-up of latent failures and organizational omissions (Reason 1997; Licu et al. 2007), the erosion of protective forms of safety (Schulman 1993), the drift of local work practices from the overall plan (Rasmussen 1997), and the reinforcing loops that move work practices further away from organizational norms (Toft and Reynolds 1994).

These two trends in incident investigation and in the study of patterns of organizational breakdown have been developed separately, with a small degree of cross-fertilization. Although both AcciMap and STAMP techniques take a systems perspective, they remain neutral with regard to the organizational structures and processes that control safety. This gap between incident investigation techniques and organizational models has left practitioners and safety investigators on their own to integrate the two strands and apply them to their specific domain.

The purpose of this chapter is to elaborate the STAMP (systems-theoretic accident model and processes) technique on the basis of a theoretical model of organizational viability. In this respect, the viable system model (VSM) seems to suit this objective as it has already been applied in several cases of incident analysis (Santos-Reyes and Beard 2006; Weir 2004; Dijkstra 2007). The STAMP technique has been adapted to identify control flaws in safety management as well as look deeper into causal patterns of organizational breakdown. To illustrate this link between STAMP and VSM, a system-wide ATM failure has been utilized as a case study in this chapter.

12.2 A Control Theoretic Approach to System Safety

Systems thinking perceives organizations as hierarchical structures with communication and control functions that operate at the interfaces between organizational levels and entail an upper level imposing constraints upon a lower one. Leveson (2004) specified several control functions and constraints that affect safety management, using the STAMP technique. Organizations that operate complex systems have to make trade-offs between conflicting goals such as safety, production, delivery times, and utilization of capacity (Marais and Saleh 2008). This brings into the fore the role of organizational models that constitute the deepest set of beliefs about how the world works, about potential hazards, and about perceptions of organizational capabilities. Safety goals are passed onto the supervisory level and are transformed into specific plans for action that are assigned to practitioners at the execution level. Plans of action are not the only constraints imposed by higher levels of control; other constraints may refer to availability of job means, resources, and degrees of freedom allowed to practitioners. At the execution level, work practices adapt safety plans to variations in the environment, making use of available resources and safety barriers. To assess the adequacy of safety plans, a feedback loop is established back to the higher management levels. Although STAMP takes a systems control approach, it remains neutral with regard to specific human and organizational models of breakdown.

12.2.1 Control Flaws and Underlying Organizational Breakdowns in Accidents

Adverse events and action failures are usually the starting point of accident investigation, while their systemic causes can be traced into the failures of system structures and safety control mechanisms. According to STAMP, problems in the structure and control of complex systems arise mainly due to control flaws in the design, enforcement, and implementation of safety constraints at different levels. The mistakes and poor decisions of practitioners and managers are usually due to violations of safety constraints or control requirements. The system should be designed to fulfill several control requirements and its operational management should ensure that controllers comply with these requirements in familiar and unfamiliar situations. In order to create a safe system design, therefore, we must understand how control requirements can be eroded or negatively influenced and what can be done to enforce safety constraints throughout the organization.

STAMP uses a safety control structure to analyze how the design, operation, and regulation of a system can work together to deliver a safe service or product. Figure 12.1 shows the interactions between four system elements listed as follows:

1. The technical system that produces a service or a product
2. The operational management that includes the procedures, policies, and team organization to deliver the system goals
3. The safety regulation that includes the safety management system, the national regulatory bodies, and cultural norms of the industry
4. The design process that describes how the system elements are created

This safety control structure shows the control and feedback relationships between the four essential system elements. The design process is used to design the technical system, operational management, and safety regulation. In fact, the design process itself is not static as it may change on the basis of information from the technical process or feedback from managers and regulators. In the design stage, the control structure is created by identifying and assigning relevant

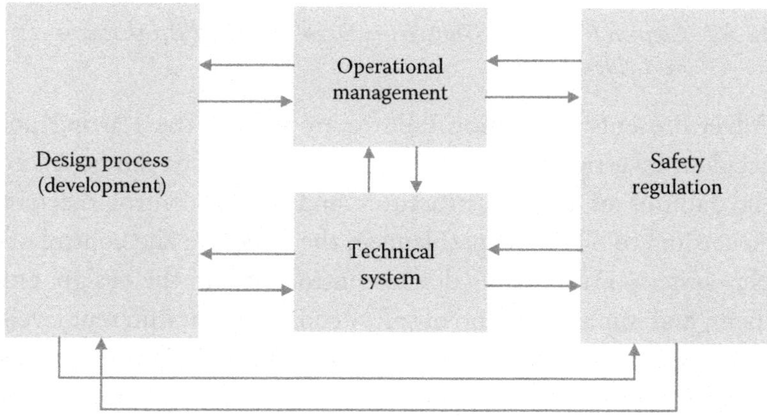

Figure 12.1 Four interacting elements in the safety control structure. (Adapted from Stringfellow, M., *Accident Analysis and Hazard Analysis for Human and Organizational Factors*, PhD Dissertation, Massachusetts Institute of Technology, Boston, 2010.)

practitioners to the operational management and the regulation functions. Subsequently, it is possible to identify the safety constraints on the design, operation, and regulation. The four elements of the safety control structure have a dynamic relationship that changes over time, as lessons are learned from experience, new technologies are introduced, or improvements are implemented.

Each element of the safety control structure can be further described in terms of a number of interacting components:

- *Goals and control inputs.* Goals may be selected from a set of options by individual controllers or may come directly from a watch supervisor or the organization's mission statement.
- *Mental models of the process under control.* This includes the current state of the process as well as a projection on how the process may evolve over time. The mental model must also contain an understanding of how the process may change in response to inputs and how the process may transform inputs into outputs. The models allow controllers to anticipate future states and use feed-forward control for faster responses.
- *Control algorithms.* They are formalized as procedures or checklists, or may be devised by human controllers as part of their experience and training. The selection of the proper algorithm is not always straightforward but may require an efficient search for data and an elaborate assessment of the situation.
- *Coordination with other controllers or decision-makers.* When multiple practitioners are responsible for a process, coordination becomes very important. For routine tasks, coordination may be achieved by organizational rules that specify the division and coordination of work, while for unfamiliar tasks coordination is better done by mutual adjustment. Other forms of coordination may include asking for advice about available options and requesting information that is not directly available from the sensors.
- *Actuators, sensors, and controlled processes.* Although controller actions may comply with safety constraints, the technical process may not implement them properly for several reasons. For instance, the actuator that executes the commands may fail to operate properly or the sensor that provides process feedback may fail. In addition, the technical process may be unable

to execute the commands or provide inappropriate feedback when its operation depends on other subsystems.

For every control component, it will be necessary to evaluate the context of work in order to understand how and why unsafe actions could occur. The context of work may include: situational demands (e.g., time pressure and uncertainty), available resources to do the work (e.g., procedures, tools, decision aids, and team support), role authorities (e.g., degree of autonomy), and policies (e.g., prevention and protection). Starting with the drawing of the safety control structure of the organizations involved in an incident, STAMP proceeds with an investigation of all the control loops and interactions between the organizations. For each control loop, STAMP looks into potential flaws in the five control components using a standard classification scheme, as shown in Figure 12.2.

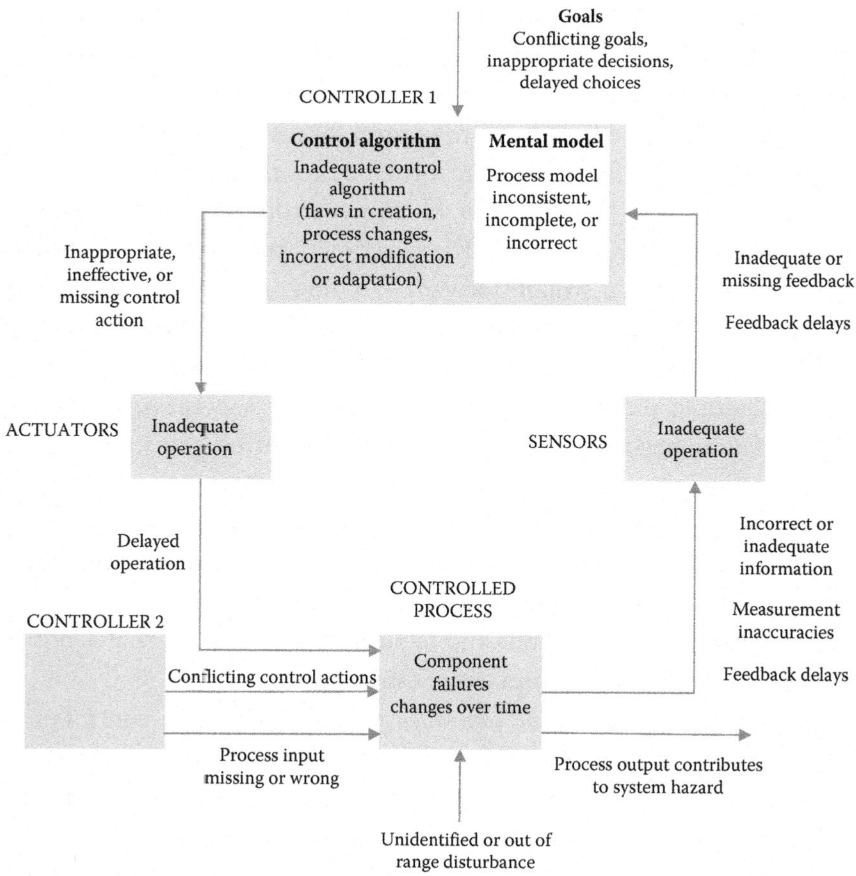

Figure 12.2 Control flaws leading to accidents according to STAMP.

The main advantage of STAMP is that it allows investigators to map out the organizational structure and control loops that regulate safety at different levels in the organizational pyramid. In addition, it allows a thorough investigation of the coordination tactics between organizations and regulatory bodies. The STAMP control structure identifies all the control loops that must be scrutinized in the process of incident investigation. This is a tedious investigation because analysts should consider all the potential flaws (Figure 12.2) for all control loops in the organizational structure. However, STAMP does not impose a theoretical model of human and organizational behavior but allows the analysts to draw on their own preferences. In the last decade, many researchers have used system archetypes in their efforts to understand flaws in the control structures of organizations (Cooke 2003; Tucker and Edmondson 2003; Marais et al. 2006).

12.3 Application of STAMP to a System Failure

12.3.1 Description of a System-Wide ATM Failure

On December 12, 2014 a failure occurred at the Swanwick operations center located southeast of Southampton airport (CAA 2016; Walmsley et al. 2015). The failure concerned a major computer system that provided information about flight data processing and distribution to air traffic controllers (ATCOs) at NATS (National Air Traffic Services). As a result, controllers did not have access to updated flight plan information, although they were still able to see the flights on their displays and provide instructions to flights with radio communications. The immediate response to the failure was to employ contingency procedures to reduce traffic and ensure aircraft separation. The affected traffic included flights arriving or departing London airports as well as flights overflying the UK airspace. At 14:55, about 10 minutes after the failure, all departures were stopped from London airports and other European airports that were planned to fly through the affected UK airspace. This was a more conservative approach than the one prescribed in the contingency plans for this type of failure. An hour after the failure, the engineering teams restored the computer system but without the normal level of redundancy; for this reason, controllers did not lift all the restrictions (i.e., air traffic flow and capacity management (ATFCM) regulations) published by Eurocontrol. Full redundancy of the computer system was restored at 20:20, which enabled controllers to lift all traffic restrictions. However, the disruption affected many airlines and thousands of passengers not only that day but well into the following day. The European Network Manager recorded a delay of 18,433 minutes and 150 flight cancellations that affected 1,900 flights and 230,000 passengers. Additionally, several airlines reported cancellations and flight disruption on the following day affecting approximately 60 aircraft and 6,000 passengers.

According to EU 2015/1018 regulation, this incident could be classified as a "failure of data processing and distribution function or service." This was not a local communication navigation surveillance (CNS) failure but a system-wide failure affecting the whole FIR. The standard response would have been to employ prescribed contingency plans in a phased approach to resume normal operations. Quite remarkably, no safety events took place during the period of fallback operations or during the recovery phase (Walmsley et al. 2015). The more conservative approach that was adopted by operational supervisors limited air traffic well below the levels prescribed in the contingency plans and played a major role in the control of safety events.

Some operational context is provided in this section in order to understand this system failure with emphasis on the architecture, function, and software of the New En-Route Centre (NERC) of the London Area Control (LAC) center. NERC is a computer system that integrates many subsystems, including radar and flight data processing, voice communications, and support information. The National Airspace System (NAS) is also served by NERC by means of a NAS computer that holds data on all flights travelling to the UK and has links to airports and control centers both in UK and abroad. It receives flight data on all flights due to travel in Europe from the Network manager in Brussels and disseminates appropriate information in a timely manner to controllers. It also receives radar data on all aircraft within range of British radars and correlates this with flight data, updating the estimated times at given points along the flight routes. NAS has also access to vast databases containing information about air sectors, routes, and aircraft characteristics. NERC operates the System Flight Server (SFS) that stores and distributes flight data to Swanwick Area Control. A simplified architecture of the systems that support controllers at LAC is provided in Figure 12.3.

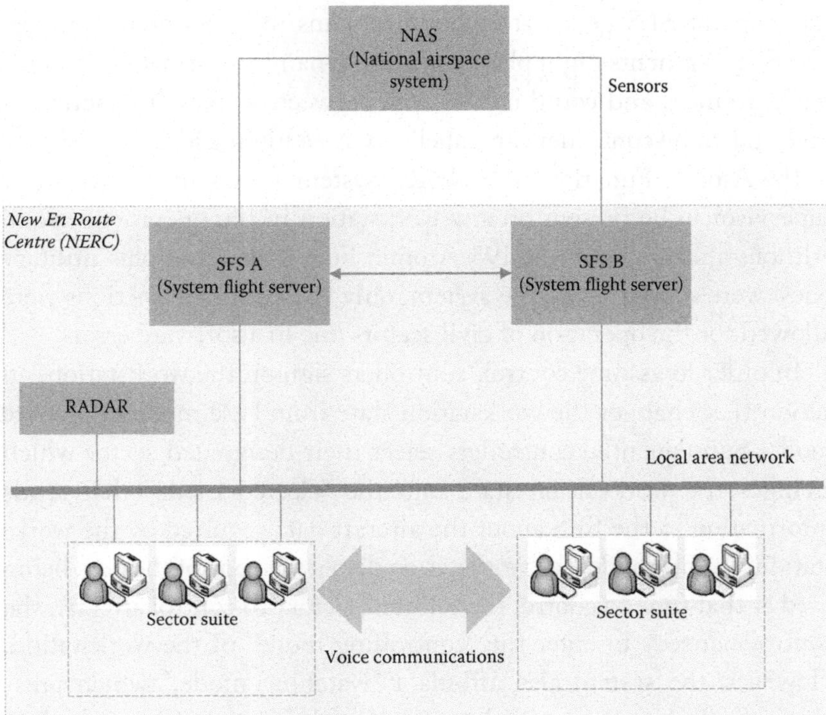

Figure 12.3 A simplified architecture of the hardware systems with two sector suites in the National Aviation Traffic Services (NATS, UK).

LAC is divided into a maximum of 32 sectors that can be combined ("band-boxed") during light traffic or separated ("split") during heavy traffic. A team of two controllers is assigned to each particular sector or combination of sectors. The Tactical or Executive Controller communicates with the aircraft while the coordinating or planning controller manages air traffic within their area of responsibility. In heavy traffic, an air traffic assistant may provide support to the two controllers. The operating teams are supervised by local area supervisors in groups of 5-8 sectors while an operations supervisor is in charge in the operations room; finally, the overall traffic in LAC is coordinated by the airspace capacity manager. At the time of the incident, there were 26 controllers in the operations room with another 42 controllers on duty elsewhere at Swanwick.

Information on civil flight plans originates from NAS and it is routed by the SFS to the appropriate controller workstation. In case of failure, the secondary SFS takes control and provides services until the primary SFS gets restored and remains in a redundancy mode. The SFS augments flight plan data with dynamic information, including clearances and coordination data between sectors. The activities undertaken by controllers are labelled with a unique identifier known as the Atomic Function. The NERC system allows any controller or supervisor to be present on any workstation in the operations room. Although a maximum of 193 Atomic Functions (civil plus military ones) were supported by the system, only 151 Atomic Functions were allowed for the operation of civil sectors due to a software error.

In order to assume control, controllers sign on the workstation, an action that changes the workstation state from base mode to prepare mode. Subsequently, controllers select their designated sector which changes the workstation state into the "elected mode" that sends information to the SFS about the aircraft data required by the workstation. In this mode, the workstation displays a copy of the data being used at that time to control the selected sector. To control aircraft, the controller needs to enter the "controlling mode" of the workstation. However, the system also affords a "watching mode," which presents a display of all aircraft data to external observers (e.g., attending training or familiarization programs). It is important to note that the watching mode was counted in the atomic functions table generated in SFS despite the fact that it has no operational capabilities.

The system failure occurred when an exception was raised in the performance of a check on the maximum permitted number of atomic functions. The software should have checked whether the limit of 193 atomic functions had been reached; instead the check was performed against a civil limit of 151 atomic functions (Walmsley et al. 2015). At the time of the incident, the total number of atomic functions in use was 153, a figure that was reached for the first time because of a change introduced on the previous day in order to include further military controller functions. This raised an exception that broke the normal flow of code execution and forced the shutdown of the primary SFS for safety reasons. Subsequently, the secondary SFS was activated that reprocessed the same list of commands, which raised the same exception in the software, resulting in the shutdown of the secondary SFS as well.

The chronology of main events and people responses are presented in Table 12.1 covering the period from the occurrence of the failure to the resumption of normal operations.

Table 12.1 A Chronology of Major Events That Took Place in the NATS Incident

TIME (UTC)	ACTOR	EVENT	DESCRIPTION
14:44	NATS	Controllers are alerted to a system failure by warnings at their workstations. The Operations Supervisor directs the teams to follow fallback on Checklist 4 for "System Flight Server Unavailable"	The SFS receives flight plans from the NAS (i.e., the central flight planning system) and distributes them to the controller workstations. An error message is issued indicating that both the primary and secondary servers failed and that operations should fall back into a reduced capability mode. Loss of SFS alone does not present an immediate safety threat as aircraft are still monitored on the radar screens and controllers can maintain radio communication with them. However, automated coordination between sectors and other flight data processing functions are not available forcing controllers to rely on voice communication to coordinate flight data manually between air traffic sectors.
14:45	NATS	Loss of the link between NAS and SFS	SFS is no longer able to receive and distribute flight plan information to the workstations.
14:55	London TMA Airports	All departures from London airports are stopped	This was a conservative response that reduced the workload of Area controllers at Swanwick but increased the workload of Tower controllers later on. Actually, the contingency plan recommended a reduced rate of departures and not a complete stop of traffic.
15:00	Eurocontrol	Zero rate regulation (ZRR) is applied	ZRR is applied in exceptionally circumstances since it sets the limit for air traffic to zero and renders an airspace unavailable (closed) to all flights.
15:25	NATS	"B" SFS is restored	Following a reset, the SFS Server B is restored and becomes ready for service.
15:41	NATS	NAS to SFS data download recovery commences	NAS to SFS download is formally activated by engineering teams.
15:43	NATS	NAS to SFS recovery is completed	SFS is repopulated with data from NAS and becomes ready to resume normal service.

(*continued*)

Table 12.1 (*Continued*)

TIME (UTC)	ACTOR	EVENT	DESCRIPTION
15:49	NATS	Resumption of ATC services and electronic coordination	Normal operation is resumed (with reduced redundancy) allowing a return to the normal operational mode from the previous reduced capability fallback mode.
15:55	London TMA and Manchester Airports	Removal of departure restrictions at Heathrow, Gatwick, and Manchester airports	First removal of some of the air traffic restrictions, allowing air traffic levels to start increasing.
16:05	Eurocontrol	ZRR is lifted and capacity is raised to 75%	The regulations applied initially are relaxed, allowing operation up to 75% of the normal capacity.
17:30	Eurocontrol	Departure restrictions are cancelled	All departure restrictions from UK airports are cancelled.
20:06	NATS	"A" SFS is made available	Engineering teams are certain that there was not a risk of a double server failure. At this point they were confident that they could re-enable the automatic standby capability.
20:30	Eurocontrol	Final restrictions are lifted	All regulations caused by the incident are lifted and normal operations are resumed.

12.3.2 STAMP Analysis of NATS System Failure

A STAMP analysis has been applied to this incident, starting with a control diagram of the wider organizational context of NATS operations together with the safety requirements and constraints (Figure 12.4). Each agent in the control structure plays a role in meeting safety requirements and ensuring viability of operations. The top half of the control structure reflects the supervision from several international regulatory bodies and local authorities (e.g., Civil Aviation Authority) that provide the context of operation for NATS. Several operational issues regarding license management are controlled by NLMCC while safety issues are considered by the Safety and Airspace Regulation Group at Civil Aviation Authority (CAA).

The bottom half of the control structure concerns the work organization of NATS and the lines of accountability for the safe and efficient regulation of traffic. Although the organization of air navigation service providers (ANSPs) may vary across different states, their control structure follows some well established principles of design and operation. The left side of the NATS En Route Ltd (NERL) organization displays the operational units (e.g. LAC) that are supported by the engineering units (shown in the right side of Figure 12.4). The safety

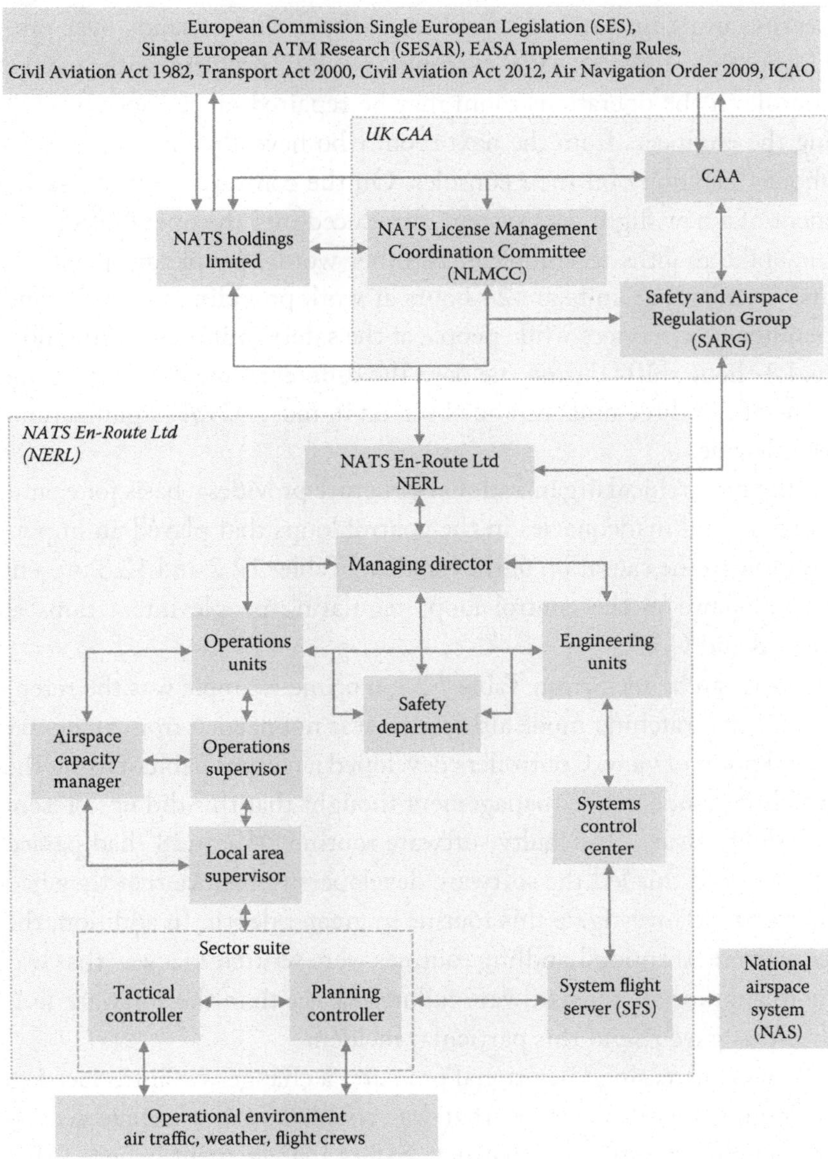

Figure 12.4 A hierarchical representation of the wider organization of NATS operations (STAMP perspective).

department interacts with the operations and engineering units in order to ensure that operations are carried out in a safe manner. However, the response times of the operational, safety, and engineering units are quite different and this has an impact on the way that safety is regulated.

In general, the safety unit interacts with the operations and engineering units in a rather asynchronous way that expands over prolonged time intervals. For example, a safety problem with a radar console in the operations room may be repaired on the spot by calling the engineers from the next room who have already been alerted about this failure on their consoles. On the contrary, a safety assessment of a new flight data system introduced into the operations room may take months to complete. In other words, people at operations and engineering units are 24 hours at work providing air traffic and maintenance services while people at the safety unit are working normal 8- hour shifts during the day. The different time scales of operations of the three units can be observed in most ANSP organizations world-wide.

The hierarchical organizational structure provides a basis for examining several inadequacies in the control loops that played an important role in the causation of the incident. Tables 12.2 and 12.3 present several flaws in the control loops regulating people interactions at NATS and CAA.

As it can be seen from Table 12.2, a prime example was the retention of the watching mode although it was not needed from an operational point of view. Controllers developed a natural habit of using the watching mode, while management thought that this did not present any safety issues. The faulty software routine "Waafu28" had passed all tests and this led the software developers to believe that they had no reason to investigate this routine in greater depth. In addition, the associated exception handling routines were written in a way that was more appropriate for hardware failures rather than the software failures that gave rise to this particular incident.

Under increasing operational pressure, a change was introduced to accommodate more sectors that was coupled with a unique sectorization configuration on that day. Safety management believed that the change to include additional military controllers was acceptably safe. In general, it was rather difficult to test the software for many

Table 12.2 Flaws in the Control Loops at the Operational Level of NATS Activities

CONTEXT	MENTAL MODEL FLAWS	INADEQUATE DECISIONS – ACTIONS
a. Controllers at the operations room habitually used the "watching mode."	a. Engineering teams believed that the number of Atomic Functions were below the maximum limit.	a. There was little testing to confirm that NERC did not perform unintended actions.
b. "Watching mode" was retained although it seemed unnecessary from early days.	b. Engineering and operations management believed that there were no safety implications from accommodating additional military controller roles.	b. Formal acceptance tests were performed with more than 130 but less than 151 Atomic Functions.
c. NERC technology traced its origins in the 1990s and required a lot of "hands-on" involvement to address amendments.	c. Safety management believed that the change to accommodate additional military controllers was acceptably safe.	c. Facility arrangements made it difficult for NATS labs to simulate system-wide failures.
d. A change was introduced in the previous day in order to accommodate more military controllers.	d. The faulty module Waafu28 had passed all tests, which led software developers to believe that they had no reason to investigate the module in any more detail.	d. Naming conventions on SFS source code (Waafu28) were difficult to follow and understand.
e. Coupled with the way that the airspace and sectors were configured that day, this change led the Atomic Functions to reach 153 for the first time.	e. Management believed that the low turnover of staff was an indicator that a comprehensive understanding existed of safety hazards at work.	e. The exception handling routines were written in a way that was more appropriate for hardware failures.
f. Relevant software test results for SFS faulty routine (Waafu28) showed six tests had been passed, with no failures.		f. Software designers did not consider the possibility that both "A" and "B" SFS could shut down on the same failure.
g. Initially the software was written for civilian use with no provisions for the inclusion of military roles.		g. Poor quality of software requirements with regard to exception handling specifications.
h. Management focused on deploying SESAR, which requires different project management, collaboration, and oversight.		h. Fixes were favored over more comprehensive inspections.
		i. Inadequate inspection of type definitions in external modules and calls on utility functions to check their correct usage.

system-wide failures since the laboratory facilities at NATS were not adequate in this respect. In addition, there was little testing to examine whether NERC could initiate different types of unintended actions. Finally, software designers did not consider the possibility that both "A" and "B" SFS could shut down on the same software

Table 12.3 Flaws in the Control Loops at the Regulation Level of NATS Activities

CONTEXT	MENTAL MODEL FLAWS	INADEQUATE DECISIONS AND CONTROL ACTIONS
The primary duty of CAA's was to maintain high standards of safety. Secondary duties included: furthering the interests of aircraft operators and passengers, promoting efficiency and economy of service, and so on.	a. CAA management believed that allowing a normal return on planned investment should give NERL a neutral investment incentive.	a. CAA provided no incentive structure for NATS to make decisions on how much to invest in any period by making NERL theoretically financially neutral to the level of investment that the company makes.
CAA regulated NERL through the enforcement of the conditions in the license and by modifying them occasionally.	b. CAA management believed that a mechanism that claws back any returns on planned investment could prevent NERL from deliberately avoiding or delaying investment.	b. CAA range of enforcement power over NERL was limited. For instance, CAA had no power to take action against past breaches of the license and levy financial penalties on NERL for significant failures of service.
CAA's approach to NATS regulation was based on the principle of giving the company strong incentives to make decisions that were in the best interest of the NERL's users.	c. CAA management believed that previous recommendations made for another system communication failure (December 2013) were closed-off and adequate provisions were in place.	c. CAA license team (NATS License Management Coordination Committee - NLMCC) did not intervene in the daily operational or investment decisions of NERL.
In general, CAA discharged its responsibility for safety by oversight of the NATS Safety Management System, operational procedures and safety cases through the Safety and Airspace Regulation Group (SARG).	d. SARG seemed to rely only on guidance material and international software assurance standards to examine practices in detecting faults in software.	d. NLMCC did not have detailed knowledge of NATS systems and was not familiar with SFS.
	e. SARG seemed unaware of the potential limitations of safety cases as a primary means of safety assessment.	e. Ineffective oversight of software safety assurance was in place.

failure. Common cause failures had been considered mainly for hardware failures since the system was built in early 1990s with little consideration for flaws in the logic of software routines.

Several flaws can also be identified in the control loops and interactions between the regulators (CAA) and the service providers, especially with regard to aspects of software safety assurance (Table 12.3).

The more the system grows in complexity, the more difficult it becomes for the engineering teams to assess how the software system performs in different situations. The increasing number of state variables and control variables results in a quick explosion of possible combinations of situations to be controlled by the software system. This combinatorial problem cannot be solved by relying solely on international software assurance standards and guidelines. It also requires a robust program of simulation of system-wide failures that many ANSPs cannot provide in their facility arrangements. Things usually get worse because instead of writing new routines software developers tend to reuse existing ones without considering all the subtleties of new situations. Software assurance remains one of the most challenging issues in safety management and safety cases have been criticized for their limited consideration of these issues (Leveson 2011). Unfortunately, the Safety and Airspace Regulation Group (SARG) at CAA overestimated the potential of safety cases as a primary means of safety assessment.

At a higher level, the NLCMCC management (NATS License Management Coordination Committee) could not intervene in the daily operations and investment decisions of NATS. In particular, NLCMCC did not provide incentives for NATS to make further investments in updating the old software system NERC in order to meet the new traffic requirements. A large financial investment in getting NERC up to the new traffic requirement would require considerable en-route charges to be brought by commercial airlines. However, the regulator NLCMCC would not provide this incentive to NATS to secure the financial resources for updating the software system. This was a matter of regulatory policy but also reflected the fact that NLMCC was not knowledgeable enough to comprehend the nuances of air traffic operations and the safety challenges of the software system.

12.4 The Viable System Model (VSM)

As systems become larger and more complex, there is an increasing need for managers and supervisors to use formal organizational models to share their understanding about the situation. The most commonly used organizational model in management is still the hierarchical model. In principle, hierarchical models represent the formal authority structures and organizational policies in the system. Hierarchical models do not provide a realistic account of fundamental things about the organization, such as its safety processes, the formal and informal structures, the communication loops, and the stakes involved in decision making. The VSM offers a more sophisticated account that can be used both for diagnosing existing problems and for designing new systems of work.

Stafford Beer (1985) sought to develop some fundamental principles that create viable organizations that exist and thrive in unpredictable and turbulent environments. The criteria of viability require that organizations are capable of adapting properly to their environment by managing the complexity of both their environment and their own activities. The VSM model unfolds complexity in a fractal structure in which systems are made up of subsystems that have the same generic organizational characteristics as those in which they are embedded.

A viable model views organizations as nested autonomous units of which each becomes a viable subsystem in its own right. System 1 is the basic unit that comprises a management and an operational element interacting with the local environment. Systems 2–5 facilitate the work of basic units and ensure the continuous adaptation of the organization as a whole. In general, organizations can make use of the five safety-related functions (Figure 12.5) presented as follows:

- *Safety policy and steering:* System 5 plays the function of policy-formulation representing the current beliefs, norms, and assumptions about the work environment as well as existing organizational capabilities. It also monitors the interaction of Systems 3 and 4 to achieve a balance between exploitation of existing safety rules and exploration of new safety concepts. Safety policies should also promote a good safety culture throughout the organization.
- *Safety development and adaptation:* System 4 plays an intelligent function that scans the environment for threats and opportunities while looking inside for internal strengths and weakness. It conducts safety research and development (R&D) and suggests changes to the safety policies for the continual adaptation of the whole system to the changes of the environment. To ensure that safety plans are grounded

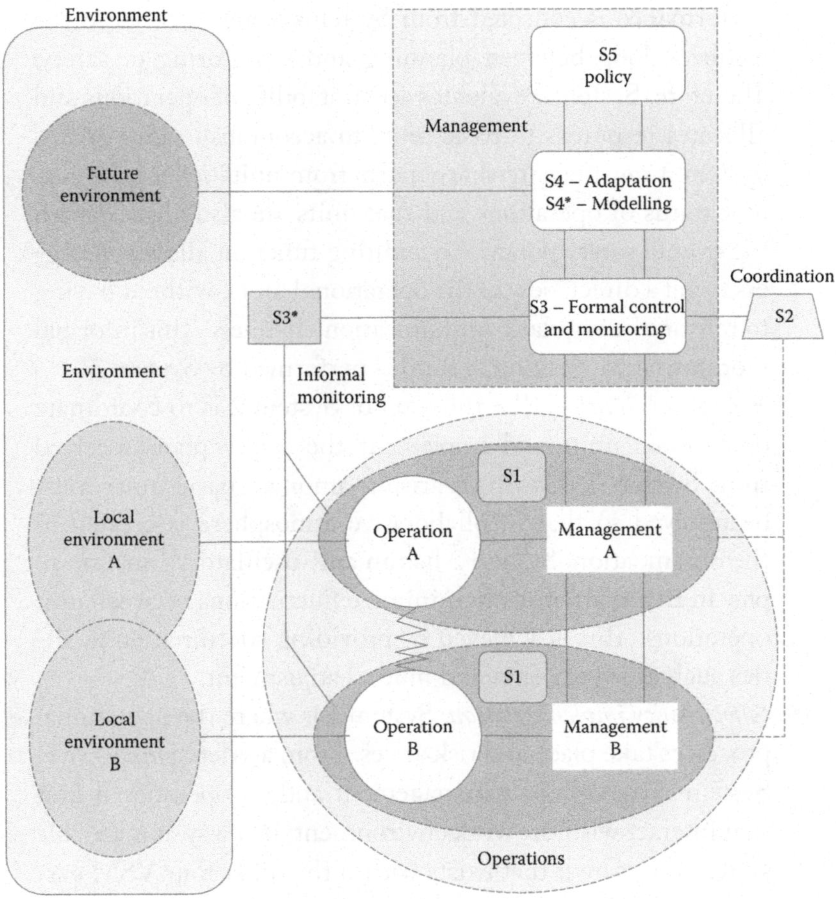

Figure 12.5 A generic recursive template for the analysis of interactions between management, operations and environment (VSM perspective).

in an accurate appreciation of the system, the intelligence function should rely on an updated model of organizational capabilities. Finally, System 4* deals with confidential or special information communicated by practitioners about near-misses and work problems to learn from actual practice.

- *Safety control loops:* System 3 is concerned with the provision of cohesion and synergy to a set of basic units. From a safety viewpoint, System 3 is responsible for maintaining risk within an acceptable level and for ensuring that units implement the organization's safety policy. Safety plans and standards are received from Systems 4 and 5 while information about safety performance is collected from Systems 1 and 2 to close the feedback loop between planning and monitoring of safety. Therefore, System 3 evaluates accountability of operations and allocates resources to basic units to accomplish safety plans. System 3 must ensure that reports from units reflect the current status of operations and that units are also aligned with the overall safety policy. An auditing function allows managers to get a direct view of the operational units, without having to rely on the regular communication channels. This informal monitoring by walking around is performed by System 3*.

- *Safety coordination:* The function of System 2 is to coordinate operational units and implement the safety plans received from System 3. Conflicts arising amongst basic units must be resolved so that a collaborative atmosphere is created in the organization. System 2 has an anti-oscillatory function to play in that it attempts to minimize fluctuations between unit operations. This is achieved by providing coordination facilities such as supervision and mutual adjustment.
- *Safety policy implementation:* System 1 is where the operational processes take place and risk arises. From a safety perspective, System 1 consists of a management and an operational unit that interact with the work environment; in a way, it is a viable system on its own that exists within the other four VSM systems. System 1 also relates to how subsystems may be grouped together to create an organizational structure.

According to VSM, the structure of an organization can be described by the way that Systems 1 and 2 are designed. *Structure* reflects the organization of basic units into higher order units as well as the type of coordination that is achieved between units. On the other hand, *strategy* refers to the managerial functions of Systems 3, 4, and 5, which determine how organizations control their processes and adapt to the environment. Finally, VSM looks at the way organizations adapt to the environment. System 4 plays an important function in scanning the environment for threats and opportunities so that new concepts of operation and safety are explored. It also addresses the process of organizational learning and change by maintaining a balance between exploration and exploitation.

The key concept in VSM is how organizations handle the complexity of their environment and their own activities. VSM deals with this complexity in two ways: (1) looking at the balance of complexity between parts of the system (i.e., the law of requisite variety) and (2) unfolding complexity in a recursive structure in which systems are made of subsystems with the same generic characteristics. Adaptation for management units involves scanning the environment for safety risks and complying with regulations. Adaptation can also be seen as a process of amplifying the variety of one's own capabilities or attenuating the complexity of the environment.

The viable system model provides an explanatory framework of the factors that can amplify the complexity of the environment or attenuate the capabilities of organizations, which can lead to a mismatch between demands and capabilities. To redress this mismatch, organizations should opt to attenuate environmental demands and amplify their capabilities. Table 12.4 presents several factors that could amplify or attenuate demands and capabilities in the context of the previous incident.

Table 12.4 Balancing the Varieties of Organization and Environment in NATS Activities

COMPLEXITY AND CAPABILITIES	DESCRIPTION
Amplification of complexity of the environment	• Coping with a wide range of interconnected regulations (EU, International, European, State) • Coping with seasonal traffic, ebbs and flows of traffic, and adverse weather conditions • Coping with some of Europe's most busy airports and airspaces • Managing incompatible demands of compliance-based and performance-based certification • Reducing delays at airport and en-route traffic • Moving into SESAR while preserving key systems and functionalities of legacy systems • Meeting EU-wide performance targets (e.g., safety, capacity, environmental and cost effectiveness) • Meeting locally imposed restrictions (e.g., noise and night curfews) • Operating in a financially neutral mode • Increased societal demands for running and changing the system
Attenuation of the complexity of the environment	• Loosening the coupling of performance targets • Establishing more realistic performance targets • Simplification of regulation • Abandoning financial neutrality for the ANSP • Slowing the progress toward SESAR

(continued)

Table 12.4 *(Continued)*

COMPLEXITY AND CAPABILITIES	DESCRIPTION
Amplification of the capabilities of the organization	• Improving sectorization techniques (i.e., the number and configuration of sectors) to meet varying demands for capacity • Introducing new systems that enhance decision making (i.e., iFACTS, the suite of tools that allow controllers to see aircraft trajectories up to 18 minutes in advance while reducing workload) • Improving service continuity plans (e.g., contingency planning) • Developing resilient CNS arrangements • Developing resilient operational plans • Searching for resilient methods for software assurance • Being ambivalent to the effectiveness of SMS and, in particular, the safety assurance function • Improving testing facilities that can simulate the full range of operationally credible scenarios and CNS systems failures • Developing technical and managerial expertise required for the deployment of SESAR projects • Searching for implicit constraints in the organization (e.g. outdated software that hinders resilient changes) • Making the system tractable
Attenuation of capabilities of the organization	• Minimizing time and resources used for software changes • Relying on regulatory compliance as a primary means of safety assessment • Safety cases have not managed to tackle communication and coordination problems in the management of safety • Strong sense of security due to low turnover of engineering staff • Inability of labs to simulate fully the operational environment and its boundary conditions (i.e., the maximum number of atomic functions with representative traffic) • Lack of consideration of potential consequence of work practices at the operational level (i.e., the widespread habit to activate the watching mode)

12.5 Mapping the STAMP Technique onto the VSM Organizational Model

Since STAMP and VSM rely on a control-theoretic view of organizations, it would be possible to map out the analytical categories of the two models by modifying some of their terminology. In this sense, the main category of control algorithms of STAMP has been extended to include aspects of steering and planning. Steering refers to the values and goals of different individuals across organizational sectors. Planning refers to the temporal and spatial constraints of control algorithms (i.e., how a plan adapts to situational changes and how it is distributed across multiple actors). A second refinement can be made to the coordination category of STAMP to create an adapting function outside the organizational boundaries (see intelligence function of VSM) and a controlling function within the organization.

At a meta-level, organizations need to adapt their collaborative work to changes in environmental demands (i.e., changing the system), while at the operational level, local coordination is needed to carry out complex tasks (i.e., running the system). As a final point, the planning and monitoring functions of STAMP correspond to the control function of VSM that is required to achieve cohesion of operational units and ensure that local coordination does not drift from the overall plan. The extended categories of STAMP (Table 12.5) can accommodate the organizational cybernetic perspective (VSM) and provide

Table 12.5 Seven Control Functions Common to STAMP and VSM

CONTROL FUNCTION	DESCRIPTION
Steering of control algorithms *(System 5)*	Control algorithms are designed with a goal in mind that should be attained and sustained over time. Practitioners face many challenges in articulating hidden goals, balancing conflicts, and seeing long term consequences. In some cases, a goal may be judged as a poor choice but a careful investigation may reveal that this is a reconciliation of conflicts at work, beyond the control of individuals. Hence, analysts should try to trace implicit goals at work that are not clearly articulated and examine tacit constraints of the organization. Steering plays a similar function to System 5 in balancing exploitation and exploration.
Adaptation to environmental demands *(System 4)*	Organizations operate in an open environment and their exchanges can be rich and dense. Increasing competition, societal pressures, and deregulation may inflict changes in technology, reforms of organizational structure, and adaptations. Organizations must adapt their structure and processes to manage these demands. Adaptation to environment and coordination between running and changing the system are functions related to System 4.
Modeling and learning *(System 4*)*	All managerial and operator interventions are associated with a mental model of what safety means. Practitioners construct their own theory of potential hazards, accident causes, and risk control strategies. Their models are vehicles for directing attention to critical signs of risk. The controllers must also have a model of how they fit into the organizational framework. For example, when a disturbance affects a process in a unique way, the controllers must have a model of the organization to know who should take charge of the problem and how to coordinate. Mental models help managers and practitioners challenge their understanding and remain vigilant to the possibility of failure.

(continued)

Table 12.5 (*Continued*)

CONTROL FUNCTION	DESCRIPTION
Monitoring and auditing *(Formal System 3)* *(Informal System 3*)*	Information handling difficulties may relate to the nature of the information itself (i.e., ambiguous data), the characteristics of the practitioner (i.e., not recognize its significance), or the environment (i.e., distractions). In studying disasters, it is important to pay attention to the distribution of information, the structures and communication networks, and the boundaries that may impede the flow of information. Organizations create assumptions about what is valued information, how it can be communicated, and what can be ignored.
Planning of control algorithms *(System 3)*	Control algorithms should be designed according to a safety plan that specifies the sequence of actions, the slack that exists, and the degrees of operator freedom. In this sense, a work practice is an algorithm with specific features, such as granularity, degree of freedom, and temporal constraints.
Coordination *(System 2)*	Cooperation of multiple units raises many important issues with regard to the delineation of responsibilities, reconciliation of different views, and communication among team members. When there are multiple controllers, decisions may be inadequately coordinated, including communication errors, unexpected side effects, and conflicting actions. When coordination crosses organizational boundaries, practitioners may not be able to see how their actions affect others or may not be motivated to do so due to a silo mentality.
Implementation of safety policies *(System 1)*	Safety policies and plans are implemented at the operations rooms where controllers interact with the internal system (i.e., displays, controls, and procedures) and the environment (i.e., adverse weather conditions and heavy traffic). To cope with economic and temporal demands, controllers often have to fall back on experience and rely on habits that seemed to work in the past. The danger is that, as habitual actions gain strength by their everyday use, they may not see certain countersigns or exceptions that make rules unsuitable to the current situation. In this sense, the balance of autonomy and control (Systems 2 and 3) are likely to influence implementation of safety policies.

a good basis for analyzing patterns of organizational breakdown presented in the literature (Busby 2006; Hoverstadt 2008; Kontogiannis 2010a). For the same reason, Table 12.6 provides a summary of ten VSM principles that are central to organizational cybernetics.

The VSM perspective highlights the recursive structure of organizations. The concept of recursion implies some sort of autonomy and self-regulation at each level of description (Table 12.6, [1]) in the sense that the same five functions apply to each individual unit to ensure viability at its level. In a sense, System 1 can be seen as a group of subunits that have relative autonomy in carrying out their tasks.

Table 12.6 Ten VSM Principles that Help Diagnose Organizational Breakdowns

VSM PRINCIPLES	DESCRIPTION
1. *Recursion/fractal structure*	Each subsystem is a viable system on its own, embedded in larger viable systems and regulated by the same five functions; complex behavior emerges from simple rules or functions that are repeated across all organizational levels.
2. *Self-regulation and autonomy*	Subsystems can remain self-regulated or autonomous as long as they do not threaten the viability of the whole organization; conflicting or hidden goals may threaten autonomy.
3. *Local coordination versus centralized control*	Local coordination can minimize fluctuations in unit interactions but may also lose sight of overall standards and plans (that is, suboptimization); hence, it must be balanced with central control.
4. *Circular processes of monitoring and planning*	Monitoring of performance is linked to accountability that, in turn, feeds to planning and allocation of resources. On the other hand, planning is used to set up performance measures for monitoring. Their interaction determines performance flexibility.
5. *Adapting organizational structure & strategy*	Organizations should adapt to the changes of the environment by changing their structure and strategy; for example, an emergency mode of operation that requires different strategy must be facilitated by appropriate changes in the organizational structure.
6. *Intelligence and vigilance*	The intelligence function of System 4 should rely on a mental model of system capabilities and remain vigilant to the possibility of failure.
7. *Exploitation versus exploration*	Exploitation that is based on existing rules and practices must be balanced with exploration or creation of new rules (that is, an important function of System 5).
8. *Prevention versus recovery*	Prevention focuses on the removal of obstacles by seeking a safe environment, specializing in a narrow niche, or minimizing spread of danger; recovery emphasizes learning from errors and mitigation by relying on teams, making use of multiple resources via many routes.
9. *Requisite variety*	The variety and competencies of the organization should match the complexity and variability of the environment.
10. *Coordination across boundaries*	Boundaries can change the exchange rate and transduction of information channels; they may affect the visibility of other units and the degree of sharing information across boundaries or sectors.

At the same time, subunits should comply with the overall requirements of the safety management system. Hence, VSM brings to the fore the balance that must exist between autonomy and centralization of control (Table 12.6, [2] and [3]). This is a delicate balance as subunits must not become isolated but, equally important, must not drift from the overall safety policy.

Two other important issues that VSM highlights regard the interactions between planning and monitoring or between strategy and structure (Table 12.6, [4] and [5]). First, it may be recognized that planning and monitoring are necessary functions for a management unit to control its operational elements. However, the two functions are coupled and create a closed control loop that often passes unnoticed and leads to several problems. For example, planning usually sets up measures of performance that drives how to monitor work in progress; in some cases, however, measures of performance may be used to make planning less challenging and lower goal aspirations (Hoverstadt 2008). Similarly, there appears to be an interaction between strategy and structure that seems to go unnoticed in many safety improvement campaigns. Strategy should be built up throughout the organizational structure where the interests of individuals at different levels are equally reflected in decision-making. Failure to do so may result in a safety campaign that may not succeed in reaching certain parts of the organization.

VSM has proposed System 4 as an intelligent function (Table 12.6, [6]) that scans the environment for threats while looking inside for internal strengths and weaknesses. However, to ensure that safety plans are grounded in an accurate appreciation of the system, the intelligence function should rely on an updated model of system capabilities; Santos-Reyes and Beard (2006) refer to this function of updating a mental model of the organization as System 4*. The functions of System 4* can be derived from earlier work on the margin of maneuver (Woods and Branlat 2010), the high reliability organization (HRO) principles of preoccupation with failure, reluctance to simplify, and sensitivity to operations (Weick and Sutcliffe 2001), and the Frame/Reframe model (Klein et al. 2007) described in Chapter 5.

Another VSM principle (Table 12.6, [8]) from general systems theory (Wildavsky 1988) refers to the tendency of organizations to move between two different safety approaches. The preventive approach relies on risk anticipation and exploitation of existing rules while the recovery approach emphasizes recovery from errors, learning, and mitigation. Prevention focuses on the removal of obstacles by seeking a safe environment, specializing in a narrow niche, or minimizing spread of danger (Wildavsky 1988); recovery emphasizes learning from errors and mitigation by relying on numerous interacting actors who are making use of multiple resources via many routes (i.e., the high flux and omnivory principles quoted in Skyttner 2005).

VSM is a systems thinking model that emphasizes the adaptation of organizational strategies and structures to the changing demands of the environment. Probably, VSM may benefit from recent sense-making and resilience approaches that provide a better insight into how practitioners and organizations make sense of changing situations and new deployments of resources and competences to handle unexpected events. It is worth noting that the principle of requisite variety (Table 12.6, [9]) addresses many aspects of how teams manage their margin of maneuver (Woods and Branlat 2010). Requisite variety is the variety of organizational resources and competences that is adequate to manage the complexity of the situation. It corresponds to the reserves of resources that an organization maintains in anticipation of new challenging events; these reserves provide a margin for maneuver when organizations operate close to their safety boundaries.

For a practical application of the joint STAMP-VSM model in a series of aviation accidents, see Kontogiannis and Malakis (2012a, b). Some general patterns of breakdown behind the control flaws are discussed in the next chapter, where the VSM organizational model is merged with a model of human performance.

12.6 Concluding Remarks

We have attempted to bridge the gap between two parallel strands in systemic accident models. On the one hand, accident investigation techniques (i.e., AcciMap and STAMP) have looked into the flaws of control functions and problems in enforcing constraints between different levels in the organization. On the other hand, a large literature of patterns of organizational breakdowns has applied organizational models to specific accidents. To help accident investigators to look at both the control flaws and the organizational breakdowns, a link was established between a control theoretic accident model (i.e., STAMP) and a cybernetic model of organizational viability (i.e., VSM).

The proposed joint STAMP-VSM framework relies on a refinement of control categories of STAMP (i.e., steering, planning, and adaptation) so that the VSM functions can be mapped onto the STAMP analysis of organizations. Second, a recursive representation of organizations was proposed that has several advantages over hierarchical representation (i.e., the same organizational principles apply at different levels). A recursive structure may help analysts to rethink the safety organization, model new information loops, identify new constraints, or see problems in the adaptation and steering functions of the organization. In addition, recursive structure can provide insights on how to unfold complexity into several organizational levels. For example, some of the complexity can be managed by intelligent behavior at the operational level and the remaining problems (sometimes called residual variety) can be managed by people at higher levels. It is argued that the recursive analysis may be more difficult to apply from the start without a preliminary analysis of the functions and interactions of the constituent subsystems. In fact, once a hierarchical analysis is made, the analysts can select particular areas and carry out a deeper recursive analysis.

Therefore, the transition from a hierarchical to a recursive representation is not so difficult to make. A recursive VSM analysis focuses on the organizational structure rather than on single actions and events; as a result, the VSM analysis is usually performed for the organization of safety management and can be used for the investigation of all accidents in the same organization. Once a VSM analysis has been performed, it can provide useful analysis for several near-misses or accidents.

Third, information channels crossing the boundaries of subsystems can be studied from the perspective of their capacity to transmit information and transform it with the use of several transducers (for example, oral instructions, written procedures, handover checklists, and so on). Fourth, the organizational cybernetic model (VSM) allows analysts to bring together several issues of organizational theory that relate to self-regulation, the autonomy-control dilemma, prevention versus recovery and the interaction of structure and strategy (see Table 12.6). The latter is very important in the management of change since strategy should be built up throughout the organization and take into account the needs and interests of all subsystems. In this context, many failures to improve a safety policy can be traced into problems of taking on board the views and interests of practitioners throughout the organization. Finally, the proposed framework addresses how organizations adapt to challenges in their environment by amplifying their own variety or by attenuating the complexity of the environment.

From this discussion, it appears that STAMP analysis provides a description of control flaws that can be explained in terms of the five organizational functions and the VSM principles of organization. In fact, the analysts have a choice with regard to the depth of analysis required; some control actions or monitoring functions can be sufficiently covered by STAMP without any need to look deeper into their underlying patterns of breakdown. Furthermore, the VSM analysis can be made proactively as part of an organizational audit on the basis of early warnings from near misses and previous incidents. In this connection, VSM can provide useful information about safety management to the STAMP analysis.

There is still, however, a dependency on the literature of organizational breakdowns that have been revealed by earlier studies (Hoverstadt 2008; Busby 2006; Kontogiannis 2010a). It may be rather difficult for an accident investigator to rely solely on the proposed framework without any prior knowledge of the literature on organizational breakdowns. Overall, the framework can help analysts to model the complex interactions across boundaries, the information structures that impede communications, and delineation of responsibilities across multiple controllers. Awareness of these traps can help organizations avoid them or at least decrease their negative impact.

Another remark should be made about the potential of the joint STAMP-VSM model to incorporate modern ideas from complexity theory. Admittedly, VSM has been one of the hard systems methodologies that offer little flexibility in describing organizations and this has been criticized in the literature (Checkland and Scholes 1999; Jackson 2003). For instance, the recursive structure of VSM assumes that the five functions are equally applicable at all levels in the organization. This may apply to safety-critical industries that rely on formal organizations and hierarchical controls; in contrast, small-to-medium size industries in lower risk domains could be more flexible with their organizational functions. In fact, some applications of complexity theory to accident analysis and crisis management (Dekker 2011; Paraskevas 2006) have used more flexible descriptions of organizations and advocated that structures and functions can emerge in different shapes. It is proposed that the predefined VSM functions are seen as high-level descriptions of the requirements of control rather than as recommendations for how organizations should control their functions.

Finally, VSM may also benefit from recent resilience approaches that provide a better insight into how practitioners and organizations make sense of changing situations and make reserves of resources and competences to handle unexpected events. Chapter 5 on the Frame/Re-Frame model (Klein et al. 2007), the concept of the margin of maneuver (Woods and Branlat 2010), and the HRO principles can be valuable in elaborating the sensemaking capabilities of System 4* and VSM in general.

13

Integrating Human and Organizational Models of Performance

13.1 Introduction

This chapter provides a joint framework for integrating models of human and organizational performance that have been discussed in the context of systems thinking. The selection of performance models should reflect both the particular domain of application and the scientific approach with its guiding principles. Air traffic control (ATC) is a domain that is increasing in interactive complexity but still maintains a manageable degree of coupling that works for replanning and error recovery. At the sharp end, ATC involves many taskwork and teamwork functions as presented in T^{2}EAM (taskwork/teamwork for effective and adaptive management). In addition, the controller's work is affected by organizational policies as well as interactions with airport facilities and airlines. ATC is part of a network of aviation stakeholders operating in a dynamic interaction as new changes are introduced by new initiatives in SESAR and NextGen. Hence, human performance should be considered within an organizational framework that addresses constraints and resources at the sharp end. This chapter aims to integrate T^{2}EAM and viable system model (VSM) to provide a common framework for analyzing the performance of people and organizations in ATC.

There are also other candidate models with similar aspects that could be used by researchers but all models are subject to a set of selection criteria. For instance, ECOM (extended control model) is another candidate model of human performance (Hollnagel and Woods 2005) that shares many similarities with T^{2}EAM. Other organizational models within the systems control paradigm may be utilized such as systems-theoretic accident model and processes (STAMP) (Leveson

2004), system dynamics (Marais et al. 2006), The Fifth Discipline (Senge 2006), dynamic decision making (Brehmer 2000), and so on. All these models of systems thinking are based on the following concepts in managing complex and dynamic situations:

1. Context of work
2. Dynamics of performance
3. Observability and change
4. Decision-making
5. Sensemaking
6. Coordination
7. Adaptation

The interplay between context and control has been thoroughly discussed in human factors (Hollnagel 1998) as well as in newer approaches to systems thinking (Leveson 2012; Hollnagel and Woods 2005). Human control should always be considered within the context of work that is, the characteristics of the situation, the constraints set by higher organizational levels, and the resources and competences made available to sharp-end controllers. Constraints at the workplace may include situational demands (e.g., time pressure, uncertainty, unfamiliarity), while the organizational level may include compliance with organizational rules and procedures. Resources refer to available tools for doing the job, support from other team members, training facilities, and safety barriers for preventing adverse events. Many applications of STAMP have presented ample illustrations of how the work context may affect human control of technical processes (Leveson 2012).

The second concept regards the ability of models to represent changes of performance as a function of time and dynamics of the situation. In dynamic environments, the situation may change, which implies that controllers should be alert to revise their thinking and approach the problem from another prespective. It may also imply that controllers should be able to maintain adequate resources to develop new plans. In most cases, dynamic environments make it difficult to apply linear models of thinking (e.g., situation assessment followed by decision making and planning). Dynamic tasks require continuous decision-action cycles so that the problem is controlled in an incremental fashion that allows early correction of unsuccessful decisions.

Control theory identifies four important conditions for control:

1. There must be a goal or decision (*the goal condition*).
2. There must be a model of the system that describes what will happen if people changed something in the system (*the model condition*).
3. It must be possible to ascertain the state of the system (*the observability condition*).
4. It must be possible to change the state of the system (*the change condition*).

The last two concepts of observability and change refer to the ability of people to control the system, that is, plan their actions and monitor/observe their work progress. For instance, System 3 in VSM controls the process by setting appropriate plans and modifying them on the basis of process feedback. Of course, planning and monitoring are not independent since the plan refers to a model of what to do, what to expect to see, and when to look for feedback. Goals are set by practitioners in a decision-making process that requires many trade-offs and dilemmas (Chapter 4). Although earlier models have addressed how people make decisions in different circumstances, less attention has been paid to the interaction between planning and situation assessment. For instance, a controller's decision may be affected by his or her degree of uncertainty in understanding a situation but also may be affected by the consequences of being wrong in planning how to control the situation.

A mental model is an important cognitive function that guides decision making and planning. Controllers' understanding of the situation is shaped by technical know how and practices in searching for critical information. A mental model is also a basis for developing sensemaking capabilities, such as awareness of one's own capabilities and resources to control the system. Therefore, research in sensemaking and reframing (Chapter 5) can be brought to bear in developing an integrated model of human performance.

The sixth condition of coordination and communication can be seen both at the teamwork level and the organization level. When there are multiple controllers, coordination problems can arise due to performance delays and lack of opportunities for recovering from errors. When coordination crosses organizational boundaries, people may not be able to see how their actions affect others and introduce side effects. T^{2}EAM has specified several behavioral markers for assessing communication and coordination, while VSM has specified System 2 for coordination across teams in the organization.

Finally, the last concept of adaptation refers to the ability of practitioners and organizations to match their modes of functioning to different contexts of work. On an individual level, adjustments have been described as sacrificing decisions or efficiency-thoroughness trade-offs (Hollnagel 2009). On an organizational level, adjustments have been described using terms as drift to the safety boundary (Cook and Rasmussen 2005) or adaptive delegation of authority (Roberts 1993). System variability usually goes unnoticed because practitioners manage to adapt their plans and control the working conditions. It is only when variability gives rise to unexpected outcomes that it becomes noticed and creates the preconditions for failure (Weick and Sutcliffe 2001). This systemic approach to safety views success and failure as the result of adaptations that organizations and practitioners perform to cope with complexity.

This chapter elaborates on T^2EAM and VSM so that human and organizational functions are better integrated within a common framework of performance. A similar effort has been undertaken by Stringfellow (2010) on the basis of an extensive literature review that resulted in a checklist of human and organizational factors that could account for control flaws as specified in STAMP. In this chapter, we take a model-driven approach where two theoretical models are modified and integrated so that a recursive pattern of performance emerges at different system levels. A model-driven framework to system safety is proposed that probes into the control functions and relationships of hierarchical subsystems in the organization. Control functions are defined by merging the functions of VSM and T^2EAM. This provides a basis for understanding the causes of many control flaws in the quality or enforcement of constraints and the quality of feedback (these causes are referred to as performance breakdowns or archetypes). A discussion follows in the last section on the benefits of this approach and the difficulties encountered in its application.

13.2 A Human Performance Model of Taskwork and Teamwork

Systems thinking perceives organizations as hierarchical structures with communication and control functions that operate at the interfaces between organizational levels and entail an upper level imposing constraints upon a lower one (Leveson 2012). This control-theoretic approach can be applied to all levels in the hierarchy of the organization, although with different emphasis on their control and timing functions. At the sharp end, control of the technical system can be described according to the taskwork and teamwork functions of T²EAM (see Figure 13.1 and Table 13.1).

T²EAM is a control model of performance that views controllers as trying to close a gap between actual and target system states. This activity of closing the gap triggers a planning process that changes the state of the system. In a socio-technical system, control is hardly ever perfect. Effective control requires a good model of the system and its operation as well as feedback about its effectiveness. The mental model helps controllers in managing uncertainty by directing them to important sources of data and by making sense of conflicting data. In this sense, monitoring of the system is guided by the mental model so that priorities are assigned to sources of information that may compete for attention. Modeling interacts with steering in a strategic mode so that modeling may reframe mental models and critique goals, hence

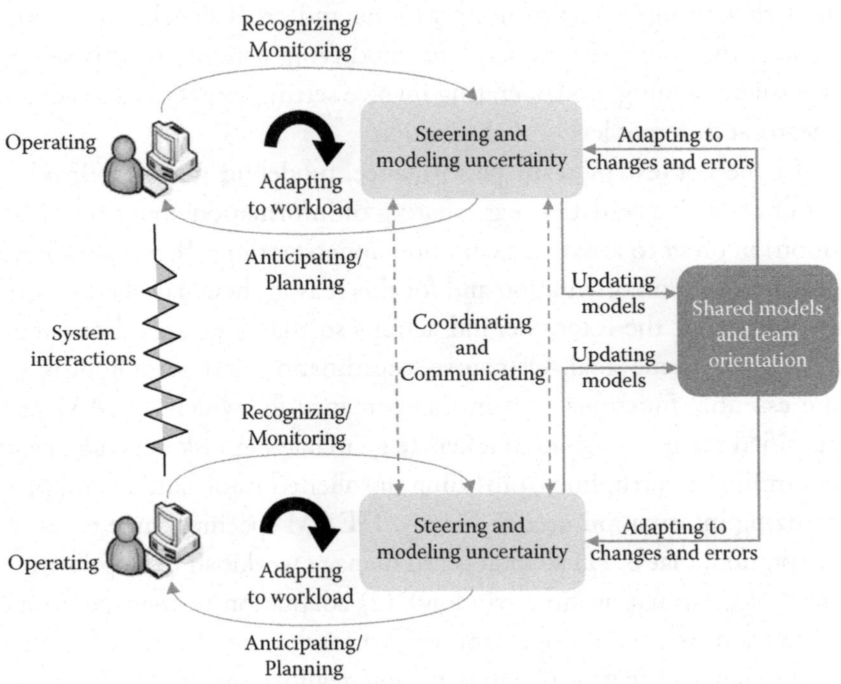

Figure 13.1 Revision of taskwork and teamwork functions (T²EAM).

Table 13.1 Revision of Taskwork and Teamwork Functions (T^2EAM)

TASKWORK AND TEAMWORK FUNCTIONS	EXAMPLES
Shared modeling and team orientation *(M)*	(e.g., shared understanding, common orientation, intent communication)
Steering *(S)*	(e.g., goal setting, specifying criteria of success, making trade-offs)
Modeling of uncertainty *(M)*	(e.g., testing data for completeness and reliability, critiquing models of the situation)
Monitoring/Recognizing *(M)*	(e.g., noticing cues, recognizing states, and receiving feedback)
Anticipation/Planning *(P)*	(e.g., acknowledging threats, staying ahead of traffic, contingency planning)
Adaptation to workload, errors, and changes *(A)*	(e.g., prioritizing tasks, reallocating tasks, detecting problems, cross-checking, supporting others in error recovery)
Coordination and communication *(C)*	(e.g., managing task dependencies and interruptions, providing unsolicited information, prioritizing data)
Operating *(O)*	(e.g., cross-checking, resuming actions, modifying actions)

steering new efforts in replanning. Finally, operating is another function that relates to performing actions and cross-checks, resuming actions that were interrupted, and modifying actions. In this sense, steering, planning, and operating involve setting targets and selecting means at different levels of abstraction.

In the context of team performance, modeling may also involve other collective activities (e.g., sharing of information about the situation) in order to arrive at a common understanding. Teams also need to set a common orientation and for this reason they may share information about the intent behind actions so that they are all oriented toward the same goal. Of course, coordination and communication are essential functions of team performance for which T^2EAM has specified many behavioral markers (e.g., managing task dependencies, controlling interruptions, providing unsolicited information, and prioritizing information needs). Finally, T^2EAM specifies three types of adaptation, that is: (1) adaptation to manage workload (e.g., prioritizing tasks, estimating time window), (2) adaptation to manage errors (e.g., team cross-checking, team support for recovery), and (3) adaptation to changes (e.g., problem detection, reallocation of tasks).

An important issue here regards the transfer of control between functions in order to match the changing nature of situations. Goals, objectives and action targets interact in order to match the context of work and the personal style of controllers. In an unfamiliar situation; for instance, controllers may spend more time in steering and modeling before planning. In a familiar situation, they may become more tactical by focusing on existing plans for controlling the situation. This transfer of control is also particular to the work styles of controllers and is referred to as vertical transfer. In contrast, horizontal transfer of control can be made between different controllers at shift handover, between controllers and automation, as well as between different teams working on adjacent sectors.

With regard to work style, a controller can adopt an explorative style that enhances interaction with the environment, or can adopt a more disciplined style and adhere to formal practices. In less busy times, for instance, en-route controllers may de-activate the plot filters to increase the coverage of radar in order to display all data derived from the radar sensors and explore the full picture of the airspace (Malakis et al. 2010a). Although deactivating the plot filters is not a standard practice, as it may give rise to radar screens cluttered by false targets, experienced controllers are in a position to discriminate valid targets and buy time for planning ahead.

It is also worth noting that a current function may be interrupted in favor of another. For instance, a controller may suspend monitoring in order to resolve a traffic conflict at the operating level. In the same way, a controller can suspend an operation in order to monitor and assess the situation or the traffic load in the next two hours. This suspension and resumption cycle may work as a switch between proactive and reactive control, depending on the context of work and the practitioner style. At any time, controllers must make judgments whether to monitor the situation and see how it develops or else make an early action that might turn out to have been needless. If they respond to every traffic conflict, they are proactive but overloaded, whereas if they respond only to definite conflicts, they are reactive and may lose the overall picture. In busy periods, controllers tend to become more conservative, resolving potential conflicts early in order to conserve attentional resources (Loft et al. 2007).

13.3 A Performance Model of Organizational Work

A similar control model can be applied to organizations with many hierarchical levels of control, according to the viable system model (Beer 1985; Espejo and Harnden 1990). Organizations that operate complex systems have to make trade-offs between conflicting goals, such as safety, production, delivery times, and utilization of capacity (System 5, policy). This brings to the fore the role of organizational models that constitute the deepest set of beliefs about how the world works, about potential hazards, and about perceptions of organizational capabilities (System 4*, mental models). Other meta-level VSM functions include an intelligence function (System 4) that helps organizations adapt to changes in environmental demands. Safety goals are passed onto the supervisory level and are transformed into resources and plans for action (Figure 13.2).

To assess the adequacy of safety plans, a feedback loop is established back to the higher management levels. The control functions of planning and monitoring correspond to System 3, which is concerned with ensuring cohesion between diverse operational units. An informal function allows supervisors to get a direct view of the system (System 3* function or management by walking around). Complex systems require cooperation of multiple controllers, which raises many important issues with regard to the delineation of responsibilities, reconciliation of different views, and communication among team members. In this sense, control is the result of coordinating multiple loops, involving different actors and artifacts (System 2, coordination). At the sharp end, operational practices adapt safety plans to variations in the environment by making use of available resources and safety barriers (System 1, operations).

According to VSM (Figure 13.2), viable organizations require that the five control systems operate properly both on an individual and a collective basis (i.e., their links and interactions work for the overall organization goals). For instance, operational units and their activities should be represented explicitly in the management structure so that shared resources, time requirements, and conflicts with other activities are carefully addressed in their operation; sometimes, activities are not done properly because they have been missed in the management structure. Similarly, System 4 supports building a mental model of a complex situation; inappropriate functioning may involve

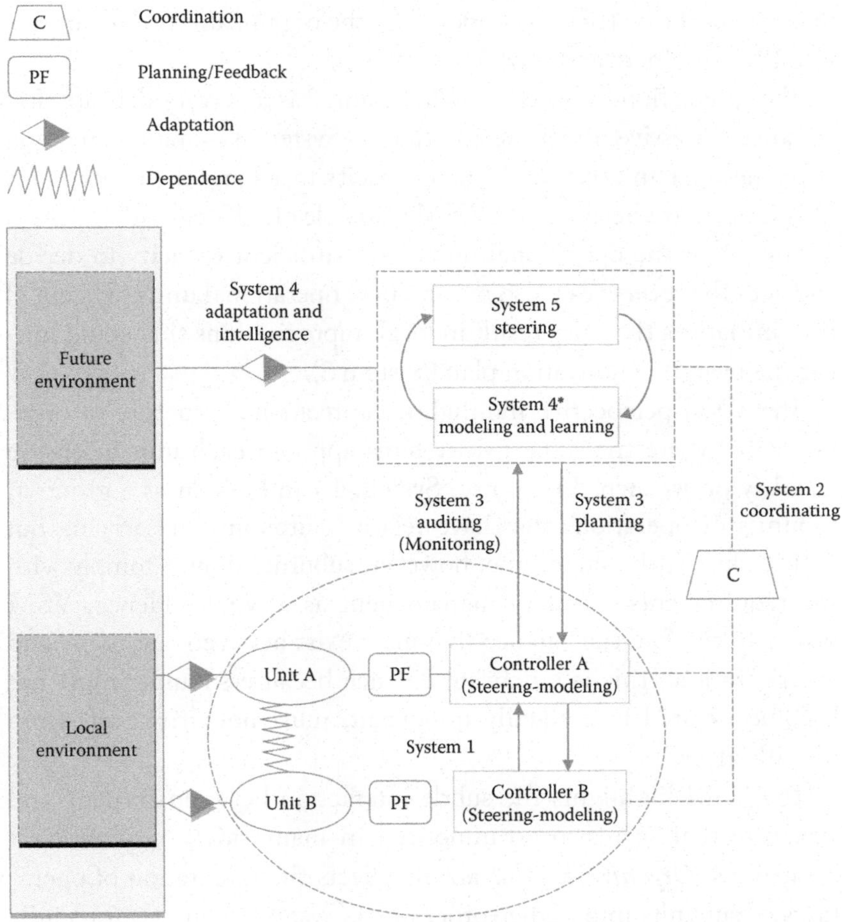

Figure 13.2 A variant of VSM for organizational processes in system safety.

squeezing the situation into an existing model instead of trying to build a new model that captures the threats and opportunities.

The interactions between the five systems are also very important in creating viable organizations. In particular the interaction between Systems 3 and 4 is considered to be a vital thermostat in organizations since this interaction regulates how novelties, new targets, and new concepts are transmitted to supervisors as well as how restrictions and incidents are communicated to higher levels. We can find a symptom of dissociation in cases where System 4 perceives System 3 as being short sighted (i.e., failing to see anything beyond the immediate), or in cases where System 3 perceives System 4 as being unrealistic and unaware of the restrictions imposed by the organization's daily operation (i.e., the "here and now").

The interaction between Systems 1 and 3 is also very delicate. For instance, excessive direct intervention by System 3 in matters relating to operational units can limit their capacity to act and their autonomy. This is in contradiction to the VSM principle of self-regulation, which requires that the operational units have sufficient capacity to decide and act. The reverse can also occur where operational units (System 1) take initiatives that may result in local improvisations that could hinder the overall organization plan (System 3).

The VSM perspective highlights the recursive structure of organizations where the same five systems apply to each unit to ensure viability on its own. In a sense, System 1 can be seen as a group of subunits or operations that have relative autonomy in carrying out their tasks. At the same time, however, subunits should comply with the requirements of safety management as a whole. Hence, VSM brings to the fore the balance that must exist between autonomy and centralization. This is a delicate balance because subunits must not become isolated but, equally important, must not drift away from overall safety.

The VSM highlights the subtle interaction between strategy and structure that seems to go unnoticed in many safety improvement campaigns. *Organizational structure* reflects the integration of operations or subunits into higher-order units (System 1) and their coordination (System 2). *Organizational strategy*, on the other hand, refers to the managerial functions of Systems 3, 4 and 5, which determine how organizations control their operations and adapt to the environment. Strategy should be built up throughout the organizational structure where the interests of individuals and teams at different levels are equally reflected in decision-making. Failure to do so may result in a safety campaign that may not succeed to reach certain parts of the organization.

Finally, the VSM looks into how organizations adapt to the environment. System 4 plays an important function in scanning the environment for threats and opportunities so that new concepts of operation and safety are explored; it addresses the process of organizational learning and change by maintaining a balance between exploration and exploitation.

Interactions between the management, the operations, and the environment can be modeled as a process of managing complexity or balancing variety. In organizational terms, managing complexity ensures that the capabilities of a management unit are sufficient to deal with the complexity of the operational problems which they have to deal with. Organizations can succeed in managing safety by amplifying their capabilities or by attenuating complexity in the environment (Figure 13.3). *Variety* is a cybernetics term referring to the number of states that a system can enter, the range of skills, and the amount of resources required to control the systems, and so on. On the other hand, variety can also refer to the capabilities of organizations, such as the range of skills, resources, and plans that can be made available in a particular situation.

Flights may encounter a range of critical conditions (e.g., adverse weather, gusting crosswinds, and ill-equipped airports) that amplify the complexity of the environment. An organization can attenuate the complexity of the environment by several such means as improving airport facilities, minimizing delays, and providing radar services. On the other hand, certain factors may amplify or attenuate the capabilities of the organization to respond to critical situations. For instance,

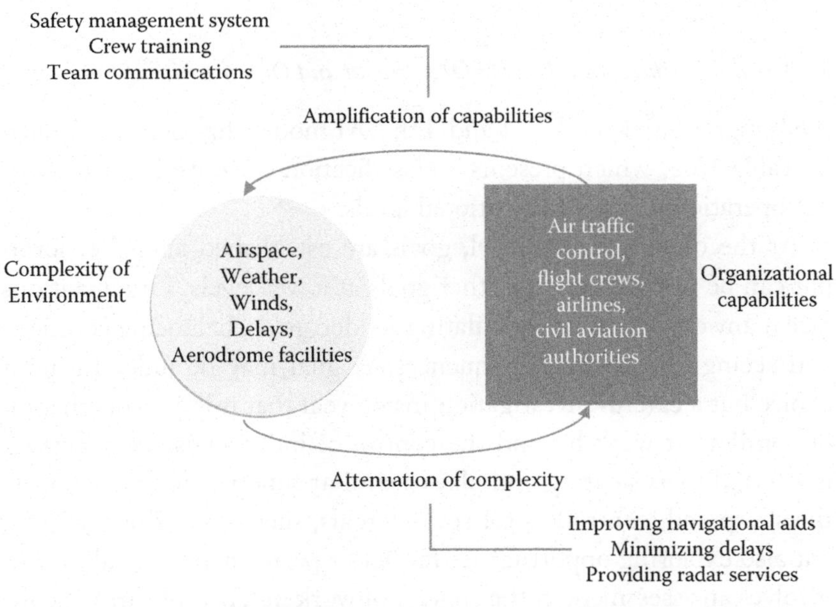

Figure 13.3 Amplifying own capability and attenuating complexity in the environment.

inadequate safety management systems and improper training may attenuate the capacity of organizations, while building on team communications can amplify their capacity and readiness for safety. In this sense, amplification and attenuation of variety are two important processes of adaptation that should be modeled in the representation of the system dynamics involved in near-misses and accidents.

13.4 Toward a Joint Model of Human and Organizational Performance

Systems thinking steers the investigation of accidents into the hierarchical control structure of organizations. According to Leveson (2004), problems in the structure and control of complex systems arise mainly due to control flaws, such as inadequate design of control algorithms, poor process models, inadequate or missing feedback, inadequate execution of control actions, and poor coordination. Although STAMP takes a systems approach, it remains neutral with regard to specific human and organizational models. A deeper analysis of causal mechanisms of control flaws could be made with reference to human and organizational models. Hence, this section considers a classification of control functions common to human and organizational performance followed by the issue of organizing performance in time.

13.4.1 A Classification Scheme of Operational and Organizational Functions

An integration of the VSM and T^2EAM models has been attempted in Table 13.2, which presents a classification of control functions at the operational and organizational levels.

At the organizational level, goals are established at higher levels but can be in conflict with other goals at lower levels. Organizations face many challenges in articulating hidden goals, balancing conflicts, and seeing long term consequences. A goal may be judged a poor choice but a careful investigation may reveal that it is a reconciliation of conflicts at work beyond the control of individuals. Goal setting is referred to as steering and plays a similar function to System 5. At the sharp end, controllers set several goals, such as controlling risks, but also exploring opportunities for better performance. Goal setting involves an assessment of the context of work in terms of threats, constraints, and resources. Sometimes, explorative goals may be taken to

Table 13.2 Control Functions at the Organizational and Operational Levels

	VSM ORGANIZATIONAL CONTROL FUNCTIONS	T²EAM TASKWORK AND TEAMWORK FUNCTIONS
Steering	Organizations face many challenges in articulating hidden goals, balancing conflicts, and seeing long term consequences. Steering plays a similar function to System 5 in balancing exploitation and exploration.	Goal setting involves an assessment of the context of work in terms of threats, constraints, and resources. Sometimes, explorative goals may be taken to see how the system reacts.
Modeling	All managerial interventions are associated with a mental model of potential hazards, barriers, and risk control strategies. Mental models help managers and controllers remain vigilant to the possibility of failure and provide a useful resources for the intelligence function of System 4 to ensure adaptation.	When information is incomplete or delayed, mental models are useful in filling gaps in understanding and testing hypotheses about causes and plausible effects. In dynamic situations, poor decisions are made when models remain outdated.
Adapting	Adaptation to environment and coordination between running and changing the system are functions related to System 4. Adaptation is also related to the ability of practitioners to learn from the past.	Adaptation usually takes the form of flexibility in changing behaviors between alternative modes of operation—for example, tight versus loose plans and feedforward versus feedback control modes.
Planning	Control algorithms or rules should be designed according to a safety plan that specifies the sequence of actions, the slack that exists, and the degrees of operator freedom (System 3).	Control algorithms may take the form of procedures or controllers' own plans. In both cases, they are evaluated by a process of mental simulation whereby their effects are projected into the future.
Feedback-Monitoring	Difficulties in information handling in System 3 may relate to the nature of the information, the observer, or the environment. Organizations often create assumptions about what is valuable information and what can be ignored.	Monitoring and auditing become more difficult in cases where feedback may be late or inadequate, warning signals are weak, noise due to irrelevant data is high, there is a large disconnect between causes and effects, and so on.
Cooperating	Cooperation (System 2) refers to the delineation of responsibilities and reconciliation of different views and decisions. When coordination crosses organizational boundaries, people may not be able to see how their actions affect others or may not be motivated to do so due to a silo mentality.	Coordination is vital for splitting difficult jobs and understanding complex situations. Reallocating tasks and plans can be useful in managing periods of high workloads. Coordination can increase cross-checking and enhance error detection and recovery.

(continued)

Table 13.2 *(Continued)*

	VSM ORGANIZATIONAL CONTROL FUNCTIONS	T^2EAM TASKWORK AND TEAMWORK FUNCTIONS
Operating	Safety policies and plans are implemented at the sharp-end where controllers interact with the process and the work environment (System 1). To cope with work demands, controllers often fall back on experience and rely on inappropriate habits that seemed to work in the past.	Implementation refers to manual and tracking activities necessary to achieve the action targets set in the plans. Tracking activities should respect the constraints of space and time in the work domain. Implementation relies on feedback control where target-outcome gaps are corrected in time.

determine how the system reacts; however, this sort of learning may appear unacceptable error to external observers.

All managerial interventions are associated with a mental model of potential hazards, accident causes, supporting affiliations, and risk control strategies. Mental models help managers and controllers challenge their understanding and remain vigilant to the possibility of failure (System 4*). They are important for the intelligence function of System 4 to ensure adaptation. A shared understanding of the situation is useful when it includes knowledge about the current system state and awareness of the information needs and expectations of others.

Organizations operate in an open environment where teams adapt their processes to meet changes in work demands, reforms of organizational resources, and updates of technology. Adaptation to environment and coordination between running and changing the system are functions related to System 4. Failure to meet these challenges could give rise to maladaptive patterns, such as unsuccessful improvisation, increased communication workload, and inability to tolerate uncertainty.

Planning refers to control algorithms or rules that usually take the form of procedures at the sharp end. In most cases, plans are evaluated by a process of mental simulation whereby their effects are projected into the future. At the organizational level, planning corresponds to a safety plan that specifies the sequence of actions, the slack that exists, and the degrees of controller freedom (System 3). In this sense, an organizational practice or procedure is seen as an algorithm with specific features such as level of granularity, degree of freedom, and temporal constraints. Monitoring refers to evaluating work progress made with regard to goals set by controllers, while auditing refers to supervision of activities by senior personnel. Monitoring and auditing become more difficult in cases where feedback may be late or inadequate, warning signals are weak, noise due to irrelevant data is high, disconnects between causes and effects, and so on. Organizations rely on System 3 to integrate planning, monitoring, and auditing in order to facilitate the control of operational units.

Coordination is vital for splitting complex jobs and understanding system goals and action constraints across units. Team members trying to follow a request have to figure out what the other team member really wants and to handle several issues not explicitly specified or explained (Klein 1998). In this sense, clarification of the intent may increase team coordination at the sharp end, without the need for further authorization from the leader. Organizations use System 2 to perform a delineation of responsibilities and reconciliation of different views and decisions. When there are multiple controllers, decisions may be inadequately coordinated, giving rise to communication errors, unexpected side effects, and conflicting actions. When coordination crosses organizational boundaries, practitioners may not be able to see how their actions affect others or may not be motivated to do so due to a silo mentality.

Safety policies and plans are implemented at the sharp end where controllers interact with the technical system and the work environment (System 1). To cope with work demands, controllers often fall back on experience and rely on habits that seemed to work well in the past. Sometimes, assumptions are made about technical or human processes that are not checked or are suddenly violated; practitioners should be alert to any countersigns or exceptions that make plans unattainable to the current situation.

13.4.2 Human and Organizational Performance in Time

In many dynamic environments, the performance of practitioners and organizations is affected by temporal changes of the situation and by delays in carrying out their control functions. In the former case, practitioners cannot wait to make decisions until they feel ready to do so since they have to respond quickly to changes in the situation. In the latter case, control functions need information, assessment, and coordination that take time to process. For instance, feedback from the flight crews may be delayed due to an evolving situation that captures the attention of crews. Reaching a decision may also take time in order to integrate information from different sources to make sense of the situation. As the situation and the performance change over time, it is important to postulate how controllers organize their cognitive functions over time. It may be argued that steering, planning, and operating may take place in a linear sequence when working in familiar situations. In unfamiliar situations, however, these functions are performed in cycles or loops, as steering could lead to planning and operations and then back to modifications in steering until the goal is achieved. To understand why practitioners and organizations operate in cyclical loops, it is important to consider how time affects performance. The literature on dynamic decision-making (see Brehmer 2010) has identified several temporal characteristics of work as follows:

1. *Feedback delays* about the effectiveness of the actions taken by controllers. This requires practitioners to be fully aware of how long things actually take and be able to translate this into a prediction of when to look for feedback. In general, systems that have long or variable information time require a feedback-based mode of control so that decisions are implemented incrementally.
2. *Decision time* to assemble all information to make sense of the situation (i.e., time to modeling) and then set some orientation and criteria how to solve the problem (i.e., time to steering). Different control modes have different decision time requirements. For instance, feedback control relies on quick decisions that allow controllers to collect additional feedback and make corrections. In contrast, "feedforward" control takes longer as controllers need to wait for extra information until they are confident of their decision.
3. *Planning time* required to think of a plan that satisfies the decision criteria set in the process of steering. At busy times, for instance, when the criterion of expedition is not viable, controllers may create aircraft sequence plans that emphasize safety over expedition. Planning time is considerably shorter at lower levels in the organization since controllers have actual contact with the operating environment. At higher levels, this

contact with the operations is more remote and this increase planning time.

4. *Coordination time* between teams working on the same or different airspace sectors. At the organizational level, different organizations and units may need to coordinate until they manage to agree on a general plan, which takes additional time (e.g., coordination in managing emergencies).
5. *Dead time* between the moment when a plan or decision has been made and that when it takes effect. This delay is usually short as applications of technology allow controllers to start acting out their plans soon after a decision has been made. In contrast, the time required to implement a plan at higher levels in the organization is much longer as it requires several preparations.

This discussion indicates that the seven control functions in Table 13.2 should be organized in a performance control loop. The dynamics of performance loops are specific to the type of problems encountered, the preferences and decision styles of practitioners, and the organizational level at which performance is studied. For instance, routine problems may be amenable to linear thinking (i.e., feed-forward control), while others may present more challenges and require an iterative processes to understand and solve the problem (i.e., feedback control). In addition, the greater uncertainty and the longer information delays at the higher organizational levels may require different decision strategies than those required at the sharp end. For these reasons, it is very difficult to propose a unified performance model that would specify not only its control functions but also its dynamics in solving different problems. In this sense, the proposed classification of the control functions caters for the structural aspects of performance, while some insights for understanding its dynamics can be gained from the presentation of systems thinking approaches (Chapter 11) and resilience engineering (Chapter 14).

13.5 Modeling Patterns of Breakdown Resulting in "Loss of Control" Events

The joint model of human and organizational performance can be used to look into potential breakdowns that create flaws in the control of events. Being in control goes beyond keeping key parameters within limits, and includes recognizing weak signals, monitoring work progress, revising plans, and adjusting coordination. In a team environment, practitioners have to coordinate in order to build a shared understanding and align their goals across different roles. Furthermore, control and coordination have to be adapted to match changes in work demands and new interdependencies between tasks. In this sense, controllers and supervisors have to maintain control of events by adapting their cognitive and organizational functions. It is worth considering, therefore,

Table 13.3 Patterns of Performance Breakdown Leading to "Loss of Control" Events

FUNCTION FAILURES	PATTERNS OF PERFORMANCE BREAKDOWN
Steering failures	• Goals may be conflicting, poorly articulated, or remain untested to particular cases. • Goals focus on firefighting while systemic solutions remain unattended • Goals may be resistant to change (e.g., failure to estimate correctly the cost of change). • Micro-managing (e.g., drifting into lower level goals).
Poor planning	• Adhering to procedures may create double binds that result in defensiveness and violations. • Plans may be overly detailed and, hence, difficult to follow or may restrict evaluation of work progress (i.e., no time slack to revise plan). • Errors in the mental simulation of a plan to respond to a threat.
Modeling failures	• Simplification of complex situations may lead to incomplete or "buggy" mental models. • Models are not updated properly (e.g., poor feedback, delays, masking effects). • Models may be resistant to change and learning.
Poor monitoring	• Information handling problems due to attentional dynamics (e.g., distractions, habituation to nuisance alarms, and problems in setting priorities for search). • Ineffective monitoring of unsafe conditions and risk factors due to flawed mental models. • Untested assumptions about what is valuable information and what can be ignored.
Adaptation failures	• Inability to transition to new modes of work and recovery. • Adaptation producing undesired consequences. • Tightly coupled plans restricting options, resources, and slack for replanning. • Restricted margins of maneuver to handle future demands.
Poor coordination	• Coordination over boundary areas remains poorly defined • Poor briefing and handover procedures. • Heedful interaction is hindered in teams, resulting in poor checking and recovery. • Intent is not clarified to allow shared understanding.

some potential breakdowns in the joint model of human and organizational performance that lead to loss-of-control events (Table 13.3).

13.5.1 Patterns of Steering Failures

A major source of complexity is the coupling of safety-related tasks that are coordinated in safety management. Although the practitioner's highest goal may be safety, there are also other goals to consider that can be less explicitly articulated. For instance, issues of cost, delayed outcome, and time pressure may encourage other types of human intervention that seem to cure only the symptoms of the problem. Some management teams may tend to diminish complexity by issuing high level and abstract goals that at first seem to cope with all eventualities but often are not well tested and contain hidden assumptions. Hence, too general or ill-specified goals may hide other subgoals that do not match up (Dorner 1996). Because there is a lot of pressure to demonstrate quick results, abstract goals often are degraded into a sort of quick fix that temporarily seems to remove obstacles at work and hence to reduce the size of the problem.

In contrast, systemic safety interventions often do not show immediate results, which creates a perception of programs being ineffective at least in the short term; this makes it likely for management to lose sight of their value in their potential to reduce accidents in the long term (Marais et al. 2006). As a result, safety efforts may be reduced to firefighting that relies on practitioners at the sharp end to undertake fast action and remove any work obstacles. Unfortunately, systemic causes may remain latent and create more work obstacles or reduce opportunities for error correction. An air crash in 1997 of a Yakovlev (YAK 42) aircraft of Aerosweet airlines near Thessaloniki, Greece, was attributed to a failure of ATC units to guide a misoriented flight crew lost in a mountainous terrain. Flight AEW-241 executed a missed approach but deviated from the published procedure in instrument meteorological conditions (Kontogiannis and Malakis 2012b). In the past, similar cases of lost crews were successfully handled by good controller communications and competent local crews who were able to compensate for the absence of an approach radar. However, the lack of a radar was an obstacle that continued to exist, providing fertile ground for human errors or lack of guidance that led to the AEW-241 flight incident.

A common source of errors in dynamic systems regards resistance to change as practitioners are required to repair a plan or come up with another plan while being under time pressure. Making changes to a plan requires understanding of decision trade-offs and costs of change (e.g., availability of resources in future, new consequences, and delays in communicating changes, etc.). In some cases, a more efficient plan may be found but teams might be reluctant to change because of time-consuming communications required to inform the affected team members. On the other hand, agreeing to a change in a plan without balancing some trade-offs may have repercussions for safety. In the American Airlines 965 incident over Cali, for instance, the crew accepted a last-minute proposal by ATC to land on a different runway without evaluating the timescale for making all necessary changes (Air Accident Investigation 1995). However, landing on the new runway increased workload and communication demands. In addition, opportunities for detecting problems were not explored (e.g., climbing with the speed brakes on), which prevented the crew from climbing up faster when they saw the obstacle and tried to change their flight route.

Micromanaging is another type of steering issue where a person gradually drifts into lower-layer activities (e.g., hands-on control), hence losing sight of his main supervisory responsibilities. Micromanaging has been a problem in many domains especially in cases where watch supervisors have progressed up the hierarchy from the operations room.

13.5.2 Patterns of Planning Failures

Aviation is a highly scripted environment where planning is carried out on the basis of formal controls (e.g., operating procedures and rules of conduct). In a sense, formal controls are systems of accountability enforced by organizations that, in some cases, are likely to produce conflicts and dilemmas. For instance, applying procedures to complex environments requires a balance between (1) under-adaptation (i.e., continuing with procedures and discovering that adapting them would have been a better choice) and (2) over-adaptation (i.e., adapting procedures in the face of unanticipated events but creating unanticipated side effects) (Dekker 2006). This policy can create a double bind because practitioners are held responsible for the final outcome but lack the authority to choose their plans and cannot control the design of their workplace and tools. The authority–responsibility bind could be a major source of hindrance in the work of practitioners (Woods and Hollnagel 2006). Hence, analysts should be cautious with reported errors (e.g., delays in making decisions, resistance in changing plans, and violation of procedures) because the authority-responsibility bind may be hidden in the accountability systems of organizations.

A common concern in planning regards its scale of detail or *granularity*. In general, planning is supported by procedures that tend to specify overly detailed plans since the usual safety practice in aviation has been to foresee and make allowances for every conceivable mishap. Overly detailed plans are difficult to rehearse and adapt because practitioners may have difficuly thinking about how plans may play out into the future. In some emergency procedures, a variation in one task can create cascade effects into other tasks because of connections and feedback loops that exist. This makes it difficult for practitioners to follow and adapt overly detailed procedures as they cannot anticipate what tasks will be affected and how they should respond. In general, mental simulation of complex plans is more prone to errors.

Marais et al. (2006) have used system dynamics to describe the procedural fix archetype where management fixes problems by specifying still more detailed procedures. Unfortunately, this increases complexity in following and revising procedures and presents fewer opportunities for improvisation. This results in failures to adapt and adaptations that fail (Dekker 2006), which does not solve the original problem and prompts management to introduce still more complex procedures; a reinforcing loop that eventually makes the problem worse. In dynamic environments, unexpected events occur, or task constraints change over time, requiring some degree of flexibility and decision latitude in the formal procedures.

13.5.3 Patterns of Monitoring and Modeling Failures

Understanding the way practitioners make sense of situations requires tracing out the cues that attracted attention or matched their expectations, the lines of thought that were triggered, and how mental models guided further exploration and action. In this sense, sensemaking involves data-driven monitoring and top-down guidance that is influenced by the mental models of practitioners.

Patterns of breakdown in monitoring appear to be of a mixed description. Some relate to the nature of the information itself and the nature of the tasks (e.g., information is ambiguous or buried in a pile of data, many activities are competing for attention, etc.). Others may relate to the mental models of practitioners (e.g., not accepting information that disconfirms expectations) or may involve the work organization itself (e.g., communication networks and organizational boundaries that impede the flow of information). Therefore, monitoring is affected by attentional dynamics and the properties of the information environment, the mental model of practitioners, and finally the organizational context of communication.

Attentional dynamics orient the mind where to find interesting events and know where to look next in the system. Attention is controlled by a perceptual cycle where salient events or data shift the focus of attention and call to mind relevant knowledge. The activated knowledge, expectations, or goals in turn guide further exploration and action. Several problems, however, can occur in the control of attention in a changing world, such as distractions, habituation to nuisance alarms, and problems in setting priorities for search (Woods and Hollnagel 2006).

Ineffective monitoring of unsafe conditions and risk factors can also be the result of flawed mental models. Practitioners need to continue to monitor for new information, look for new information to support or disconfirm their expectations, and follow up checking their actions. Therefore, mental models guide attention to search for interesting events or weak evidence that may be related to the current understanding of the problem. For instance, practitioners may miss weak signals of early threats, especially when signals are not related to their mental models. The remainder of this section focuses on patterns of breakdown in making sense of the problem due to flawed mental models.

In studying disasters, it is also important to pay attention not only to attentional dynamics and mental models but also to the distribution of information in the organization, to the communication networks, and to the boundaries that impede the flow of information (Turner and Pidgeon 1997). In this respect, organizational processes can affect the use of information by creating assumptions about what is given value as information, how it is to be communicated, and what is to be ignored.

Many patterns of breakdown relate to problems in managing complexity and flawed mental models. Errors in mental models are not uncommon as the complexity of the situation increases. Some simplification in terms of heuristics and workarounds may actually be beneficial in handling complex situations. However, there is a risk of oversimplification, where controllers generalize conditions in which plans apply, get stuck with plans that were successful in the past, or become unable to revise their understanding. Oversimplification usually results in failure to learn from earlier unsuccessful attempts as the situation unfolds.

In the context of aviation, Sarter and Woods (1995) found that "buggy" mental models contributed to problems in using cockpit automation. For example, the pilot might believe that the aircraft was in speed mode but the computer has changed the mode to "open descent" and as a result the pilot issues inappropriate commands. Unfortunately, areas of buggy knowledge remain hidden from practitioners because they have the capability to work around these areas by resorting to heuristics and workarounds. However, some unusual or novel situations may arise that reveal gaps or bugs in the mental models of how automated systems function in actual practice. For systems thinking, the critical issue is whether practitioners can recognize that their models are simplified for a particular situation and have the ability to use more complex models or integrate knowledge from different agents.

Apart from failing to start with an accurate model, problems in processing feedback, lack of feedback, or inaccurate feedback may give rise to incorrect mental models. Specifically, human-computer interfaces and artifacts may aid or hinder the process of model updating and learning. In particular, when the response time of the artifact is too long, it is likely that other people or supervisory systems may take additional actions whose outcomes can mask the consequences of the first action. The design of a device may obscure important states, or create the appearance of linkages between states that are not in fact linked. This makes it difficult to correct flawed mental models, especially when available time is limited.

Mental models can also be resistant to change or difficult to revise as new evidence becomes available. The initial situation assessment may seem appropriate, given the information available at an early stage, but practitioners or organizations may be reluctant to revise their mindsets in response to new evidence. Unfortunately, resistance to change may lead to failure to learn from previous experiences. An example can be taken from helicopter emergency medical services (HEMS) where the aviation regulatory authorities failed to revise their mental model of the way that HEMS services were managed by HELITALIA, a private company in Greece. In the period 2001–2003, three fatal accidents occurred involving HEMS helicopters carrying patients from remote Greek islands to Athens. HEMS services had a very safe record prior to 2001 because they were undertaken by a consortium comprising the Hellenic Army, Navy, and Air Force in close cooperation with the national airline Olympic Airways (OA). The aviation regulatory authorities in Greece failed to see the new evidence from HELITALIA operations suggesting a degraded HEMS capability because their crews had no experience in adverse weather conditions in the Aegean sea, human resources were inadequate, the fleet did not include any jets that could fly above bad weather, while helicopters were ill-equipped in their capabilities for HEMS services. At the regulator level, no one anticipated the speed and severity of the escalation pattern of the three accidents; each one was perceived as a single nonrepeatable case of unwanted events (Kontogiannis and Malakis 2012a).

13.5.4 Patterns of Adaptation Failures

Adaptation breakdowns refer to difficulties in adapting modes of functioning when control breaks down. Hollnagel (2009) proposed that adaptability can be seen as the ability of practitioners to match their modes of functioning to different contexts of work. Modes of functioning can be seen as a space of fundamental trade-offs such as efficiency-thoroughness, feedforward-feedback control, centralization, local improvisation, implicit-explicit coordination, and so on. Woods and Branlat (2010) summarized the main challenges to practitioners as follows: Do practitioners know where they are positioned in the trade-off space? Can they assess whether they have adopted the right mode of functioning? How do they know what cues provide early indicators of the need to shift to another mode? Failure to meet these challenges can give rise to maladaptive patterns such as local improvisation, increased workload in communications, and errors of fixation.

Tightly coupled tasks work well for efficiency but allow little scope for absorbing disturbances and revising plans. In high-tempo situations, practitioners may have to switch to more modular or looser plans that afford more opportunities for recovery (e.g., more redundancies, more options, and more time to recover). Woods and Branlat (2010) referred to a similar adaptive behavior that has to do with how practitioners manage the capacity to handle future demands or contingencies (e.g., sustain resources, maintain redundancies, redistribute resources, and so on). Being in control is the ability to assess how margins of maneuver change relative to uncertainties and the potential for surprise. Adapting margins for maneuver requires the ability to anticipate when the system is exhausting its adaptive capacity before it collapses in a failure. Problems in anticipating and managing the margin for maneuver can give rise to what has been termed as going solid in a system or all hands tied up (Woods and Branlat 2010).

13.5.5 Patterns of Poor Coordination

Other patterns of breakdown are observed in complex systems that require cooperation of multiple controllers, sometimes at different scales in the organizational hierarchy. In cooperative systems, control is the result of multiple interacting actors who try to share information, reconcile different points of view, and align their subgoals to avoid any side effects from working together. Loss of control and poor coordination can be the result of inadequacies in technological, organizational, and social systems. For instance, an analysis of the shoot-down of U.S. Army Black Hawk helicopters over Northern Iraq by U.S. fighter aircraft (Leveson 2004) indicated that the Air Force had upgraded their radio technology while the Army had not; in other words, there was an asynchronous evolution in communication media whereby changes in one part of a system were not followed by changes in other related parts. In the same incident, coordination over boundary areas was poorly defined, leading to further confusion over who was actually in control. This risk of unclear boundary areas may increase as the team structures are joined higher in the hierarchy (Leplat 1987).

Another pattern of breakdown refers to the transfer of control between teams in different sectors. A poor ATC briefing at transfer of control between air sectors was one of the causes that led to the late detection of the problem facing the flight crew of Helios Airways in the crash of flight HCY522 (AAIASB 2006). The Cypriot ATCOs were suspicious of a problem facing flight HCY522 because the flight crew would not reply to their repeated attempts to make radio contact with them; additional attempts to make contact with the flight crew via another aircraft also failed. Although the Cypriot controllers were very concerned about this problem, they did not communicate their suspicion and assessment of the problem to the Greek controllers during their briefing at the transfer of control from the Cypriot to the Greek airspace. Instead, their briefing focused on some hints that the crew did not answer their call, which may well have indicated that the crew was busy with their own flight actions. No information was given at the transfer point about the persistent efforts of the Cypriot controllers to make contact with flight HCY522 in the past. The Cypriot controllers were very concerned about this issue and continued to make additional contact with both the flight crew and the Greek controllers, although the aircraft was outside their area of responsibility. This failure to communicate the team's stance and suspicion at the transfer point was a contributory factor to the failure of the Greek controllers to detect the problem facing flight HCY522 at an early stage. The controllers did not suspect any problems as flight HCY522 was programmed to pass precisely through all waypoints of the Greek airspace.

Finally, coordination can be supported when intent is communicated to team members so that misunderstandings and assumptions are managed. When intent is not clarified, it becomes very difficult for controllers to understand how hard others are trying to control a problem to provide assistance. Moreover, difficulties in deducing the "intent" of others from observable actions may deprive controllers from any opportunities to detect errors of others in a team environement (Kontogiannis and Malakis 2009).

13.6 Concluding Remarks

The proposed joint VSM-T^{2}EAM framework relies on a recursive representation of the safety organization that helps analysts model new information loops, identify new constraints, or see problems in the adaptation and steering functions of the organization. Patterns of performance breakdowns have been derived from the integration of two models at the human and organizational levels (i.e., T^2EAM and VSM). In particular, T^2EAM can help us examine how controllers coordinate, build a shared understanding, and align their goals across different roles. In a supplementary role, the Viable System Model can help us go deeper into the control functions of the safety organization, such as by recognizing weak signals, managing goal conflicts, and monitoring progress toward goals.

The recursive representation of the joint VS –T^2EAM framework is worthwhile because it can be used in a generic fashion for the analysis of many adverse events within the same safety organization. A recursive VSM analysis focuses on the organizational structure rather than on single actions and events; as a result, the VSM analysis is usually performed for the organization of safety management and can be used for the investigation of many accidents in the same organization. Once a VSM analysis has been performed, it can provide a useful basis for the analysis of many near-misses and accidents. Likewise, Santos-Reyes and Beard (2009) presented a systemic SMS that can be used for the analysis of several incidents in the oil and gas industry. With the shift toward systemic models of safety, there is also likely to be a greater degree of transfer of knowledge across different industries that are assigned to the same field of practice.

The proposed VSM-T^2EAM framework allows safety analysts to bring together several organizational and control issues that relate to self-regulation and the interaction of structure and strategy. The latter is very important in the management of change since many failures to improve a safety policy can be traced to problems of taking on board the views and interests of practitioners throughout the organization. The joint framework also addresses how practitioners and organizations adapt to challenges in their environment by adapting their modes of functioning and recovery when control breaks down. Modes of functioning can be seen as a space of fundamental trade-offs, such as efficiency, thoroughness, centralization, local improvisation, tightly-loosely coupled plans, and so on. Finally, adaptation can be seen as a form of organizational learning from previous near-misses and incidents.

An important issue in the elicitation of performance breakdowns in incidents regards the standards relative to which performance falls short in some way. External observers and analysts usually make reference to formal systems of work (i.e., safety rules and operating procedures) as standards to spot deficiencies in the actual problem-solving process. Unfortunately, standard practices and procedures provide very weak criteria for defining errors and performance breakdowns because they are underspecified, cannot cope with all eventualities and tend to underestimate conflicts and constraints (Woods et al. 2010). Assessing poor practices is a difficult task for safety analysts because a variety of standards and criteria should be consulted. Dekker (2006) has proposed that neutral-practitioner criteria can also be used to examine what other practitioners would have thought or done in similar situations. The purpose of peer expert opinion is to help define the envelop of appropriate behaviors and capture the goal conflicts and trade-offs present in the actual workplace. This may help analysts understand why it made sense to people to do what they did in a specific situation.

In this sense, the performance breakdowns (Table 13.3) should be seen as the result of local rationality that is bounded by system constraints, goal conflicts, and trade-offs between alternative ways of performing the control functions. For example, commercial pressures and conflicts may cause problems in steering, while trade-offs in planning may affect work scheduling. In addition, typical ergonomic problems with the CNS systems may deteriorate system monitoring while past expectations and perceptions of threats may influence the modeling of critical situations. This implies that analysts should examine whether controllers take past successes as a guarantee of future safety, whether they keep a discussion about risk alive even when everything looks safe, and whether they remain open to fresh perspectives on a problem (see also Dekker 2007).

Another aspect of adaptation is the possibility of recovering from unsuccessful actions and learning from earlier performance. This is very important because complex systems require many efforts to manage the problem. Therefore, analysts should not take for granted that an unsuccessful plan is an error because controllers may experiment with the system and learn from their experience; it is important that analysts look into the opportunities for error recovery created by the controllers themselves.

Overall, the joint framework focuses on the control functions in the safety organization that are necessary to adapt to changes in the environment and disturbances in the technical system.

14

Organizational Decision Making in Managing Work Trade-offs: A Resilience Approach

14.1 Introduction

A central theme of human performance has been how practitioners balance work demands and capabilities or resources (e.g., available tools, manpower) in different ways that match the characteristics of work situations. In today's hectic workplaces, very often work demands exceed resources so practitioners have to do their best and manage their work by adjusting their practices. In this sense, they are trying to maintain a continuous balance between demands and resources. This theme has received particular emphasis in resilience engineering by Hollnagel's (2009) proposition of the efficiency-thoroughness trade-off (ETTO) principle. The effort to tailor human performance to work demands can be described as if it involved a trade-off between efficiency and thoroughness.

The present chapter discusses a contemporary view of organizational safety that emphasizes the decision dilemmas or performance trade-offs faced by practitioners and managers in everyday work. A number of studies have recognized practitioners' dilemmas in balancing productivity and safety requirements at the same time (Morel et al. 2008; Gomes et al. 2009; Amalberti 2013). Practitioners have to reach their safety and productivity goals, neither of which should be achieved at the expense of the other. Organizations are also seeking ways to preserve their level of economic performance without degrading their safety margins. As industrial systems grow in complexity, their work demands also increase; for example, do more, faster, cheaper, and better. Organizations try to amplify their own capabilities to control complexity (the viable system model [VSM] principle of variety) by engaging more practitioners in the system (that is, ensure a multiplicity of perspectives), by balancing tasks between practitioners and automation, and by delegating authority to multiple levels in the system. However, this increase in the capabilities and the operating modes of the organization may also create several decision trade-offs, such as multiplicity of views vs. coordination cost, manual vs. automated control, and centralized vs. decentralized control. As a result, many decision trade-offs are created at both the operational and organizational levels that should be resolved in the context of the particular situation. Organizational life is full of small or large decision trade-offs that must be balanced so that the final decision suits the requirements of the problem at hand.

Studies in naturalistic decision-making have shown that the same heuristics used by experts to adapt performance to work demands can be seen by external observers as biases that result in errors. Trading off formal procedures and heuristics involves a comparison along a number of criteria (e.g., resources, time constraints, cost, rewards, and so on) some of which may not be well articulated (e.g., supervisory pressure to get the job done). In hindsight, when the outcome is undesirable, a specific resolution of a trade-off may be labeled human error but the real problem may rest with hidden or tacit criteria for choices, managerial pressures, or peer rewards. The implication for safety management is that unsuccessful outcomes are not always due to errors and malfunctions, but mostly due to hidden criteria, work constraints, unavailability of resources, and conflicting requirements that are built into the design and operation of the work system. Hence, safety management should focus on how to make the system more controllable either by making it less tightly coupled and complex, or by ensuring that more resources and means are available to support practitioners in their jobs.

This chapter attempts to apply this resilience approach to human and organizational performance. To this end, the seven cognitive functions described in T^2EAM and VSM models are revisited in order to consider the decision trade-offs that make performance challenging.

14.2 Human and Organizational Performance in Balancing Work Trade-offs

Managing trade-offs can be seen as a critical aspect of performance in the way that practitioners monitor the situation, make decisions, develop plans, and organize their authority delegation and coordination.

In monitoring a situation, for instance, controllers have to decide how long they should keep gathering information to understand the problem before making a decision on how to act. For example, how long to keep gathering information about a communication navigation surveillance (CNS) system failure before applying contingency plans. In the same way, developing a plan requires controllers to trade-off resources spent on preparing for an activity versus resources spent on doing it. In work organization, supervisors may face similar trade-offs with regard to authority structures. For example, a decentralized structure may increase flexibility but comes at a high cost of training and coordination. Maintaining a balance between different authority structures could also introduce other side effects. For instance, a study of authority structures in the cockpit found that captains who adopted a flexible style of centralized and decentralized authority caused some confusion as the rest of the crew were unable to predict which style the captain would adopt next (Helmreich et al. 1998); additional training was necessary so that the flight crews could improve their coordination skills.

The ETTO principle is a useful framework for looking at how practitioners manage their trade-offs in the workplace. To cater for efficiency, practitioners generally try to achieve their goals by keeping their efforts and resources as low as possible. Hollnagel (2009) asserts that the decision how much effort to spend is usually not explicit but rather a result of perceived demands, habits, social norms, and established practice. A controller, for instance, may think that a plan may be good enough for a particular traffic scenario since it meets certain requirements. An operational supervisor may face a similar dilemma in the decision for band-boxing or splitting a sector given certain traffic conditions. Thoroughness, on the other hand, requires that more time and resources are spent on ensuring that the necessary work conditions are in place, so that production goals are achieved without any risks. Thoroughness implies that practitioners spend more time thinking about whether preconditions for their activities are met, execution conditions are right, and preparations for contingencies are made in advance. In most cases, it is not possible to maximize efficiency and thoroughness at the same time. Hollnagel (2009) argues that a work activity usually requires a blend of efficiency and thoroughness in order to succeed. In this sense, balancing goal trade-offs may involve blending the two properties, or choosing one and then reverting to the other when conditions permit.

A similar approach has been proposed by Kontogiannis (2010a) in explaining the variability observed in complex organizations. In this approach, organizational control usually takes the form of adaptation of organizational processes to a continuum between optimization and agility. Optimization focuses on hierarchical structures that opt for efficiency, optimized planning, and economical use of resources while agile structures opt for thoroughness in assessing problems and using resources. The two poles can also be described as minimizing uncertainty versus handling uncertainty (Grote 2009), since the former focuses on how to predict uncertainties and minimize their effects while the latter focuses on how to enable organizations to cope with uncertainty locally.

In any organization, practitioners adjust what they do to match the conditions of work (e.g., a work practice or an authority structure is given priority over other options). In most cases, balancing performance trade-offs may lead to successful outcomes but the same cognitive functions may lead to adverse outcomes in other circumstances. It is only in hindsight, when the outcome is wrong, that the resolution of trade-offs is labeled human error. The advantage of recognizing the trade-offs that practitioners face is that safety management can identify potential flaws in the system and consider more efficient ways of working.

Recent developments in the management of trade-offs have expanded the ETTO approach beyond the mismatch between demands and resources. Hoffman and Woods (2011) have specified five principles in managing trade-offs in view of a wider range of system requirements, such as adapting to unexpected events, incorporating multiple perspectives, managing goals at local and global levels, and delegating authority to many organizational levels. The five trade-off principles can be briefly described as follows:

1. *Optimality versus resilient adaptive capacity.* Complex systems cannot always predict their work environment and meet the situational demands fully; in some cases, there are gaps in fitness or abilities to respond. As a result, systems require a capability to adapt to rare or surprising events that may call for additional resources. In contrast, optimization requires that less time and fewer resources are spent on maintaining an adaptive capacity for unforeseen events. For example, an approach sector

is normally designed around standard arrival and departure routes that serve one or more airports. The structure of routes, standard instrument departures (SIDs), standard arrival routes (STARs), and instrument approach procedures (IACs), are optimized to cater for standard traffic flows. Work practices of controllers take into account these structures in normal day-to-day operations. However, sometimes the same sector may accommodate a large military exercise or serve as an en-route sector for a large formation of slow-moving visual flight rules (VFR) traffic that crosses the sector in nonstandard patterns. No terminal maneuvering area (TMA) structure can be optimized for both traffic flow conditions and this calls for an adaptive capacity to accommodate standard and nonstandard flows efficiently and safely into a sector.

2. *Efficiency versus thoroughness of work plans.* Practitioners often engage in a process of testing fit between plans and situations to decide whether they should follow a well-established plan or become more thorough and improvise. Thoroughness expands the scope of plans but constrains the ability to put plans into action in a timely fashion. For example, in all air traffic control (ATC) units there is a standard regulatory requirement to establish contingency plans in case of CNS failures. Given the complexity of the CNS systems and the range of possible failures, there is sometimes a dilemma between following up the contingency plan or becoming more thorough regarding the number of aircraft that can be accepted and the arrangements to be made on the sectors (sectorization practices).
3. *Impulsiveness versus reflection on perspectives.* Complex systems require coordination of different specialties, each bringing its own perspective on the problem. Reflection involves stepping out of one's own perspective and looking at the problem in new ways. However, reflection comes at a cost because it requires sharing perspectives and bridging contrasting ones. In contrast, impulsiveness relies on one's own perspective and may provide faster responses to imminent problems. The challenge in controlling complex systems is to strike a balance between the two opposing approaches according to the characteristics of the situation.

4. *Global versus local goal responsibility.* Complex systems divide up roles and responsibilities to manage multiple goals. Hence, systems may be simultaneously cooperative over shared global goals and potentially competitive when local goals may be in conflict at different units. Work systems must devote additional resources to manage responsibility across units and ensure reciprocity. For example, tower controllers may have to change the runway in use due to local noise restrictions. However, this may be in conflict with the approach controllers since the new runway configuration may increase the complexity of handling the new traffic flows.
5. *Concentrated versus distributed action* (i.e., the centralization-decentralization dilemma). Complex systems require coordinated efforts of many practitioners at the sharp end. The type of authority structure is a usual managerial dilemma since centralized control may effectively coordinate individual efforts but it is not very flexible. In contrast, distributing authority to many levels or units increases flexibility but comes at a cost in coordinating efforts across units. An example in air traffic management (ATM) regards coordinating efforts between flow controllers who handle strategic aspects of traffic and local controllers who handle operational aspects of air traffic. At the operational level, an additional aircraft may give rise to a large increase in workload of a sector which is not perceived by the flow controllers.

These performance trade-offs cover a wide range of organizational life from individual and team performance (e.g., plans and adaptive capacity) to organizational performance (e.g., multiplicity of perspective, responsibilities across levels, and distribution of authority).

Managing trade-offs is similar to choosing operating modes in organizations; that is, alternative forms of work practices, coordination, and work organization. In terms of safety management, human errors and management failures are seen as inappropriate decisions in managing trade-offs or as poor adjustment of operating modes. This approach emphasizes that the process of balancing trade-offs (or switching between operating modes) is made in the context of a particular work system to achieve the best match between demands and resources. In this sense, human errors and management failures are seen as inappropriate adjustments of operating modes rather than as limitations or weakness of human behavior. The same decision that was successful in one case may be inappropriate in another. So the challenge for practitioners is to recognize new problem features that call for an adaptation of established plans and behaviors. Balancing trade-offs or operating modes may imply several strategies, such as choosing the best one for the circumstances, recognizing the need to switch to another operating mode, or even blending alternative operating modes. Table 14.1 shows several trade-offs or alternative

Table 14.1 Controller Performance as Balancing Trade-offs at Multiple Levels

FUNCTIONS	EXAMPLES OF OPERATIONAL TRADE-OFFS	EXAMPLES OF ORGANIZATIONAL TRADE-OFFS
Steering	Trading efficiency and thoroughness in selecting goals and options.	Staying away from the safety boundary minimizes risks but deprives organizations from learning opportunities.
Modeling	Achieving a balance between confirmation and mindfulness.	Organizational culture may imbue controllers with common approaches and priorities but may also be turned into "collective blindness."
Adapting to changes	The cost of change may be higher than continuing with an ineffective plan.	First-order performance (or firefighting) may produce immediate results but fails to remove work obstacles and hidden organizational problems (i.e., a failure of second-order performance).
Planning	Tightly coupled plans may be efficient but make it difficult to cope with feedback delays, changing priorities, and interruptions.	Nonlinear interactions require a balance between risk assessment and intuition as information load and task complexity increase.
Monitoring	Narrowing attention on plan progress (focused attention) versus keeping an eye for weak signals and subtle events at disparate times (divided attention).	Thorough safety reporting may create an impression of being ineffective while others who report less may be rewarded.
Cooperating	Switching between standard and proactive coordination.	Differentiation increases efficiency but creates different orientations and conflicting priorities, which combine to defeat responses in complex systems.
Operating	As habitual actions gain strength by their everyday use, controllers may not see certain countersigns or exceptions that make habits unsuitable to the situation.	Formal versus informal means of supporting operations (for example, formal procedures vs. job rotation and teamwork).

operating modes that are at the heart of making decisions how to collect data, how to interpret data and create a model, how to develop plans, how to coordinate, and how to adapt to changes.

The discussion on performance trade-offs is structured along seven human and organizational functions that have been described in the joint T²EAM-VSM framework (Chapter 13). The emphasis of the analysis in the following sections has been on maladaptive patterns of performance stemming from inappropriate balances of trade-offs at the individual, teamwork, or organizational levels.

14.3 Balancing Trade-offs at the Organizational Level

14.3.1 Steering

Steering usually involves balancing trade-offs between safety and production that is challenging because organizational processes have different time responses, dependence, and long-term effects. For instance, safety efforts often do not show immediate results, which creates a perception of their being ineffective, at least in the short term; this makes it likely for managers to lose sight of the value of their efforts which ultimately leads to downward adjustments of safety priorities (Marais et al. 2006).

Another trade-off that safety managers face regards the distance from the safety boundary that an organization should maintain. Organizations that choose to stay well away from the safety margin may minimize their risks but do not learn how to cope with unexpected events (Amalberti 2013). Others may choose an operating point much closer to the safety margin and increase their learning potential; however, there is often a risk regarding the magnitude of migrations or transgressions they undertake. Cook and Rasmussen (2005) found that high reliability organizations (HROs) manage small transgressions inside the margin of safety without losing sight of the safety boundary (see Chapter 3). However, it is not easy to judge what constitutes a small or large transgression and how to bring the operating point back within the safety margin. This implies that a performance failure due to relaxation of rules, or circumvention of rules, should not be seen as deviant behavior but as a poor judgment of the distance and dynamics of transgression of the safety margin.

14.3.2 Modeling

Mental models that are shared in organizations may create a multiplicity of views that provides the requisite variety necessary to cope with a spectrum of safety critical situations. However, this requisite variety comes at the initial cost of resolving certain conflicts that emerge from the multiple models or views in the organization. For a flow controller, for example, adding another aircraft to the flow stream may not be considered a risk factor. In contrast, a tower controller may perceive several risks in this action, since extra aircraft may have to stay on the taxiway, block standard taxiway routes, or increase the runway occupancy time. Organizational culture plays an important role in imbuing controllers with common approaches and reconciling different models. A balance should be achieved in the degree of harmonization of views because converging mental models brings with it a danger of "collective fixation" (i.e., a danger that some vital factors may be left outside the bounds of organizational perception).

14.3.3 Adapting to Change

Authority structures are usually associated with particular styles of problem solving (i.e., first- vs. second-order performance). In the real world, many teams try to resolve work obstacles by themselves without bringing them to the attention of supervisors (i.e., first-order problem solving); this tendency seems to be more prominent in autonomous or decentralized teams. As a result, teams are less likely to be searching for systemic organizational shortfalls that gave rise to the work obstacles (e.g., failure to engage in second-order problem solving). For example, a short closure of the main taxiway in an airport, due to urgent maintenance, could be managed locally without having to comply with formal safety procedures. People at the safety department may come to know about this event only through the tabulation of safety occurrences, although they may not have all the facts available (e.g., how controllers managed the traffic and what hazards emerged). Tucker and Edmondson (2003) asserted that the flip side of empowerment and decentralization is the removal of supervisors from daily work activities, leaving the local teams on their own to resolve problems that may stem from other parts of the organization with which they have limited interaction. Hence, some well-intended efforts to create empowerment can generate side effects in the long term.

14.3.4 Planning

Many safety critical systems are characterized by nonlinear interactions between units or tasks whereby outcomes are not proportional to inputs, connections are unexpected or hidden, and cause-effect relationships are cyclical (e.g., feedback loops). Nonlinear interactions make the work system less predictable, especially in emergency situations. Unexpected connections and cyclical effects create problems that cascade across multiple areas and affect the work of many practitioners. This makes it difficult to set up a safety planning system with clearly defined roles because the safety problem cannot be broken into simple individual tasks that can be combined at a later stage. However, safety planning has been based mainly on predictive techniques for assessing risks. In Chapter 3, a critique was presented for formal risk assessment techniques that cannot factor in the uncertainties involved in assessing complex scenarios. For instance, managing risks does not always mean reducing them; it may also imply exchanging them for other smaller risks in the short term or even postponing risks for later on when more capabilities would be available. It is now increasingly recognized that expert intuition can play an important role in making sense of nonlinear situations where risks can be exchanged for smaller local risks so that safety is maintained for the whole organization (Weick and Sutcliffe 2001; Amalberti 2013). In managing traffic in adverse weather conditions, for instance, controllers may accept higher separation minima and provide more vectoring options for the flight crews in order to account for unexpected deviations due to bad weather. In this sense, they replace the risk of aircraft converging from opposite directions due to a sudden weather change with the risk of a conflict geometry with no safety implications due to traffic diverging in parallel routes. These subtle ways of managing risks are difficult to handle in formal risk assessments as they require significant inputs from expert intuition.

14.3.5 Monitoring

Another trade-off has to do with the amount of risk information to be reported by a department, or a subcontractor, to the higher organization. Thoroughness is associated with openness and reporting of even minor mishaps because the absence of a report may be seen as everything is well. In this sense, both critical reports and weak signals should be monitored to improve safety. On the other hand, thorough reporting may create an impression of being ineffective, while others that report less may be rewarded. In particular, subcontractors and suppliers often feel under pressure to comply with the standards of the organization for reporting. At the same time, they may believe that they will be punished if they have too many things to report, while a competitor who reports less may be rewarded (Hollnagel 2009). Hence, they may report enough to sound credible but not so much that one loses the contract.

14.3.6 Cooperating

Specialists are essential people in organizations that operate complex systems, as they increase the effectiveness with which tasks are performed. However, as organizations become increasingly differentiated or specialized, the likelihood that an unforeseen and adverse event falls into the gaps between highly specialized personnel increases. Even in the best organizations, high levels of job specialization can result in gaps of expertise or situations for which no one seems to have the right skills. In these cases, no actions are taken to attend to a particular concern or problem, with tragic results. In addition, highly specialized teams may have different perceptions of the problem or different orientations on how to solve the problem. When faced with ambiguous or unusual events that do not fit current models of work division, the very same organizational process that accomplishes many ordinary tasks effectively can work to defeat appropriate responses in unusual events.

14.3.7 Operating

Operating refers to the timely and safe execution of safety programs and plans. A well-known challenge here regards the degree of compliance of programs/plans to official systems of work. Compliance with procedures enables controllers to perform fast but, unfortunately, several performance aspects cannot be codified in procedures. For instance, many patterns of coordination and mutual adjustment are not easily reproduced or described in words. For example, not all the handover procedures can be described in detail in the letters of agreement (LoAs) between two adjacent units. *Ad hoc* coordination appropriate for certain types of traffic situations are introduced at the very instant when they are needed. Controllers build, test, and eventually adapt their work practices through operational practice but they know that these cannot be transferred into a detailed coordination plan.

The challenge for organizations is to balance formal and informal means of transferring knowledge to controllers. Organizations may select different forms of knowledge transfer, according to the context of work. Instead of documenting all practices, they may choose to use other means to transfer knowledge, such as relying on demonstration, rotating controllers between units, and embedding knowledge in tools and technology (Argote and Ingram 2000).

14.4 Balancing Trade-offs at the Operational Level

14.4.1 Steering

An important aspect of steering involves the balance between efficiency and thoroughness. An emphasis on efficiency may simplify the problem and set more manageable goals. In the long term, however, this may increase workload or introduce side effects because insufficient time would have been spent on looking into all factors that are likely to affect decisions. On the other hand, a thorough consideration of alternative options would take time and cause delays. One way to balance efficiency and thoroughness is through a planscape of goals, tactics, goal conflicts, and windows of opportunity (Klein 2007). When practitioners are not sure how to recover from a problem, they may be simultaneously preparing for a few goals or tactics. They may have a preferred tactic but, as they are not sure if it will work out, they prepare some backups. In this sense, controllers may be juggling several goals at the same time; finally, as the window of opportunity gets smaller and smaller, they are forced to choose one option.

In a complex arrival traffic approach, for instance, controllers may delay sequencing arrival aircraft until a point when the cost of replanning would be too high or even unsafe. Similar cases can be drawn from crew dilemmas to divert or fly into adverse weather at the destination airport; another option may be to choose an alternate airport where the chances of bad weather are lower (Batt and O' Hare 2005).

Flight crews may try to delay their decision to the last moment in the hope that their preferred option would fall into place (e.g., continue to destination). But at the same time, preparations should be made for the diversion possibility (e.g., after a certain distance traveled into the destination airport, fuel may not be sufficient to go back to a diversion airport). This type of steering allows practitioners to conclude with a workable plan after completing many of the required preparations.

14.4.2 Modeling

A common problem in dynamic systems is failure to revise a mental model or mindset as new evidence becomes available. The initial situation assessment may seem appropriate, given the available information, but people may fail to revise their mindsets. Several causes of fixation have been cited in the literature (Gaba and DeAnda 1989; De Keyser and Woods 1990), such as personality factors, incomplete models, and dependency on salient cues. A modern approach attributes failure to revise mental models to a poor balance between confirmation and mindfulness (Kontogiannis 2011).

Abnormal situations usually generate an overload of information that makes practitioners resort to some sort of data filtering in order to build a coherent mental model of the situation. This creates a reinforcing loop which is useful at the beginning because it boosts confidence (Kontogiannis 2011); however, as this loop grows up continually it can cause fixation on the initial assessment. A possible limit to confirmation is staying mindful; for example, taking regular action to test the leading assessment or provoking a reaction from the system to generate more data. Mindfulness creates a counter-loop that cultivates new hypotheses about the problem and helps practitioners mentally run through several system states; hence hidden assumptions and missing data can be discovered and corrected. Therefore, practitioners must maintain a balance between the confirmation and mindfulness loops.

14.4.3 Adapting to Change

Adapting plans requires an understanding of the cost of change (e.g., availability of resources in the future, time to communicate changes, and new impacts from changes). In some cases, a more efficient plan may be found but teams could be reluctant to change because of time-consuming communications required to inform the affected team members. On the other hand, a change of plan without balancing these trade-offs may have repercussions for safety. Some incidents in the aviation domain have resulted from flight crews accepting a last-minute offer by air traffic control to land on a different runway, without evaluating the workload and timescale for making all necessary changes. For instance, Croft (2009) reported that one aircraft landed on a taxiway after having accepted a switch to another landing runway via a sidestep maneuver after sighting the runway visually on the approach (it was presunrise darkness). The approach lights would have prevented the crew from landing on the taxiway but these were inoperative due to maintenance, while the instrument landing system was not operating as it was not the baseline runway for the approach on that day. Reason (2008) used the term error wisdom to refer to the mental skills required to recognize and avoid situations with a high error potential.

14.4.4 Planning

Time is an important aspect of control and planning. It is not enough for controllers to make a correct decision and implement it in the appropriate order; they also have to make a decision in a timely fashion and implement it within the time window available. To maintain control under time pressure, controllers often resort to a working style that is more efficient but less thorough and make plans that are tightly coupled. While this mode of control may work for the initial planning of activities, tightly coupled plans become difficult to adapt to the unfolding situation. For instance, unanticipated events and errors may introduce interruptions of tasks and changes in priorities. As dependency on time becomes a feature of many dynamic situations, plans should make provisions for adaptations, including when and how to interrupt tasks or resume tasks when opportunities arise. In this sense, loosely coupled plans can tolerate more disruptions and become more robust.

14.4.5 Monitoring

An abnormal situation can create a data-overload problem, which increases monitoring requirements. Controllers may continue monitoring for new information, look for new priorities for action, and follow up checking their actions. Attentional dynamics orient the mind where to find critical events and know how to accomplish recovery plans without getting absorbed in the situation (Woods and Hollnagel 2006). This raises an important challenge regarding how people accomplish their plans while remaining sensitive to subtle events occurring at disparate times in the environment.

Complex plans provide a large amount of information to consult to accomplish tasks which diverts attention from subtle events that may appear at disparate times in the environment. For instance, controllers could miss cues and events that might even be of little apparent relevance to their tasks but could help them assess the overall situation and prepare for adaptation. In this sense, controllers should be able to narrow attention on plan progress but also keep an open eye for weak signals and subtle events at disparate times. Hollnagel (1992) presented several tactics for coping with the trade-off between focused and divided attention, such as reducing the accuracy of the main task to avoid missing any important cues and reducing the amount of information processing (e.g., filtering information and reducing the level of discrimination by noting only large variations or extreme values).

14.4.6 Coordination

Coordination involves team members exchanging information to articulate their plans, which requires sufficient time and cognitive resources to be accomplished. Proactive coordination is a strategy that minimizes workload and requires knowing when to interrupt colleagues and when to offer information that has not been requested. Several studies have found that providing unsolicited and proactive information can make team communication more efficient, especially at high tempos of work. In a field study by Malakis et al. (2010b), expert controllers were able to communicate effectively without unnecessary elements that prolonged and garbled communications. They were able to appreciate major attributes of information (i.e., criticality and timelines) and were able to judge the level of workload and interruptibility of other team members. Some en-route controllers reported that, in high-workload situations, they changed their speaking habits (e.g., giving stricter and shorter instructions or raising their voice). This was not only to save frequency time but also to signal to flight crews that they should listen carefully. Therefore, controllers should be able to know when to switch from standard to proactive coordination.

14.4.7 Operating

Execution of plans faces similar trade-offs between speed and accuracy, or between efficiency and thoroughness. Controllers tend to rely on plan and action habits that proved to be efficient in past situations and apply them cautiously to new problems. This transfer of habits and action patterns relies of a process of drawing analogies between old and new experiences. Unfortunately, as habitual actions gain strength by their everyday use, controllers may not see certain countersigns or exceptions that make them unsuitable for new situations. An urgent situation may prompt controllers to rely on old habits and workarounds that turn to be unsuitable for a new situation. Therefore, controllers have to make several speed-safety trade-offs, while the results of one experience may not be transferable to a new one.

Another trade-off in the execution of plans regard the level of specificity or detail in the actions or instructions given by controllers. Adjusting the level of specificity enables controllers to adjust their own workload as well as adjust the degrees of freedom allowed to the flight crews. It is apparent that specificity of instructions should be tailored to the demands of the situation and the working styles of controllers.

14.5 Challenges in Managing Performance Trade-offs

Balancing trade-offs, or operating modes, requires a great deal of experience and competence in several strategies that may have their relative strengths and weaknesses. Kontogiannis (2010b) proposed four strategies for managing performance trade-offs, namely:

1. Choosing the best option or operating mode for the circumstances
2. Recognizing the need to switch to another operating mode
3. Blending alternative operating modes
4. Developing a mindset for adaptation and change

The next section provides a short review of these challenges in managing performance trade-offs.

14.5.1 Developing Competence to Operate at Both Sides of the Spectrum to Choose the Right Option

Effective management of trade-offs or operating modes implies that controllers and organizations are competent in operating at both ends of the spectrum despite the fact that the different goals have their own requirements. Trading off goals or options requires a deep knowledge of their relative strengths and weaknesses as well as an ability to discern the range of applicability of these options to a variety of situations. Developing this capability, however, comes at an increased cost of training so that controllers can acquire redundant skills for a variety of domains. Broadening the bandwidth of competences may be a good strategy to increase flexibility, for instance, but it also leads to increased demands for training.

A typical example in the ATM system is the dilemma facing an area control center (ACC) when it comes to the training of controllers in different sectors. An ACC may operate with 15 different sectors that may be quite dissimilar in traffic demands, complexity, conflict resolution, coordination requirements, weather patterns, and so on. The training section of the ACC has to make a tough decision whether to train all controllers for all sectors or to tailor training to dedicated sectors for selected controllers. The first option creates a lengthy progression of controllers toward acquiring ratings and sector endorsements but provides rostering flexibility as all controllers can work satisfactorily in any sector. In the second case, controllers may develop in-depth expertise, efficient work practices may flourish, but the margin of maneuver could be significantly lower as rostering becomes more difficult to adjust. In addition, system-wide failures and contingency plans can be better managed with the first option while day-to-day operations can become smoother with the second option.

14.5.2 Switching between Modes and Evaluating the Cost of Change

Switching between operating modes in the spectrum is an essential capability for revising earlier options and changing decisions. In many cases, it is unlikely that the same operating mode will be suitable for many situations; controllers should be able to set revision steps to revise options and evaluate alternative ones. In addition, changing to another option may also require a special method so that the transition is smooth and widely acknowledged (e.g., several controllers should be informed of this change in options and plans). A typical example regards the decision how to split or ban-box sectors to accommodate changes in traffic demands.

Sometimes mode changes may require extensive communication to avoid confusion and minimize risks resulting from the new operating mode; this may be hard to achieve under time pressure. In some cases, a more efficient operating mode may be found but teams could be reluctant to change because of time-consuming communications. The process of change to a new mode may be complicated by the complexity of the precautions and safeguards required to bring technological systems to stable states before aborting the existing mode of operation.

14.5.3 Blending Alternative Options or Operating Modes

An alternative to mode changes could be to create a synthesis of alternative modes. Self-organizing behaviors, for instance, require decentralized planning so that controllers are able to make rapid decisions without the need to notify and get agreement from their watch supervisors. However, an authority structure that is too decentralized may have trouble synthesizing data from different sources to develop an accurate picture of the situation (Klein et al. 2005). Is the team appropriately structured to permit self-organizing behavior and, at the same time, synthesis of different inputs? Unfortunately, there are very few concrete suggestions on how to bridge the gap between different modes and provide a safe integration of modes.

14.5.4 Developing a Mindset for Adaptation and Change

An adaptive organization is one that expects to find problems with the current assessment of the situation and therefore expects to make changes in operating modes in the course of problem-solving. In the context of military organizations, Klein and Pierce (2001) have raised several questions regarding this mindset for adaptation. Examples include: Does the team try to preserve or challenge the current understanding of the situation? Does the team expect to find weaknesses in the current plans? Is the team orientation to dismiss weaknesses or take them seriously? This mindset for adaptation can be expressed as sustaining an ambivalent and critiquing stance toward the problem.

In complex systems, practitioners are faced with situations that are partly familiar and partly novel. For these cases, Weick and Sutcliffe (2001) argued that "people should retain a model of situation created by their past experience but also watch for unfamiliar and novel cues in the interest of building a comprehensive story or account of events." Maintaining ambivalence requires controllers, on the one hand, to retain well tried and proved operating modes but, on the other hand, to remain vigilant to the possibility of changing to another mode if the situation takes an unexpected turn. Engaging in simultaneous belief and doubt is admittedly a difficult exercise but this stance of ambivalence may be required in order to exploit the valuable experience of controllers and, at the same time, leave more opportunities for improvisation, and error recovery.

14.6 Concluding Remarks

This chapter has argued that the work of practitioners and organizations involve some sort of decision trade-offs, not only in planning and doing, but also in other functions such as steering, modeling, and adaption (see Table 14.1). Making trade-offs as part of everyday activities is necessary because modern systems have become intractable and difficult to predict; as a result, practitioners have to adjust their performance and switch between different operating modes. For instance, organizations may switch from a hierarchical style to a more autonomous one that grants practitioners greater decision latitude when uncertainty and time pressure increase. However, adjusting performance in terms of balancing trade-offs brings to the fore three important questions:

1. How can one recognize the conditions in the organization that would require an adjustment of performance?
2. When is it the right moment to start making this adjustment?
3. How can practitioners manage this adjustment without introducing any side-effects or coordination problems?

The first question is important because practitioners need to know how to match the conditions in the organization to alternative operating modes. Although several studies have commented on the right conditions for different modes, there is still a lack of a systematic framework in the literature. For instance, organizations should strive for thoroughness, collective mindfulness, and decentralization when time pressure increase and opportunities for recovery are restricted. However, this match between operating modes and conditions of work will be influenced by the practitioner styles. In this respect, controllers who are granted more authority should be well trained and willing to take additional roles, while supervisors and managers should be willing to accept greater uncertainty with regard to actions of their subordinates.

The other two questions regard the time frame and the preconditions for making adjustments safely. In switching between operating modes, there is a risk that the switch may be delayed or may introduce additional difficulties to other teams involved in the same situation. In addition, the technological system may impose further time constraints and preconditions for changing operating modes. These preconditions for change most often escape the attention of accident investigators who perceive the lack of change in leadership or teamwork as cognitive fixation or problem solving rigidity. In some cases, a reluctant stance to adopt a new operating mode may be the result of careful thinking of the costs of change rather than any fixation on previous modes.

A new way to look at the balance of trade-offs would be in terms of a synthesis of options and views that are seemingly at odds or in some form of conflict. For example, constrained autonomy refers to cases where controllers decide on the constraints of their own autonomy through their involvement in the design of procedures. A synthesis of alternative operating modes can be facilitated with the use of training simulators, which allow controllers to experiment with the safety boundaries and learn without risk. Training simulators may also facilitate teams in creating multiplicity of views and finding ways to solve conflicting events. Additionally, training simulators may provide practice opportunities for self-regulation so that the workload in combining task execution and self-regulation becomes lower. This synthesis of alternative operating modes presents new challenges for future studies in cognitive engineering.

Finally, performance adjustments may involve making trade-offs at multiple levels, such as steering, modeling, planning, and coordinating. In this sense, a poor trade-off at one level of performance may be compensated by a skillful trade-off at another level. For instance, a tight plan may be a wrong choice because it restricts error detection but proactive team communication may compensate by making available other team members to pick up and recover errors. When performance is seen at multiple levels, the problem of decision trade-offs becomes more difficult to tackle as it requires a better understanding of the interactions of the seven functions depicted in Table 14.1. Resilience engineering offers valuable insights in addressing the trade-offs at individual levels of performance but further work should be done in the mechanisms that regulate the cognitive and organizational functions.

15

Effective Safety Risk Management

15.1 Introduction

In this book, an effort has been made to show how to operationalize or utilize theory in cognitive engineering and safety organization in order to provide practical guidance in the areas of safety management, training, personnel assessment of cognitive and technical skills, evaluation of traffic complexity, and system design. This chapter revisits the safety requirements of regulatory authorities and the challenges to safety management (see Chapter 3) in order to demonstrate how to meet them with the use of models and methods presented throughout the book. *Safety organization* is a general term that refers to both theoretical models and techniques for managing safety in organizations. *Safety risk management* is more concerned with the organizational processes and, in particular, the risk management activities for assessing operational hazards, quantifying risks, and applying measures for mitigating risks.

Safety risk management can address system safety at three interacting organizational levels. At the strategic level, organizations should be able to specify their safety requirements, as well as their internal or external challenges to safety in order to generate safety intelligence. Chapter 14 is particularly useful in regard to managing strategic trade-offs at this level of analysis. At the supervisory level, organizations should be able to specify their safety functions and processes with the use of organizational models and methods such as VSM (viable system model) and STAMP (systems-theoretic accident model and processes) (see Part IV). Finally, at the operational level, the cognitive engineering perspective (Parts II and III) can provide a good basis for modeling the interactions of practitioners such as controllers, pilots, and airport staff. In the present chapter, the focus is mainly on the supervisory and operational levels of safety risk management.

The content of this chapter is organized so that traditional models and techniques of risk management are contrasted with new proposals derived from the material presented in earlier chapters. Since existing risk management models have been promoted by regulatory agencies (e.g., Eurocontrol and European safety aviation agency [EASA]) and applied by many organizations in the aviation domain, an effort has been made to show how to elaborate them rather than propose radically different methods that could present a problem of integration with what has been current practice in risk assessment. In this sense, this chapter is not a field guide to risk assessment but a general guide how to improve on existing practices in risk management.

15.2 An Overview of Safety Risk Management

Safety risk management requires a robust approach to modeling risks, assessing risks, and applying risk mitigation measures. In this respect, Eurocontrol has developed the integrated risk picture (IRP) modeling approach (Eurocontrol 2006), which has been applied with varying degrees of success to the air traffic domain. This model-based approach to risk management requires a description of the organizational processes, the workflow activities, the risks associated with performing work under different conditions, and the barriers or defenses that can control or mitigate consequences. A model-driven approach to risk management includes three types of models:

1. A *system model* that describes workflow activities, organizational processes, the work constraints, communication of risk information, and practitioner strategies for controlling their work
2. A *risk model* that identifies all hazards related to a specific work activity or aviation scenario, the underlying causes (i.e., organizational and human factors), the minimal routes to failure (i.e., minimum combination of causes), and the mitigation means for minimizing the consequences of hazards
3. An *influence model* that examines how workplace and organizational factors influence the likelihood of risks and the choices of risk mitigation measures.

The models are usually built in a sequential fashion but some degree of elaboration can be achieved by revisiting them briefly. For instance, the system model identifies the safety processes, risk communication channels, and organization of work in different operational units. This provides a good basis for building a risk model to identify prominent hazards, quantify their risks, and design risk mitigation measures. Subsequently, the influence model can tailor risk quantification to the work characteristics of a particular organization. This analysis entails some degree of uncertainty about the completeness and reliability of available data and hence a new round of analysis may be required to obtain additional risk information and elaborate the previous models.

The next stage of risk management requires analysts to develop safety barriers and ergonomic interventions that mitigate risks and reduce their consequences. Examples of risk mitigation measures may include: system design, computerized support, controller training, task reallocation, and systems for disseminating risk information. It is anticipated that the theoretical models and methodologies presented in this book can provide valuable assistance in making the risk management process more effective, as discussed in the following sections.

15.3 System Models for Risk Management

Effective risk management requires not only collection of historical data on system operation but also the design of models that describe how the system works and how its operations function to achieve safety. In this context, system models are developed to examine the safety organization of the departments and the operational units as well as the coordination with other aviation stakeholders. At a more technical level, system models examine the operations and plans of the practitioners at the sharp end, the factors that are taken into account to make decisions, the allocation of work, and the coordination demands with other sectors.

A prominent model for system analysis has been the structured analysis and design technique (SADT), which describes the inputs, outputs, controls, and resources required in many work processes in the air traffic management (ATM) system (Eurocontrol 2005). Figure 15.1 shows a SADT description of the functions of en-route conflict management, such as surveillance, traffic synchronization, tactical separation, conflict resolution (short term conflict alert), inter-sector coordination, and ATC communications. Their connections are as follows:

- *Inputs and outputs* (horizontal arcs) show the flow of information among several functions such as, aircraft data and sequences, traffic pictures, information from flight crews, ATC coordination information, etc.
- *Controls* (downward arcs) are the aviation rules and procedures that specify the goals, means, and constraints for achieving the functions. Controls also include the reference business trajectories that specify the client requirements.
- *Resources* (upward arcs) are the practitioners and the technical or computer means that are required for achieving the functions.

This functional description of the system is useful for understanding how information travels through the organization, how work is organized in terms of goals and requirements, and how resources can be allocated to functions. Hence, it is possible to examine how functions can be reallocated to different human or computer resources and

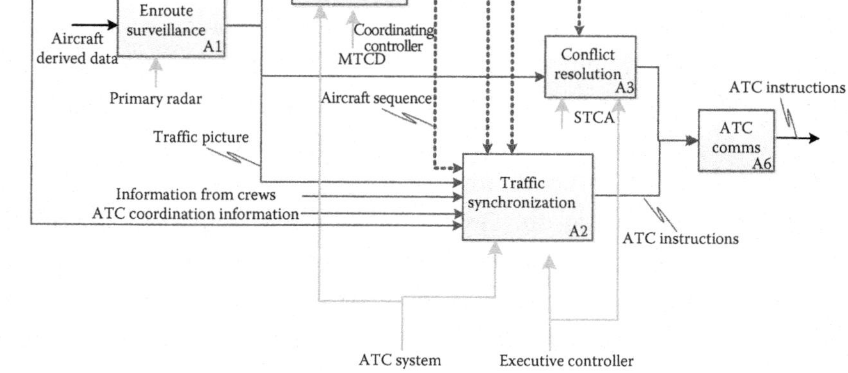

Figure 15.1 A SADT description of the functions of en-route conflict management. (Adapted from Eurocontrol, *ATM Process Model SADT Diagrams*, Eurocontrol, Bretiny, 2005.)

how information requirements can be met in order to improve the functioning of the system.

However, this functional description of the system does not enable us to understand how supervisors and practitioners make critical decisions, how they assess the situation and monitor traffic, how they adapt to changes in their work, or how they coordinate their work. Chapters 12 and 13 present two cybernetic approaches (i.e., STAMP and VSM) for understanding the interdependencies and dynamics of systems in terms of control loops that comprise seven processes:

- Goal steering
- Modeling the situation
- Planning
- Operating
- Monitoring
- Coordination
- Adaptation

As can be seen from Figure 15.2, the links and interactions between the departments and units that are engaged in air traffic operations can be described in terms of a similar pattern of analysis.

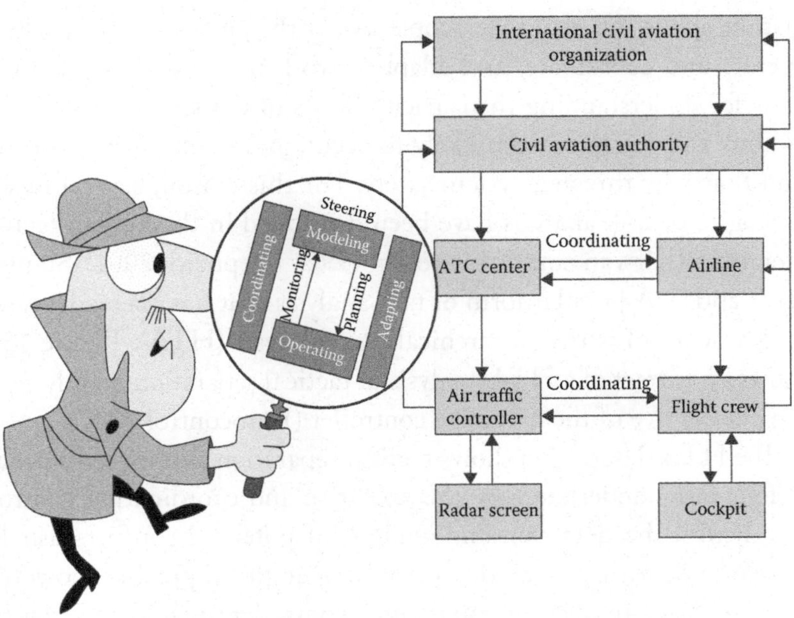

Figure 15.2 A system description of the safety organization in terms of control loops. (STAMP analysis.)

The identification of the control loops can enable analysts to examine feed-forward and backward effects, time lags of effects, and nonlinear interactions as they propagate throughout the system. A system description in terms of STAMP can provide supplementary information to SADT analysis for a number of issues such as system interactions, dynamic relationships, and control flaws that may threaten the safety of the system. The advantage of cybernetic models (i.e., STAMP and VSM) is that the aviation system can be described in general terms that correspond to "how-the-system-works" in a range of circumstances. Hence, it can be used to analyze particular incidents (Chapter 12) and to identify system hazards and risks (Chapter 13). The safety organizational structure (Figure 15.2) can also provide insights about how to unfold complexity into a number of control loops at different levels in the organizational hierarchy. Each control loop relies on a similar organization of seven functions, although differences may exist in the time available to make plans, the types of feedback available, the nature of constraints, and the degrees of freedom in making decisions.

At a more detailed level of analysis, system models provide valuable information for the particular plans and operations that supervisors and practitioners use to achieve the system functions. Detailed information about the way that people assess the situation, make plans, operate and coordinate, and adapt to work pressures can provide a basis for understanding the capacity limits of the system, the sort of failures and errors that are likely to occur, and the hazards of operation that may threaten system safety. For this reason, several methodologies of task analysis have been developed in the human factors literature (Kirwan and Ainsworth 1992; Shepherd 2001; Stanton et al. 2005). A popular form of task analysis that has been advocated by Eurocontrol is the hierarchical task analysis (HTA). Figure 15.3 shows an extract of a HTA analysis of tactical separation mainly from the perspective of the executive controller (Eurocontrol 2007).

The HTA description shows tactical separation as a sequential plan of four tasks undertaken by the executive and coordinating controllers; hence, the detection and analysis of potential conflicts usually proceeds the actual separation plan. In operational practice, however, the two tasks may be ongoing since controllers may detect several conflicts with different dynamics and mentally play out a couple of

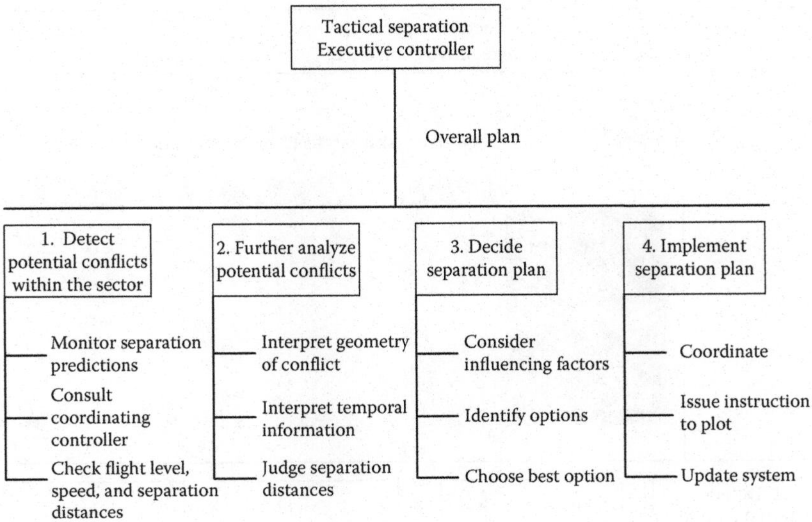

Figure 15.3 An extract of HTA analysis of the tactical separation function of the executive controller.

candidate separation plans. This may result in a decision to intervene late that makes an external observer believe that planning follows detection. On other occasions, however, an early intervention is made to avoid imminent conflicts in the near future but this early resolution of converging traffic cannot be captured by external observers. A potential concern with HTA is that the task descriptions usually match routine situations where a sequential order of tasks is the most efficient one. When there are variations or disturbance in the environment, different HTA descriptions should be drawn to capture the different ways of organizing work.

In general, the HTA presents a layout of tasks and subtasks that are required to achieve certain functions. Little information is provided about the difficulties that practitioners have in doing their work, the ways that they assess situations and make decisions, the errors that are likely to occur, or the ways that the operating teams work to detect and recover errors. For this reason, a cognitive task analysis (see Chapter 8) may offer supplementary information for these human factors issues (Figure 15.4).

The cognitive task analysis in Figure 15.4 corresponds to the ABCDE method (Chapter 8), which relies on a set of probes for putting in practice the T^{2}EAM model of human performance. A

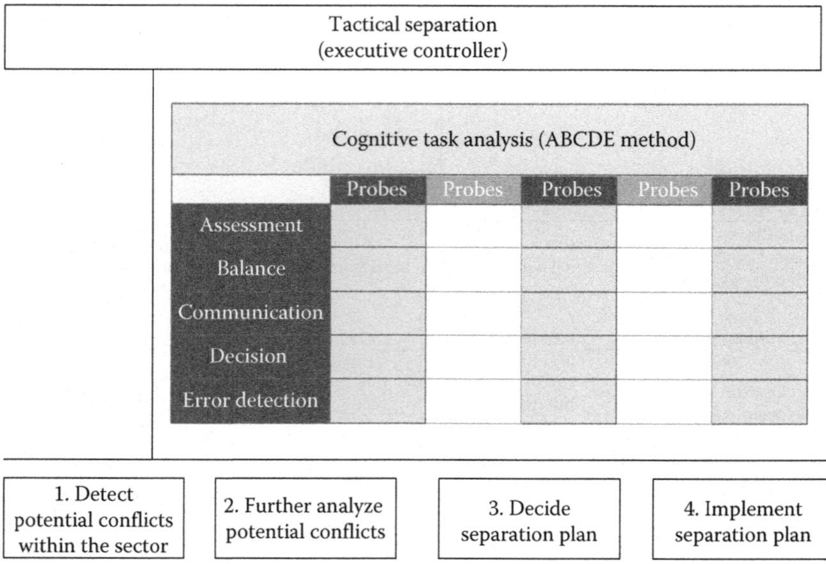

Figure 15.4 A cognitive task analysis of the tactical separation function.

workshop with practitioners may be required in order to make an introspection of their work and record data about how they assess the situation, how they balance different options, how they communicate, how they make decisions, and how they correct previous errors. Usually the analysis focuses on the overall task of tactical separation rather than the individual steps the task comprises. If the task is complex, the analysis can focus on each of the four lower level tasks (Figure 15.4) but not on the individual steps. That is, instead of focusing on a detailed description of "how-the-job-is-done," the cognitive task analysis explores different ways of doing the job under different circumstances. Chapter 8 provides an elaborate description of the tasks of the executive and coordinating controllers for en-route sectors on the basis of the ABCDE method.

If there is a requirement, however, to develop a detailed description of all operations and plans that are required by the two controllers in en-route separation tasks, the results of the CTA can be used to make the HTA description more versatile over a wider range of circumstances that might be encountered on the job. In any case, CTA provides a wealth of human performance data that can provide valuable input to the risk models that follow in the next stage of risk assessment. Finally, CTA can be helpful in designing courses of controller training or refresher training because it can describe the cues, challenges, decisions, and strategies used by experienced controllers in the course of a complex scenario.

15.4 Risk Models

The analysis of system functions, practitioner tasks, and organizational processes feeds into the next stage of risk modeling of the system. This stage entails an analysis of potential system weaknesses, practitioner errors, coordination problems, and barrier failures that may lead to critical hazards. The literature has presented many types of risk models and techniques including bow ties, fault trees, event trees, cause consequence diagrams, and so on. A simple risk model that has been extensively used in aviation is the bow-tie analysis that is briefly presented in Chapter 3. Bow ties consider the causal paths to particular hazards and the potential consequence stemming from hazards. Bow ties provide a good basis for building more accurate representations of failures of preventive barriers with the use of fault trees as well as more extensive representations of consequences with the use of event trees of protective barriers. The quantification of event trees and fault trees relies on influence models that show the effects of workplace factors, environmental factors, and safety organization that influence the likelihood of system failures and human errors.

15.4.1 Fault Tree Analysis

Fault tree analysis (FTA) has been extensively used in aviation for the analysis of mechanical systems and control systems that involve routine human actions. Fault trees consider failures of components that have distinct categories of operation and failure; states that cannot be split into simple failure/operational states are represented through the influence model. Fault trees are relatively easy to learn and apply in practice; however, there are some weaknesses that have already been mentioned in Chapter 3. For instance, fault trees cannot model accidents where the components remain functional but their design does not cover extreme conditions of operation.

What is most important, it becomes very difficult to model interactions between components or human activities. In modeling loss of separation events, for instance, fault trees tend to model human activities in a sequential fashion where the detection of a potential conflict usually precedes the development of a separation plan. In the fault tree in Figure 15.5, inadequate separation instruction can be due to poor system information, undetected conflicts, or inadequate separation planning. Poor information and failures in detection and planning are added through an OR gate to estimate the overall separation instruction failure.

As argued in Chapters 4 and 5, detection and tactical planning are not static events, as assumed in fault trees, but dynamic events that interact in complex ways. Controllers can adjust their attention and planning from long-term predictions of traffic flows to short-term

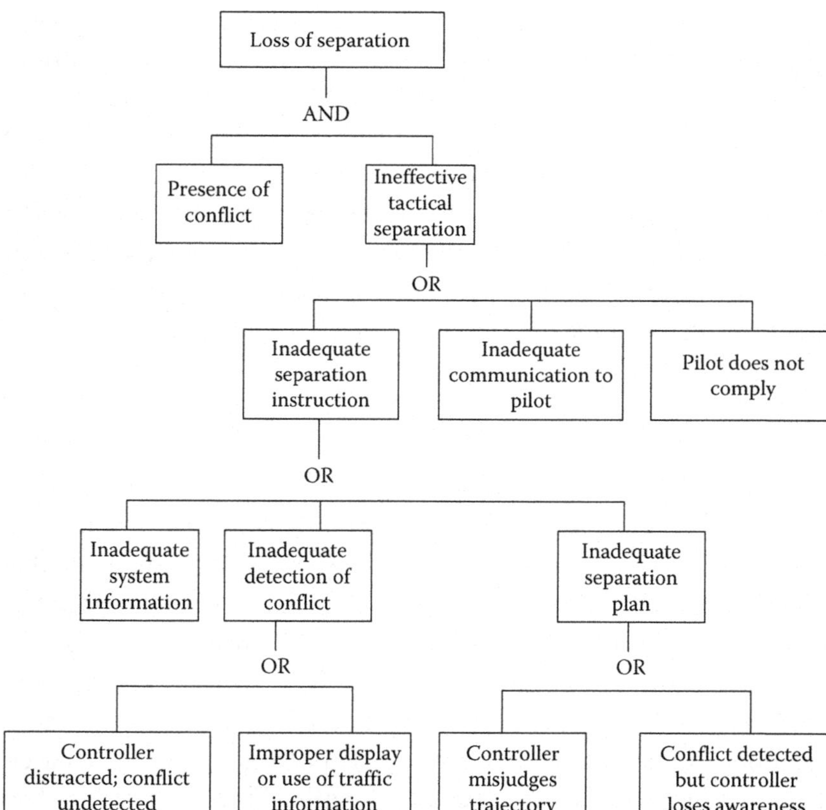

Figure 15.5 A conventional fault tree of a "loss of separation" event.

predictions of imminent conflicts. On many occasions, the initial traffic plan may frame subsequent phases of monitoring and planning. Hence an initial traffic projection and plan may lead to a tight traffic pattern that increases the chances of conflicts later on.

The proposed T^{2}EAM model enables us to understand that conflict detection and planning are dynamic events with complex patterns. For instance, two aircraft descending at different speeds may be safely separated in one sector but come in conflict as they cross sectors; this is more likely to occur when the next sector has lower separation minima. On the other hand, separation planning is also a dynamic event in the sense that a successful plan may go astray later on because of unexpected events or an unsuccessful plan may improve the situation in the near future:

- *A plan starts well but remains incomplete due to other interruptions.* For instance, separation planning may be tightly coupled, leaving little scope for crew diversions or unexpected events. A tight traffic pattern may not be recognized by crews, who may wish to change their flight to a continuous descent from a stepwise one, giving rise to other conflicts later on. In a similar way, a correct separation plan may be interrupted by other crews blocking the radio frequency for too long (e.g., taking too long in initial contact formalities).
- *An unsuccessful plan may improve the situation in the near future.* In other cases, separation planning may be recorded officially as unsuccessful yet it may cause no harm. For instance, a separation plan may result in two aircraft violating the separation minima but the conflict geometry may be such that the conflict is resolved soon after its recording by the system; in this case, there may be no complaints by the crews nor by the safety managers. This may happen because avoiding a temporary conflict may require tremendous effort while its tolerance may improve traffic separation in the near future.

Once we understand the nature of dynamic events in conflict detection and planning, we may be in a better position to extend the conventional fault trees. This effort is undertaken in Figure 15.6, which presents an extended fault tree where the top event of loss of separation incorporates another branch on the right side that refers to the case of an unsuccessful plan that is easily resolved by pilots in the near term. In fact the separation plan may be judged unsuccessful if it creates even a minor loss of separation event; however, the intention of the controllers could have been to improve the traffic situation and let the pilots resolve it in the near future. The other case of a plan that

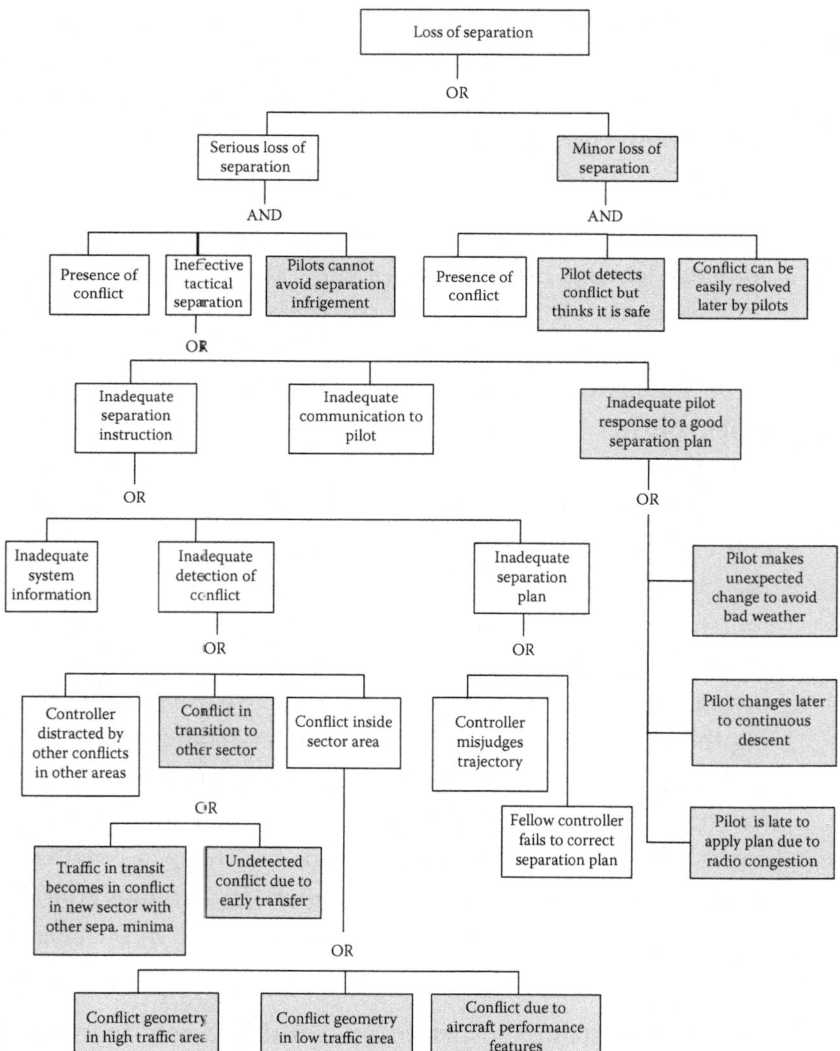

Figure 15.6 An expanded fault tree of a "loss of separation" event represented in gray boxes.

starts well but remains incomplete is also modeled in the lower-right hand of the fault trees as inadequate pilot response to a good separation plan.

A fault tree is a convenient format for quantitative estimates of human errors and system failures. This requires a good knowledge of the context of work and the situations that may lead to ineffective performance. In this respect, the analysts will have to identify many situations that could potentially lead to inadequate detection of conflicts. A similar point is made in Chapter 3 (Table 3.2) where five new conflict situations were presented. The lower-left part of the extended fault tree (Figure 15.6) shows five new gray boxes that provide a wider range of situations leading to unsuccessful conflict detection.

15.4.2 Event Tree Analysis

Event trees is a convenient format for modeling the consequences of a particular hazard when several safety barriers fail to provide the required protective functions. Most aviation organizations would rely on conventional event trees that, although not very sophisticated, can easily be applied by practitioners on a regular basis. Figure 15.7 shows a generic event tree where an initial event or hazard occurs followed by a number of human actions to mitigate safety consequences. For simplicity reasons, we assume that there are no technical or computer barriers apart from the actions of controllers. An example could be

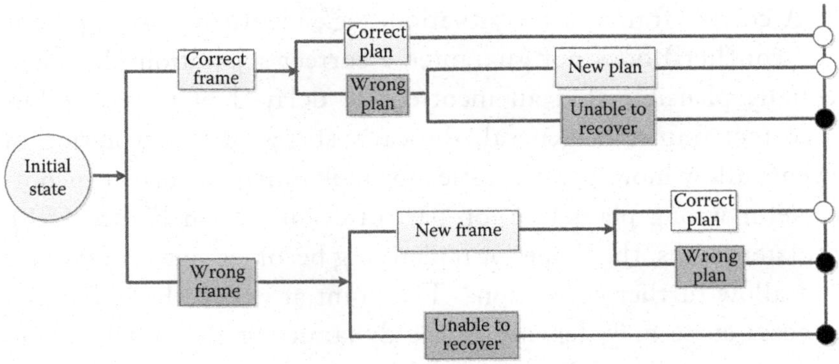

Figure 15.7 A conventional event tree of the response of controllers to an initial disturbance.

the response to a loss of separation event where the controller may develop a correct frame of the problem and proceed with a separation plan; if the plan was wrong, there may be another opportunity later to recover the problem and develop a new separation plan. However, it is also possible that the controller may be late in recognizing this event (i.e., wrong frame at the bottom part of event tree). In this case, there may be another opportunity for developing a new frame and an appropriate separation plan. In its simplest format, an event tree can model both failures and recovery opportunities at a later stage.

On the basis of the proposed models of sense-making, error recovery and T^2EAM (Chapters 4, 5, and 6) it may be possible to propose an extended event tree that could provide a better risk model. For reasons of simplicity, we may focus only on two interacting elements of human performance, such as sensemaking and planning. The dynamic cycle of sensemaking and planning can produce a more elaborate event tree in Figure 15.8. For instance, the processes of identifying and questioning a frame may produce at least three outputs: (1) a correct frame, (2) a wrong frame, and (3) an incomplete frame due to lack of appropriate or reliable data. The same three performance gradients can be adopted for the process of planning, which may result in a correct, incomplete, or wrong plan. It is anticipated that more complex gradients can be proposed including finer levels of detail. For instance, a plan may be in the right direction at a high level but it could turn out to be an inefficient one when it comes to its implementation details at a lower level of specification. For reasons of simplicity, the branching rule in Figure 15.8 includes only three levels of performance gradient.

A correct frame of the situation (e.g., identification of potential conflicts) does not guarantee a correct separation plan; it is equally plausible that an incorrect or delayed plan may follow a correct frame. In general, the early stages of the evolution of events allow more opportunities for replanning to correct incomplete or wrong plans (see top part of event tree in Figure 15.8). At later stages, the scope for replanning becomes smaller and may not allow further corrections. The point at which the branching process stops may depend on the dynamics of the situation and the amount and quality of data that exist for making quantification judgements for further corrections.

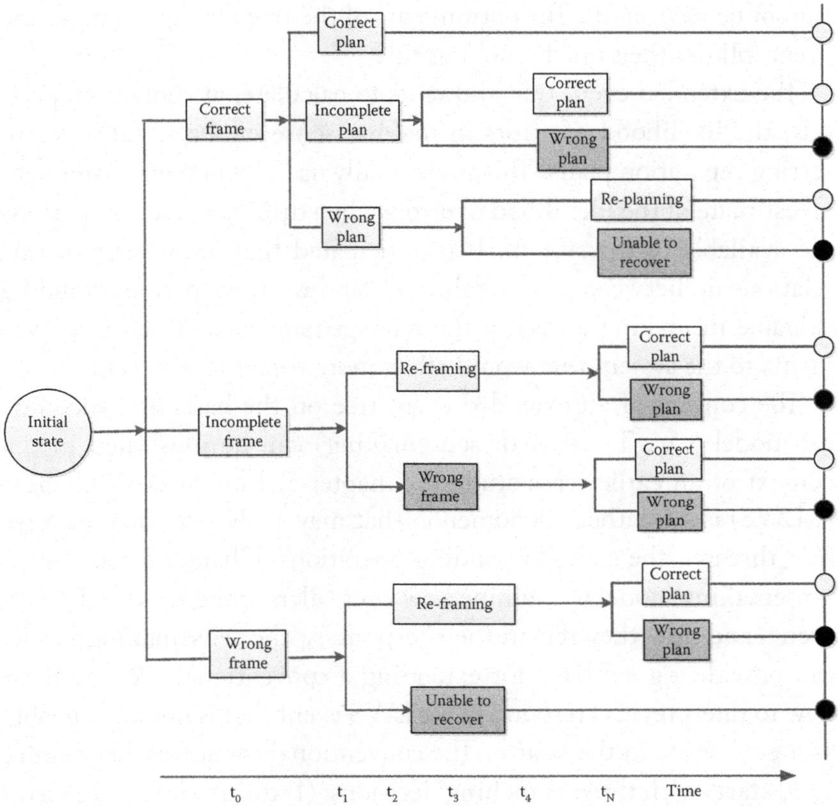

Figure 15.8 An expanded event tree of the response of controllers to an initial disturbance.

The middle part of the event tree (Figure 15.8) refers to the development of an incomplete frame or assessment of the situation. As the situation progresses, the level of analysis should have to become smaller, including a smaller number of branches. This is because the number of alternative branches gets smaller as more information becomes available to the practitioners regarding the nature of the situation and the results of earlier frames and decisions. Gradually, the number of branches will fall into one or two output states and the analysis will be concluded.

Figure 15.8 uses two gradients for the analysis of incomplete and wrong frame into two future states (i.e., reframing or unable to recover). The same logic is used for the branching of events related to planning and replanning; that is, a two-gradient analysis of replanning results in two outputs (i.e., a correct plan or a wrong plan that cannot be recovered). The bottom part of the tree for the wrong frame event follows the same branching rule.

The extended event tree allows us to calculate, at short time periods, the likelihood of errors in making sense of the situation or in setting separation plans. This is especially useful when we wish to get an estimate of the likelihood of errors when different recovery periods are available in the system. It is anticipated that knowledge of this relationship between error likelihoods and recovery periods could be valuable in making a case to the management for making improvements to the system that would allow more time for recovery.

The concept of an extended event tree on the basis of a theoretical model (e.g., T^2EAM or sensemaking) can be illustrated in the context of an earlier case study in Chapter 5. Low level wind shear (LLWS) is a weather phenomenon that may evolve in different ways that threaten the safety of landing operations. Chapter 5 has used a sensemaking model to examine how controllers make sense of LLWS events and how they replan their responses. The sensemaking model can provide a good basis for extending a conventional event tree for how to interpret and respond to a LLWS event that is not manageable as it gets worse in the course. The conventional event tree (see Figure 15.9) starts with three branching decisions: (1) divert aircraft, (2) send aircraft to a holding pattern, hoping that the weather gets better, and (3) allow aircraft to land. The last decision is not desirable since the exacerbating LLWS may put flight crews in a difficult situation during the landing phase and lead to runway excursions. The second decision is wiser since controllers may buy time until they collect more data about the LLWS event and replan their behavior. Nevertheless, the flight crews will make the final call and may even insist on landing despite the severity of the LLWS event. This is shown at the top of the event tree (Figure 15.9) where a crew could insist on landing despite of controller advice to divert to another airport.

Putting in action the T^2EAM model, the analysts may look deeper into the interaction cycle of reframing and replanning of the executive controller in response to a LLWS event and produce an extended event tree (see Figure 15.10). In the initial assessment of the situation, the analyst will have to take into account two factors with two levels of results: (1) whether the LLWS is manageable or not at the initial stage and (2) whether the LLWS event will get worse or not in the

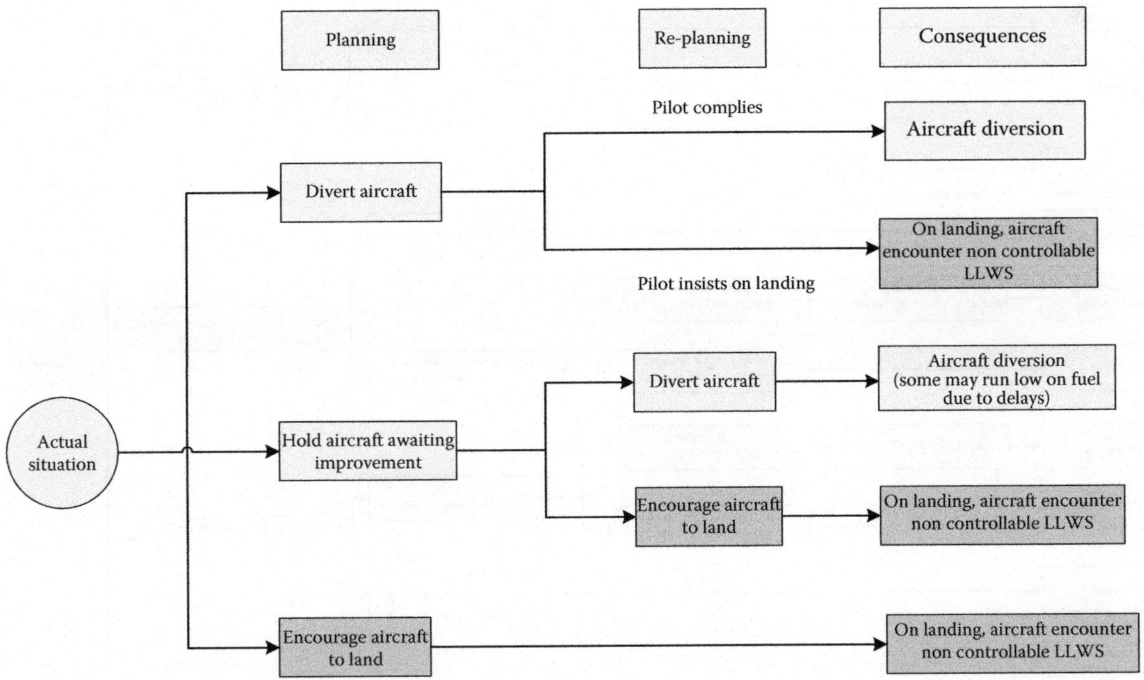

Figure 15.9 A conventional event tree of a response to a severe and deteriorating LLWS phenomenon.

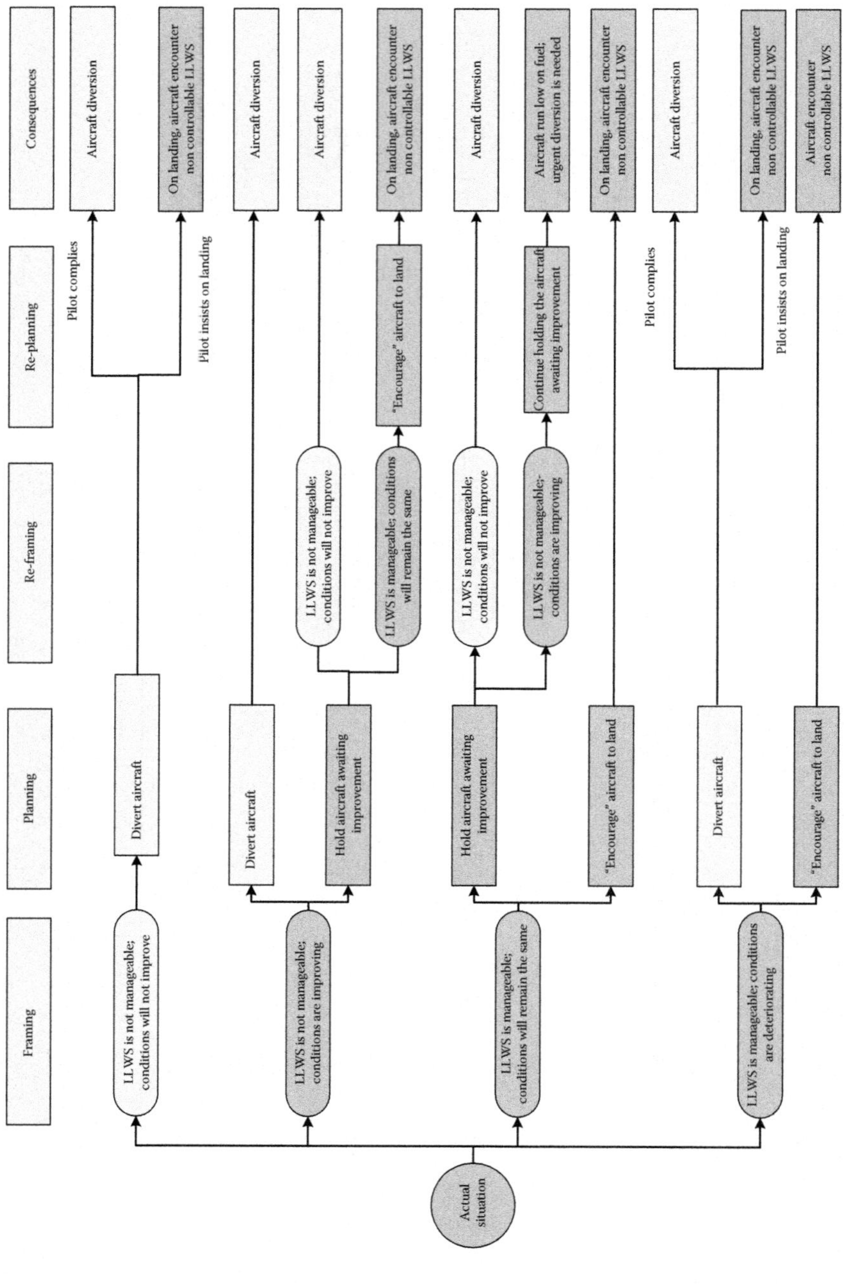

Figure 15.10 An expanded event tree of a response to a severe and deteriorating LLWS phenomenon.

near future. Although the actual case may refer to a LLWS event that is not manageable at present or in the future, the analyst will have to consider four possibilities or branches at the start of the event tree (Figure 15.10):

1. *LLWS is not manageable and the phenomenon will not get better.* In this case, the correct frame has been identified and the controllers have only one option to consider and divert aircraft to other airport. However, there is a possibility that some crews may insist on undertaking a landing on the premise of reliable assisting technology onboard the aircraft.
2. *LLWS is not manageable but the weather may get better later.* A safe decision would always be to ask flight crews for a diversion to another airport. However, some controllers may choose to put aircraft on a holding pattern, awaiting weather improvement. Subsequently, there is a chance that they realize that the weather will not improve and they may divert aircraft; there is also another chance that they expect the weather to improve and allow aircraft to attempt an unsafe landing.
3. *LLWS is manageable and the weather will remain stable.* This reflects a wrong frame of the present and future state of the LLWS event and may result in a decision to allow aircraft to enter an unsafe state of landing as crews may encounter a strong LLWS event. The most likely decision in this case would be to allow landing in conditions of strong LLWS. However, the cautious controller may decide to buy time and see how the weather evolves; so the controller may reframe the situation and make the correct decision to divert aircraft. A more likely outcome may be that the controller may realize that LLWS is not manageable at present but keeps aircraft on holding in the hope that things will get better. The risk here is that the holding pattern may last too long, resulting in fuel consumption and in urgent pilot requests to abort this pattern. This may create high workload since a large number of aircraft may be requesting diversion to different airports.
4. *LLWS is manageable but the weather may get worse.* The belief that things will get worse precludes the option of putting aircraft on holding. Hence, the controller may have only two

options: either to encourage aircraft to undertake landing or ask aircraft to divert to another airport. In the second case, the pilots can still make the final call and insist on landing.

In this way, the original event tree can be extended to include many additional branches that are likely to capture the wide variety of choice made by different controllers. The stopping rule of the analysis depends on the characteristics of the situation and the quality of data that are available to make further speculations about human performance. The extended event tree may produce more consequences, more accurate quantitative estimates, and better relationships between error likelihoods and recovery periods.

15.5 Influence Models

To tailor a risk model to the characteristics of a particular organization and derive quantitative estimates of human performance and hazards, it is important to model the influence of workplace, environmental, and management factors. The integrated risk picture (IRP) model of Eurocontrol (2006) uses an influence model that tailors generic estimates of failure probabilities in the event trees and fault trees to the characteristics of the organization. The influence model shows how a set of factors (e.g., training, interface design, procedures, traffic complexity) influence a specific event (e.g., the detection of a conflict) in a critical scenario (e.g., the occurrence of loss of separation). In general, influence models should provide information about how to perform a number of critical activities in risk quantification, such as the following:

1. The set of factors that influence an event controlled by a practitioner
2. The type of effect on human performance (e.g., direction, size, and duration of effects)
3. The additive effect of all factors on a specific event
4. A quantitative estimate of total influences on a specific event

Several techniques for building and quantifying influence models have been developed in the literature, including expert judgement, influence diagrams, Bayesian belief networks (BBN), and so on. These techniques mainly focus on the fourth activity and rely on general human factors knowledge to identify the set of influencing factors and their types of effect on performance. In this sense, the models of human and organizational performance that are presented in earlier chapters may be useful in understanding how a network of influencing factors interact together and affect performance of a particular event.

Many influence models assume a linear relationship between workplace factors (e.g., traffic complexity) and human performance (e.g., conflict detection). However, this is rarely true since human experts are able to sustain performance even when a traffic factor gets worse. Chapter 9, for instance, presents many examples showing that the relationship between traffic complexity and performance is nonlinear (see Figure 9.2). This implies that we should not expect a drastic deterioration of performance as complexity rises up to a certain threshold; beyond this tipping point performance can fall significantly as complexity rises to its highest levels. Although, the precise complexity-performance curve may be difficult to prescribe, an approximate relationship with different regions of effects could help analysts proceed with risk quantification.

Another example of nonlinear relationships regards the effects of procedures and safety rules on human performance. There is usually an assumption that an increase in the use of procedures may also enhance performance as practitioners may benefit from further rules and guidelines. Again this assumption may simplify the effort of quantifying the influences on performance but that is not always the case. In Chapter 7, for instance, it was shown that controllers usually adapt their practices to traffic situations in ways that are prescribed in procedures. Practitioners may consult procedures to develop their own practices but these may be developed in a different direction than the original procedures. Without an understanding of how procedures are incorporated into the work practices of controllers and how they are modified with the benefit of further experience, it is very difficult to estimate the effects of procedures on human performance.

Another important issue in the development of influence models regards the examination of the sources of origin or the root causes of influential factors. For instance, the root causes of traffic complexity are usually traced into the design of airspace, the original planning of traffic and the imbalance between demands and capacity of the sectors. The review of performance breakdowns (see Chapter 13) can provide further assistance into other forms of traffic complexity that are generated in a dynamic fashion during the interaction of controllers and flight crews.

Dysfunctional and unexpected interactions between controllers and crews can modify complexity as practitioners may transfer risks to others since the solution of one's own concerns may create problems elsewhere. For instance, adverse weather is a safety hazard for all flight operations. When weather cells are encountered, flight crews may request permission to circumnavigate cells, which could increase traffic complexity for controllers particularly in congested airspaces. Hence, granting a cell circumvention to aircrews may increase the risk of separation minima infringement and reduce the margin of maneuver for controllers. On the contrary, flight operations hazards are effectively reduced. Further increases in traffic complexity are generated as several flight crews may require changes to their routes to circumvent the weather cell.

In general, many influence models assume a simple model of human performance and influencing factors. In a typical risk assessment, analysts assume that there is an optimal standard of performance that all controllers should follow. This assumption, however, fails to address the trade-offs and decision dilemmas that controllers face in their environment (see Chapter 14). In many cases, safety may be the primary goal of controllers but this is often in conflict with efficiency of operations. For instance, controllers may maintain a high safety record at the expense of efficiency, forcing airlines to spend more mileage and fuel on their sectors, increasing aircraft delays or issuing a large number of route changes to aircraft. It is anticipated that this balance between safety and efficiency, may sometimes turn toward efficiency where controllers may create a tight traffic pattern that improves efficiency but leaves little scope for recovering problems. A better understanding of how controllers trade-off their decisions is necessary to make estimates of the pattern of effects of workplace and organizational patterns.

Finally, influence models should also address the issue of error detection and recovery since influencing factors differ in the way that affect the processes of error production and error recovery. In this regard, Chapter 6 provides useful knowledge in the processes of error detection and recovery that can be used in obtaining a better estimate of how practitioners recover their errors before any critical consequences are observed.

15.6 Risk Mitigation Measures

The previous stage of risk assessment has identified system failures and human errors that may lead to critical hazards as well as safety barriers that failed to protect the system from risk consequences. In particular, information about inadequate safety barriers and performance influencing factors is very useful for specifying risk mitigation measures in the form of training, procedural support, interface design, new allocation of functions, and technical barriers. Although the influences of risk mitigation measures have been considered in the influence models, introducing new measures or modifying existing ones requires a thorough human factors knowledge. In other words, the assumptions and simplifications used during the analysis may facilitate the generation of failure estimates but weaken the process of designing risk mitigation measures. For this reason, a good knowledge of human factors is needed to propose mitigation measures for reducing the risk level. In this section, a brief discussion is made of three risk mitigation measures that have been presented in Chapters 7, 8 and 10.

15.6.1 Aspects of System Design

The most common measure of risk mitigation regards changes in system design and task allocation between stakeholders. For instance, human performance may be increased by reducing workload that is assigned to other human or computer agents. An automated decision aid may enhance human performance by helping controllers to visualize aircraft trajectories and detect or resolve conflicts early. Changes in system design bring to the fore many research challenges in relation to the interaction between humans, artifacts and teams. Chapter 10 provides several examples, such as changes in authority and areas of regard, transfer of control between flight crews and controllers, issues of trust on automated aids, shared understanding between practitioners, and finally, coordination between people with different decision criteria and planning styles.

Any changes in system design will have to consider not only what tasks to automate but also how the jobs of practitioners change as a result of automation. Major changes introduced by NextGen may have significant implications for the roles of controllers since NextGen technologies would inflict the following changes in the tasks of controllers (see Chapter 10):

- A change in the way that controllers structure information in their airspace and cluster aircraft into categories for action
- A change in the pattern-recognition strategies of controllers
- A change from tactical to strategic roles since controllers would supervise the entire trajectory of aircraft instead of resolving local conflicts

In addition, changes in system design may be associated with different allocations of tasks between controllers, flight crews, and automation. In this case, attention should be paid not only to the newly allocated tasks but also to the coordination cost to manage the interactions between the new roles. As discussed in Chapter 10, the coordination cost may involve new arrangements, such as the following:

- Managing interactions between agents that have different working styles and priorities
- Managing the transfer of control from one agent to another and supervising the whole operation
- Sharing understanding of goals and intentions between agents so that they can anticipate the behavior of others and plan ahead for potential contingencies

The paradigm of cognitive engineering and the T^{2}EAM framework have been used in Chapter 10 to address these challenges in taskwork and teamwork in the extended team of controllers, artifacts, and flight crews required to work together in the new airspace environment advocated by SESAR and NextGen. In this respect, T^{2}EAM can be useful in postulating hypotheses about changes in controller and flight crew strategies to match the demands of the new human–artifact interaction.

15.6.2 Aspects of Controller Training

The design of refresher training can provide valuable knowledge how to reduce risks by enhancing the skills of practitioners, particularly in handling emergency and abnormal situations (EAS). As argued in Chapter 8, refresher training programs provided limited opportunities to practice EAS scenarios in the context of real world demands. Time constraints and cost considerations restricted the range and depth of EAS training to only the most common system malfunctions and emergencies.

In addition, a review of a sample of EAS reports from the Aviation Safety Reporting System (ASRS) found that most aviation incidents were not reflected in textbook emergencies (Burian and Barshi 2003). Also, the field study in Chapter 8 showed that the training scenarios did not simulate many elements of team interaction, system degradation (e.g., the STCA aid always alerted controllers), and workload management that were the major sources of error in incidents. In real events, controllers had made more errors in anticipating threats, planning traffic and their own workload and, finally, managing their own errors.

Real-life scenarios should go beyond normative if–then rules that signify well-rehearsed activities and require skills in cognitive functions such as

- Recognizing subtle cues in the environment
- Synthesizing patterns of cues
- Adapting to new constraints of work
- Replanning earlier actions
- Judging the timelines of interventions
- Coordinating actions
- Communicating intentions behind actions

Although modern training approaches offer more opportunities for hands-on learning and practice of realistic scenarios, they still suffer from:

- Inadequate representation of real-world demands (i.e., "cognitive fidelity")
- Lack of integration of cognitive strategies and technical skills.

Meeting the challenge of cognitive fidelity in training requires a clear understanding of the cognitive functions involved in the tasks that controllers undertake in realistic conditions. Many instructional features could be manipulated during training to allow practitioners to practice their cognitive functions, including adjusting the timing of events, adjusting the rate of update of information, introducing interruptions, late transfer of aircraft from other sectors, and so on. In addition, cognitive and technical skills could be integrated within the same training program with the use of cognitive task analysis. In Chapter 8, the ABCDE method was proposed for studying how cognitive strategies emerge in a complex domain and how they can be used to design refresher training that provides practice conditions to integrate technical skills with cognitive functions.

The ABCDE method can be helpful for specifying the cues, the challenges, the decisions, and the strategies used by controllers in the course of events of a complex scenario. It can also help instructors to enhance the cognitive fidelity of training by identifying events in a scenario that would provide opportunities to controllers to practice specific cognitive functions. The findings of Chapter 8 can help ANSP organizations to diagnose weaknesses in their training and seek advice how to overcome them.

15.6.3 Communities of Practice and Safety Knowledge

A large body of knowledge about how-the-system-works in risk assessment comes from official descriptions of work, standard operating procedures, safety rules, and regulatory documents. In most cases, these formal descriptions of work are usually taken at face value as descriptions of how-the-job-is-done in actual practice. As argued in Chapter 7, actual practices on the job may deviate from formal procedures as practitioners try to cope with unanticipated events, goal conflicts not addressed in formal rules, time pressure, and changes in technology that require modifications to existing procedures.

This gap between formal and informal work descriptions has been identified by many studies (Kontogiannis 1999a, 1999b; Dekker 2006) as a potential weakness in risk assessment. In general, safety audits performed on industrial systems usually exclude work practices from consideration or perceive them as workarounds to be avoided. As a result, the risk assessment relies on formal information that does not reflect the current state of affairs in the organization. It may be proposed then that informal work practices should be identified and brought into the foreground as a potential means of improving work procedures.

This opens up several critical questions that have already been examined in Chapter 7, that is, how informal practices should be identified, how they can be tested for their reliability, and whether they should be improved and documented to act as a reference material for practitioners. Although this is the formal way of testing and documenting work practices, there are still other avenues for facilitating transfer of knowledge between practitioners in organizations. For instance, organizations may rely on communities of practice of experienced controllers who communicate their knowledge to each other by less formal means such as weekly workshops, organizational journals, professional conferences, informal discussions, chats on professional websites, and so on. Between these formal and informal avenues, there may be other ways to communicate practical knowledge, such as relying on demonstration or show-how, rotating workers between units, and embedding knowledge in tools and technology (see Chapter 7).

Communities of practice require cultivation if organizations are to fully exploit their benefits; they will not flourish in inhospitable organizational environments. Organizations must seek to harness communities of practice in order to fully leverage their knowledge capacities. Sharing practices within an organization requires an atmosphere of mutual respect and trust. The presence of a trusty relationship indicates an ability to share a high degree of mutual understanding that is built on a common appreciation of a shared work context. Trust, familiarity, and mutual understanding are prerequisites for the successful transfer of practices. Indeed, empirical evidence suggests that trust leads to higher levels of openness between partners, thereby facilitating transfer. A fundamental purpose of managing knowledge is to build some degree of shared context so that practitioners exchange their assumptions and align their different perspectives.

15.7 Concluding Remarks

This chapter presents several cases how to apply the theoretical models of earlier chapters to provide practical guidance in safety risk management. The aim is not to present a field guide to risk management but rather to show how the book material can be used to expand existing approaches and techniques that constitute current practice in ATM. Emphasis has been given to elaborating system and risk models that could be tailored to particular organizations by paying attention to a range of workplace and organizational factors.

The second stage of safety risk management concerns the development of safety barriers and risk mitigation measures. This area requires a good knowledge of human factors in order to address a wide variety of barriers and measures including interface design, job aids, and computerized support, controller training, task reallocation, automation of tasks, and systems for disseminating risk information. The material in this book has covered only three areas of human factors interventions regarding system design, controller training and communities of practice that could disseminate information about work practices. Other areas of risk mitigation measures require additional knowledge of human factors.

The application of the techniques of safety risk management requires the recruitment of experienced personnel in safety departments since the collection, analysis and dissemination of risk information requires human, technical and financial resources. However, the safety record of organizations is not proportional to the amount of work and the number of safety people recruited in the safety department. Many safety analysts focus on the bureaucratic aspects of safety assurance mandated by an over-regulation of safety. As Dekker (2014) put it, "organizations are in need of more safety people not so much to manage safety but to feed a series of regulators with data sliced and parceled in particular ways." Safety departments are often organized around safety targets and indicators, risk assessment techniques, and compliances to regulatory safety systems.

This view of safety is unlikely to put in good practice the models and techniques that are presented in this book. Safety intelligence requires a mixture of effective safety models or techniques and active engagement of practitioners who do the dangerous work. The safety department must have an inquisitive mindset and an honest desire to learn what makes the work frustrating or dodgy at the sharp end. Safety should be the responsibility of all people at the organizational hierarchy and especially the practitioners who are actually exposed to risks. This view of total safety should be able to create valuable safety intelligence where: (1) hidden system causes are examined together with human errors at the sharp end, (2) unexplored work practices are compared with official procedures, and (3) proposed system changes and risk mitigation measures are continually evaluated and improved on the basis of operational experience.

The three safety approaches that have been discussed in Chapter 11 may be appropriate for organizations at different levels of safety maturity. For instance, a "defense-in-depth" approach may be useful to design appropriate barriers and close the "holes" in the organizational processes that may set the latent conditions of failure. As safety maturity grows up, error prevention can be supplemented with error recovery and mitigation so that adverse events that form a dangerous chain are detected and recovered in time. This systems thinking approach can explore better the dynamics of human and organizational performance so that a more proactive approach is adopted. As safety becomes more mature, more emphasis can be placed on resilience qualities and adaptive capacities required of organizations to withstand and recover from complex situations. Finding one's own place in the space of safety maturity and creating a blend of different safety approaches for a particular organization still remains an important challenge for many safety practitioners and managers.

References

AAIASB. 2006. *Helios Airways Flight HCY522 Crash at Grammatiko*. Report 5/2002. Athens: Air Accident Investigation and Aviation Safety Board, Hellenic Ministry of Transport and Communications.

Ahlstrom, U. 2005. Work domain analysis for air traffic control weather displays. *Journal of Safety Research*, 36, 159–169.

Air Accident Investigation. 1995. *Controlled Flight into Terrain American Airlines Flight 965 Boeing 757–233, N651AA near Cali, Colombia, December 20, 1995*. Aeronautica Civil of the Republic of Colombia.

AIRBUS. 2007. Adverse Weather Operations Windshear Awareness. FLT_OPS – ADV_WX – SEQ 02 – REV 03 – OCT. 2007. Flight Operations Briefing Notes. Blagnac Cedex France: Airbus.

Alange, S., Jacobson, S., and Jarnehammar, A. 1998. Some aspects of an analytical framework for studying the diffusion of organizational innovations. *Technology Analysis and Strategy Management*, 10, 3–12.

Ale, B. J. M, Bellamy, L. J., Cooke, R. M., Goossens, L. H. J., Hale, A. R., Roelen, A. L. C. and Smith, E. (2006). Towards a causal model for air transport safety—an ongoing research project. *Safety Science*, 44, 657–673.

Amalberti, A. and Wioland, L. 1997. Human error in aviation. In Soekkha, H. (Ed.), *Aviation Safety: Human Factors, System Engineering and Flight Operations*. CRC Press, 91–108.

Amalberti, R. 1992. Safety in process control: An operator-centered point of view. *Reliability Engineering and System Safety*, 38, 99–108.

Amalberti, R. 2001. The paradoxes of almost totally safe transportation systems. *Safety Science* 37, 109–126.

Amalberti, R. 2013. *Navigating Safety: Necessary Compromises and Tradeoffs*. New York, NY: Springer.

Amalberti, R., Vincent, C., Auroy, Y., and De Saint Maurice, G. 2006. Violations and migrations in health care: A framework for understanding and management. *Quality and Safety in Health Care*, 15, 66–71.

Amaldi, P. and Leroux, M. 1995. Selecting relevant information in a complex environment: The case of air traffic control. In L. Norros (Ed.), *5th European Conference on Cognitive Science Approaches in Process Control.* Espoo, Finland: VTT Automation, 89–98.

Andersen, S. and Mostue, B. A. 2012. Risk analysis and risk management approaches applied to the petroleum industry and their applicability to IO concepts. *Safety Science*, 50, 2010–2019.

ANSV. 2004. *Final Report. Accident Involved Aircraft Boeing MD-87, Registration SE-DMA and CESSNA 525-A, Registration D-IEVX Milano Linate Airport October 8, 2001.* Agenzia Nazionale per la Sicurezza del Volo.

Argote, L. and Ingram, P. 2000. Knowledge transfer: A basis for competitive advantage in firms. *Organizational Behavior and Human Decision Processes*, 82, 150–169.

Athenes, S., Averty, P., Puechmorel, S., Delahaye, D., and Collet, C. 2002. Complexity and controller workload: Trying to bridge the gap. *Proceedings of the 2002 International Conference on Human-Computer Interaction in Aeronautics.* Cambridge MA: Massachusetts Institute of Technology, 56–60.

Bainbridge, L. 1975. *Working Memory in Air Traffic Control.* Unpublished manuscript. London: University College, Department of Psychology.

Bainbridge, L. 1983. Ironies of automation. *Automatica*, 19, 775–780.

Batt, R. and O'Hare, D. 2005. Pilot behaviors in the face of adverse weather: A new look at an old problem. *Aviation, Space and Environmental Medicine*, 76, 552–559.

Battiste, V., Johnson W., Brandt, S., Dao, A. Q., and Johnson, N.H. 2008. Assessment of flight crew acceptance of automated resolutions suggestions and manual resolution tools. *Proceedings of the 26th International Congress of the Aeronautical Sciences.* (ICAS), Anchorage, AS.

Baumard, P. 1999. *Tacit Knowledge in Organizations.* London: Sage.

Beer, S. 1985. *Diagnosing the System for Organizations.* Chichester: Wiley.

BFU, 2004. *Investigation Report.* AX001-1-2/02 May 2004. German Federal Bureau of Aircraft Accident Investigation. Bundesstelle für Flugunfalluntersuchung.

Blavier, A., Rouy, E., Nyssen, A.S., and de Keyser, V. 2005. Prospective issues for error detection. *Ergonomics*, 48, 758–781.

Bolic, T. and Hansen, M. 2005. User request evaluation tool (URET) adoption and adaptation, three center case study. *6th USA/Europe Air Traffic Management Research and Development Seminar*, Baltimore, MD, 27–30 June 2005.

Bourrier, M. 1996. Organizing maintenance work at two nuclear power plants. *Journal of Contingencies and Crisis Management*, 4, 104–112.

Bove, T. 2004. *Development and Validation of a Human Error Management Taxonomy in Air Traffic Control.* Unpublished PhD Thesis. Roskilde, Denmark: Riso National Laboratory.

Branlat, M., Morison, A. M., Finco, G. J., Gertman, D. I., Le Blanc, K., and Woods, D. D. 2011. A study of adversarial interplay in a cybersecurity event. In S. M. Fiore and M. Harper-Sciarini (Eds.), *Proceedings of the 10th International Conference on Naturalistic Decision Making,* 31 May–3 June 2011, Orlando, FL: University of Central Florida.

Brehmer, B. 2000. Dynamic decision making. In C. McCann and N. Pigeau (Eds.), *The Human in Command: Exploiting the Modern Military Experience*, New York, NY: Kluwer Academic/Plenun Publishers, 233–248.

Brooker, P. 2003. Control workload, airspace and future systems. *Human Factors and Aerospace Safety*, 3(1), 1–23.

Brooker, P. 2005. ATC automation: For humans or people? *Human Factors and Aerospace Safety*, 5(1), 23–42.

Brooker, P. 2007. Air Traffic Safety: Continued Evolution or a New Paradigm? *Transport Risk Management Lecture*, Imperial College, Lloyds Register Educational Trust, 17 October 2007.

Burian, B. K. and Barshi, I. 2003. Emergency and abnormal situations: A review of ASRS report. *Proceedings of the 12th International Symposium on Aviation Psychology*, 14–17 April 2003, Dayton, OH: Wright State University Press.

Burian, B. K., Barshi, I., and Dismukes, K. 2005. *The Challenge of Aviation Emergency and Abnormal Situations.* Technical Report No. NASA/TM—2005-213462. Moffett Field, CA: NASA Ames Research Center.

Busby, J. S. 2006. Failure to mobilize in reliability-seeking organizations: Two cases from the UK railway. *Journal of Management Studies*, 43, 1375–1393.

CAA. 2005. *Aircraft Emergencies Consideration for Air Traffic Controllers.* CAP 745. Gatwick Aerodrome: Civil Aviation Authority.

CAA. 2016. *Implementation of the Recommendations from the Independent Enquiry into the NATS Systems Failure on 12th December 2014.* CAP 1480. Gatwick Aerodrome: Civil Aviation Authority.

Cannon-Bowers, J. A., Tannebaum, S. I., Salas, E., and Volpe, C. E. 1995. Defining competencies and establishing team training requirements. In R. A. Guzzo and E. Salas (Eds.), *Team Effectiveness and Decision Making in Organizations.* San Francisco: Jossey-Bass, 333–381.

Cardosi, K. M. 1993. Time required for transmission of time critical air traffic control messages in an en route environment. *International Journal of Aviation Psychology*, 3, 303–314.

Checkland, P. and Scholes, J. 1999. *Soft Systems Methodology in Action.* Chichester: John Wiley.

Clarke, D. M. 2005. Human redundancy in complex, hazardous systems: A theoretical framework. *Safety Science*, 43, 655–677.

Cohen, M. S., Freeman, J. T., and Thomson B. 1998. *Critical Thinking Skills in Tactical Decision Making: A Model and a Training Strategy*. Arlington, VA: Cognitive Technologies Inc.

Cohen, M. S., Freeman, J. T., and Wolf, S. P. 1996. Meta-recognition in time stressed decision making: Recognizing critiquing and correcting. *Human Factors*, 38, 206–219.

Colville, I. and Pye, A. 2010. A sensemaking perspective on network pictures. *Industrial Market Management*, 39, 372–380.

Commission Decision (EU) of 7.7.2011a on the nomination of the Network Manager for the air traffic management (ATM) network functions of the single European sky.

Commission Implementation Regulation (EU) 2016/1377 of 4 August 2016 laying down common requirements for service providers and the oversight in air traffic management/air navigation services and other air traffic management network functions, repealing Regulation (EC) No 482/2008, Implementing Regulations (EU) No 1034/2011 and (EU) No 1035/2011 and amending Regulation (EU) No 677/2011.

Commission Implementation Regulation (EU) No 1035/2011 of 17 October 2011 laying down common requirements for the provision of air navigation services and amending Regulations (EC) No 482/2008 and (EU) No 691/2010.

Commission Regulation (EU) 2015/340 of 20 February 2015 laying down technical requirements and administrative procedures relating to air traffic controllers' licences and certificates pursuant to Regulation (EC) No 216/2008 of the European Parliament and of the Council, amending Commission Implementing Regulation (EU) No 923/2012 and repealing Commission Regulation (EU) No 805/2011.

Commission Regulation (EU) No 1018/2015 of 29 June 2015 laying down a list classifying occurrences in civil aviation to be mandatorily reported according to Regulation (EU) No 376/2014 of the European Parliament and of the Council.

Commission Regulation (EU) No 216/2008 of 20 February 2008 on common rules in the field of civil aviation and establishing a European Aviation Safety Agency, and repealing Council Directive 91/670/EEC, Regulation (EC) No 1592/2002 and Directive 2004/36/EC.

Commission Regulation (EU) No 255/2010 of 25 March 2010 laying down common rules on air traffic flow management.

Commission Regulation (EU) No 677/2011b of 7 July 2011 laying down detailed rules for the implementation of air traffic management (ATM) network functions and amending Regulation (EU) No 691/2010.

Cook, R. I. 2006. Being bumpable. In D. D. Woods and E. Hollnagel (Eds.), *Joint Cognitive Systems: Patterns in Cognitive Systems Engineering*. Boca Raton, FL: CRC Press.

Cook, R. I. and Rasmussen, J. 2005. "Going solid": A model of system dynamics and consequences for patient safety. *Quality and Safety in Health Care*, 14(2), 130–134.

Cooke, D. L. 2003. A system dynamics analysis of the Westray mine disaster. *System Dynamics Review*, 19, 139–166.

Corver, S. and Grote, G. 2016. Uncertainty management in en-route air traffic control: A field study exploring controller strategies and requirements for automation. *Cognition Technology and Work*, 18, 541–565.

Cox, T. 1987. Stress, coping and problem solving. *Work and Stress*, 1, 5–14.

Crandall, B., Klein, G. A., and Hoffman, R. R. 2006. *Working Minds: A Practitioner's Guide to Cognitive Task Analysis*. Cambridge, MA: MIT Press.

Croft, J. 2009. *Complex Error Chain Preceded Delta 767 TaxiwayLanding*. NTSB Report. http://www.flightglobal.com/articles/2009/12/29/336663/ntsb-complex-errorchain-preceded-delta-767-taxiway.html.

D'Arcy, J. F. and Della Rocco, P. 2001. *Air Traffic Control Specialist Decision Making and Strategic Planning—A Field Study*. DOT/ FAA/ CT-TN01.05. Atlantic City International Aerodrome, NJ: DOT/FAA William J. Hughes Technical Center.

Davoudian, K., Wu, J. S. and Apostolakis, G. 1994. Incorporating organizational factors into risk assessment through the analysis of work processes. *Reliability Engineering and System Safety*, 45, 85–105.

Degani, A. and Wiener, E. L. 1994. Philosophy, policies, procedures, and practices: The Four "P"s of flight deck operations. In N. Johnston, N. McDonald, and R. Fuller (Eds.), *Aviation Psychology in Practice*, Hants: Avebury Technical, 44–67.

De Keyser, V. and Woods, D. D. 1990. Fixation errors: Failures to revise situation assessment in dynamic and risky environments. In A. G. Colombo and A. Saiz de Bustamante (Eds.), *Systems Reliability Assessment*, Amsterdam: Kluwer Academic, 231–251.

Dekker, S. W. 2003. Failure to adapt or adaptations that fail: Contrasting models on procedures and safety. *Applied Ergonomics*, 34, 233–238.

Dekker, S. W. 2005. *Ten Questions about Human Error: A New View of Human Factors and System Safety*. Mahwah, NJ: Lawrence Erlbaum Associates.

Dekker, S. W. 2006. *The Field Guide to Understanding Human Error*. Aldershot: Ashgate.

Dekker, S. W. 2011. *Drift into Failure: From Hunting Broken Components to Understanding Complex Systems*. Aldershot: Ashgate.

Dekker, S. W. 2014. *The Field Guide to Understanding Human Error*. Second Edition. Aldershot: Ashgate.

Dekker, S. W., Dahlstrom, N., van Winsen, R., and Nyce, J. M. 2008. Crew resilience and simulator training in aviation. In E. Hollnagel, C. P. Nemeth, and S. Dekker (Eds.), *Resilience Engineering Perspectives: Remaining Sensitive to Possibility of Failure*, Vol. 1. Aldershot: Ashgate, 119–126.

Dekker, S. W. and Woods, D. D. 1999. To intervene or not to intervene: The dilemma of management by exception. *Cognition Technology and Work*, 1, 86–96.

Dijkstra, A. 2007. Cybernetics and resilience engineering: Can cybernetics and the viable system model advance resilience engineering? In *Proceedings of the Resilience Engineering Workshop*, 25–27 June 2007, Vadstena, Sweden.

Dismukes, R. K., Berman, B. A., and Loukopoulos, L. D. 2007. *The Limits of Expertise: Rethinking Pilot Error and the Causes of Airline Accidents.* Aldershot: Ashgate.

Dorner, D. 1996. *The Logic of Failure.* New York, NY: Metropolitan Books/Henry Holt.

Dunbar, M., McGann, A., Mackintosh, M., and Lozito, S. 2001. *Re-examination of Mixed Media Communication: The Impact of Voice, Data-link, and Mixed Air Traffic Control Environments on the Flight Deck.* NASA/TM-2001-210919. Moffet Field, CA: NASA Ames Research Center.

Dwyer, J. P. and Landry, S. 2009. Separation assurance and collision avoidance concepts for the next generation air transportation system. In M. J. Smith and G. Salvendy (Eds.), *Human Interface*, Part II, HCII, LNCS 561, Berlin: Springer, 748–757.

EASA. 2011. *The ARMS Methodology for Operational Risk Assessment in Aviation Organisations.* http://www.easa.eu.int/essi/documents/Methodology.pdf

EATMP. 1999. *Controller Training in the Handling of Unusual Incidents.* HUM.ET1. ST12.3000-GUI-01. Brussels: Eurocontrol.

EATMP. 2000. *Validation of the Human Error in ATM (HERA) Classification Schema.* Brussels: Eurocontrol.

Edmonds, B. 1999. *Syntactic Measures of Complexity.* PhD Thesis. Manchester: University of Manchester.

Embrey, D., Kontogiannis, T., and Green, M. 1994. *Guidelines for Reducing Human Error in Process Operations.* New York: Center for Chemical Process Safety.

Endsley, M., Mogford, R., Allendoerfer, K., Snyder, M., and Stein, E. 1997. *Effects of Free Flight Condition on Controller Performance, Workload and Situation Awareness.* Technical Report No. DOT/FAA/CT-TN97/12. Atlantic City, NJ: Federal Aviation Administration.

Entin, E. B. and Entin, E. E. 2000. Assessing team situation awareness in simulated military missions. *Proceedings of the Human Factors and Ergonomics Society 44th Annual Meeting.* San Diego, CA: Human Factors and Ergonomics Society Press, 73–77.

Entin, E. E. and Serfaty, D. 1999. Adaptive team coordination. *Human Factors*, 41, 321–325.

Erzberger, H. 2006. Automated conflict resolution for air traffic control. *Proceedings of the 25th International Congress of the Aeronautical Sciences*, 3–8 September 2006, Hamburg, Germany.

Espejo, R. and Harnden, R. 1990. *The Viable System Model: Interpretations and Applications.* New York, NY: Wiley.

Eurocontrol. 2005. *ATM Process Model SADT Diagrams.* Bretiny: Eurocontrol.

Eurocontrol. 2006. *Main Report for the 2005/2012 Integrated Risk Picture for Air Traffic Management in Europe*. Bretiny: Eurocontrol.

Eurocontrol. 2007. *First ATC Support Tools Implementation (FASTI) –Task Analysis*. Bretiny: Eurocontrol.

Eurocontrol. 2009. *Risk Analysis Tool*. ESP/2009-81. Brussels: Eurocontrol.

Eurocontrol. 2016. *Air Traffic Flow & Capacity Management Operations: ATFCM User's Manual*. Twentieth Edition. Brussels: Eurocontrol.

Falzon, P. 1982. Display structures: Compatibility with the operator's mental representation and reasoning process. *Proceedings of the 2nd European Annual Conference on Human Decision Making and Manual Control*. Bonn, 297–305.

Feltovich, P. J., Spiro, R. J., and Coulson, R. L. 1997. Issues of flexible flexibility in contexts characterized by complexity and chance. In P. J. Feltovich, K. M. Ford, and R. R. Hoffman (Eds.), *Expertise in Context: Human and Machine*. Cambridge, MA: MIT Press, 125–146.

Fletcher, G., Flin, R., McGeorge, P., Glavin, R., Maran, N., and Patey, R. 2004. Rating non-technical skills: Developing a behavioral marker system for use in anaesthesia. *Cognition Technology and Work*, 6, 165–171.

Flin, R., Martin, L., Goeters, K., Hoermann, J., Amalberti, R., Valot, C. and Nijhuis, H., 2003. Development of the NOTECHS (Non-technical skills) system for assessing flight crews' CRM skills. *Human Factors and Aerospace Safety* 3, 95–117.

Flin, R., Salas, E., Strub, M., and Martin, L. 1997. *Decision Making Under Stress: Emerging Themes and Applications*. Aldershot: Ashgate.

Gaba, D. M. and DeAnda, A. 1989. The response of anesthesia trainees to simulated critical incidents. *Anesthesia and Analgesia*, 68, 444–451.

Garud, R., Kumaraswamy, A., and Karnøe, P. 2010. Path dependence or path creation? *Journal of Management Studies*, 47(4), 760–774.

Gomes, J. O., Woods, D. D., Carvalho, P. V. R, Huber, G., and Borges, M. R. S. 2009. Resilience and brittleness in the offshore helicopter transportation system: The identification of constraints and sacrifice decisions in flight crews' work. *Reliability Engineering and Systems Safety*, 94, 311–319.

Gordon, R., Kirwan, B., and Perrin, E. 2007. Measuring safety culture in a research and development centre: A comparison of two methods in the air traffic domain. *Safety Science*, 45, 669–695.

Gronlund, S. D., Dougherty, M. R., Durso, F. T., Canning, J. M., and Mills, S. H. 2005. Planning in air traffic control: Impact of problem type. *International Journal of Aviation Psychology*, 15, 269–293.

Grote, G. 2009. *Management of Uncertainty: Theory and Application in the Design of Systems and Organizations*. London: Springer.

Hale, A. R. and Swuste, P. 1998. Safety rules: Procedural freedom or action constraint? *Safety Science*, 29, 163–177.

Hammond, K. R., Hamm, R. M., Grassia, J. and Pearson, T., 1987. Direct comparison of the efficacy of intuitive and analytical cognition in expert judgment. *Proceedings of IEEE Transactions on Systems, Man, and Cybernetics,* SMC-17, 753–770.

Hansman, R. J. and Davison, H. J. 2000. The effect of shared information on pilot / controller and controller / controller interactions. *3rd USA/Europe Air Traffic Management R&D Seminar.* Napoli, Italy.

Helmreich, R. L, Hackman, J. R., and Foushee, H. C. 1998. *Evaluating Flight Crew Performance: Policy, Pressures, Pitfalls and Promises.* NASA Technical Memorandum. Moffett Field: NASA Ames Research Center.

Helmreich, R. L., Klinect, J. R., and Wilhelm, J. A. 1999. Models of threat, error, and CRM in flight operations. *Proceedings of the 10th International Symposium on Aviation Psychology.* Columbus, OH: The Ohio State University, 677–682.

Helmreich, R.L., Klinect, J.R., and Wilhelm, J.A., 2001. System safety and threat and error management: The line operational safety audit (LOSA). In: *11th International Symposium on Aviation Psychology.* Columbus, OH: Ohio State University.

Henneberg, S.C., Mouzas, S., and Naudé, P. 2006. Network pictures: Concepts and representations. *European Journal of Marketing*, 40, 408–429.

Hilburn, B. 2004. *Cognitive Complexity in Air Traffic Control: A Literature Review.* Brétigny-sur-Orge CEDEX: Eurocontrol.

Histon, J. M. and Hansman R. J. 2002. *The Impact of Structure on Cognitive Complexity in Air Traffic Control.* Report No ICAT-2002-4. Cambridge, MA: MIT International Center for Air Transportation.

Hoffman, R. R. 1998. How can expertise be defined?: Implications of research from cognitive psychology. In R. Williams, W. Faulkner, and J. Fleck (Eds.), *Exploring Expertise.* New York, NY: Macmillan, 81–100.

Hoffman, R. R, Crandal, B., and Shadbolt, N. 1998. Use of the critical decision method to elicit expert knowledge: A case study in the methodology of cognitive task analysis. *Human Factors*, 40, 254–276.

Hoffman, R. R. and Woods, D. D. 2011. Simons's slice: Five fundamental tradeoffs that bound performance of human work systems. *Proceedings of the 10th International Conference on Naturalistic Decision Making*, Orlando, FL: University of Central Florida.

Hollnagel, E. 1992. Coping, coupling and control: The modeling of muddling through. *Proceedings of 2nd Interdisciplinary Workshop on Mental Models*, 23–25 March 1992, Cambridge: Robinson College.

Hollnagel, E. 1998. *Cognitive Reliability and Error Analysis Method.* London: Elsevier.

Hollnagel, E. 2004. *Barriers and Accident Prevention.* Aldershot: Ashgate.

Hollnagel, E. 2007. Flight decks and free flight: Where are the system boundaries? *Applied Ergonomics*, 38, 409–416.

Hollnagel, E. 2008. Investigation as an impediment to learning. In E. Hollnagel, C. P. Nemeth, and S.W. Dekker (Eds.), *Remaining Sensitive to the Possibility of Failure.* Aldershot: Ashgate.

Hollnagel, E. 2009. *The ETTO Principle: Efficiency-Thoroughness Trade-Off: Why Things that Go Right Sometimes Go Wrong.* Aldershot: Ashgate.

Hollnagel, E. and Amalberti, R. 2001. The emperor's new clothes: Or whatever happened to human error. In S. W. A. Dekker (Ed.), *Proceedings of the 4th International Workshop on Human Error, Safety and System Development*. Linkoping, Sweden: Linkoping University, 1–18.

Hollnagel, E., Leveson, N., and Woods, D. D.. (Eds.). 2006. *Resilience Engineering: Concepts and Precepts*. Aldershot: Ashgate.

Hollnagel, E., Paries, J., Woods, D.D., and Wreathal, J. 2011. *Resilience Engineering in Practice*. Aldershot, UK: Ashgate.

Hollnagel, E. and Woods, D. D. 2005. *Joint Cognitive Systems. Foundations of Cognitive Systems Engineering*. London: Taylor and Francis.

Holyoak, K. J. 1991. Symbolic connectionism: Toward third-generation theories of expertise. In K. A. Ericsson and J. Smith (Eds.), *Toward a General Theory of Expertise: Prospects and Limits*. Cambridge, MA: Cambridge University Press, 301–335.

Hopkin, D. V. 1995. *Human Factors in Air Traffic Control*. London: Taylor & Francis.

Hopkins, A. 2009. Thinking about process safety indicators. *Safety Science*, 47, 460–465.

Hormann, H. J. 1995. FOR-DEC: A prescriptive model of aeronautical decision making. In R. Fuller, N. Johnston, and N. McDonald (Eds.), *Human Factors in Aviation Operations. Proceedings of the 21st Conference of the European Association for Aviation psychology (EAAP)*. Aldershot: Avebury, 17–23.

Hoverstadt, P. 2008. *The Fractal Organization: Creating Sustainable Organizations with the Viable System Model*. Chichester: John Wiley.

HSE. 2006. *Developing Process Safety Indicators: A Step-by-Step Guide for Chemical and Major Hazard Industries*. Sudbury: Health and Safety Executive books.

Hudson, P., Reason, J., Wagenaar, W., Bentley, P., Primrose, M. and Visser, J., 1994. Tripod-delta: A proactive approach to enhanced safety. *Journal of Petroleum Technology*, 40, 58–62.

ICAO. 1984. *Air Traffic Services Planning Manual*. Document 9426. First Edition. Montreal: International Civil Aviation Organization.

ICAO. 2001. *Air Traffic Services: ANNEX 11 to the Convention of International Civil Aviation*. Thirteenth Edition. Montreal: International Civil Aviation Organization.

ICAO. 2002. *Line Operations Safety Audit (LOSA)*. Document 9803, AN/761. First Edition. Montreal: International Civil Aviation Organization.

ICAO. 2005a. *Rules of the Air: ANNEX 2 to the Convention of International Civil Aviation*. Tenth Edition. Montreal: International Civil Aviation Organization.

ICAO. 2005b. *Threat and Error Management (TEM) in Air Traffic Control*. Preliminary Edition. *Montreal: International Civil Aviation Organization*.

ICAO. 2005c. *Manual on Low-level Wind Shear*. Document 9817, AN/449. First Edition. Montreal: International Civil Aviation Organization.

ICAO. 2006a. *Aircraft Operations, Flight Procedures*. Document 8168. Vol. 1, Fifth Edition. Montreal: International Civil Aviation Organization.

ICAO. 2006b. *Aircraft Operations, Construction of Visual and Instrument Flight Procedures*. Document 8168. Vol. 2, Fifth Edition. Montreal: International Civil Aviation Organization.

ICAO. 2006c. *Aeronautical Telecommunications: ANNEX 10 to the Convention of International Civil Aviation. Radio Navigation Aids*, Vol. 1, Sixth Edition. Montreal: International Civil Aviation Organization.

ICAO. 2006d. *Convention on International Civil Aviation Organization*. Document 7300/9. Ninth Edition. Montreal: International Civil Aviation Organization.

ICAO. 2007a. *Air Traffic Management. Procedures for Air Navigation Services*, Document 4444, ATM/501. Seventh Edition. Montreal: International Civil Aviation Organization.

ICAO. 2007b. *Manual of Radiotelephony*. Document 9432, AN/925. Fourth Edition. Montreal: International Civil Aviation Organization.

ICAO. 2007c. *Manual on the Prevention of Runway Incursions*. Document 9870, AN/463. First Edition. Montreal: International Civil Aviation Organization.

ICAO. 2008. *Normal Operations Safety Survey (NOSS)*. Document 9910, AN/473. First Edition. Montreal: International Civil Aviation Organization.

ICAO. 2010a. *Aircraft Accident and Incident Investigation: ANNEX 13 to the Convention of International Civil Aviation*. Tenth Edition. Montreal: International Civil Aviation Organization.

ICAO. 2010b. *ANNEX 6 to the Convention of International Civil Aviation. International Commercial Air Transport — Aeroplanes*, Vol. 1, Ninth Edition. Montreal: International Civil Aviation Organization.

ICAO. 2013a. *Aerodrome Design and Operations: ANNEX 14 to the Convention of International Civil Aviation*, Vol. 1, Sixth Edition. Montreal: International Civil Aviation Organization.

ICAO. 2013b. *Safety Management: ANNEX 19 to the Convention of International Civil Aviation*. First Edition. Montreal: International Civil Aviation Organization.

ICAO. 2013c. *Safety Management Manual (SMM)*. Document 9859, AN/474. Third Edition. Montreal: International Civil Aviation Organization.

IRGC. 2010. *The Emergence of Risks: Contributing Factors*. Geneva: International Risk Governance Council, IRGC Report.

Isaac, A., Shorrock, S.T., Kennedy, R., Kirwan, B., Andersen, H. B. and Bove, T. 2001. *The Human Error in ATM (HERA) Technique*. Edition 0.2 - Second Draft. Internal document. Eurocontrol Experimental Center, Eurocontrol.

Ivancevich, J. M. and Matteson, M. T. 1980. *Stress and Work*. Glenview, IL: Scott Foresman.

Jackson, M. C. 2003. *Systems Thinking: Creative Holism for Managers*. Chichester: John Wiley.

Jentsch, F. and Bowers, C. A. 1998. Evidence for the validity of low-fidelity simulations in air crew coordination research and training. *International Journal of Aviation Psychology*, 8, 243–260.

Jha, P. D., Bisantz, A. M., and Drury, C. G. 2011. Air traffic controlllers' performance in advanced air traffic control systems: Part II Workload and trust. *Journal of Air Traffic Control*, 51(2), 46–52.

Jha, P. D, Bisantz, A. M, Parasuraman, R., and Drury, C. G. 2011. Air traffic controllers' performance in advance air traffic management system: Part I—Performance results. *International Journal of Aviation Psychology*, 21, 283–305.

Johnson, C. W. 2003. *Failure in Safety Critical Systems: A Handbook of Incident and Accident Reporting*. Glasgow: Glasgow University Press.

Johnston, N., Seamster, T. L., Redding, R. E., and Kaempf, G. L. 1997. *Applied Cognitive Task Analysis in Aviation*. Aldershot: Ashgate.

Kahneman, D., Slovic, P. and Tversky, A. (Eds.) 1982. *Judgment under Uncertainty: Heuristics and Biases*. Cambridge: Cambridge University Press.

Kallus, K. W., Van Damme, D., and Dittman, A. 1999. *Integrated Job and Task Analysis of Air Traffic Controllers: Phase 2. Task Analysis of En-route Controllers*. European Air Traffic Management Programme Report No HUM.ET1. ST01.1000-REP-04. Brussels: EUROCONTROL.

Kanki, B. G. and Palmer, M. T. 1993. Communication and crew resource management. In E. L. Wiener, B. G. Kanki, and R. L. Helmreich (Eds.), *Cockpit Resource Management*. San Diego: Academic Press, 99–136.

Kanse, L. and van der Schaaf, T. 2001. Recovery from failures in the chemical process industry. *International Journal of Cognitive Ergonomics*, 5, 199–211.

Kennedy, R. and Kirwan, B. 1998. Development of a hazard and operability based method for identifying safety management vulnerabilities in high risk systems. *Safety Science*, 30, 249–274.

Kirwan, B. 2013. *Safety Intelligence for ATM CEOs: A White Paper*. Bretigny: Eurocontorl.

Kirwan, B. and Ainsworth, L. K. 1992. *A Guide to Task Analysis*. London: Taylor & Francis.

Kirwan, B. and Flynn, M. 2002. *Towards a Controller-based Conflict Resolution Tool: A Literature Review*. European Air Traffic Management Program Report No ASA.01. CORA.2.DEL04- A.LIT. Brussels: EUROCONTROL.

Klein, G., Pliske, R., Crandall, B., Woods, D.D., 2005. Problem detection. *Cognition Technology and Work*, 7, 14–28.

Klein, G.A. 1989. Recognition-primed decisions. In: W.B. Rouse (Ed.) *Advances in Man-Machine Systems Research*, Vol. 5. Greenwich, CT: JAI Press, 47–92.

Klein, G. A. 1998. *Sources of Power: How People Make Decisions*. Cambridge MA: MIT Press.

Klein, G. A. 2004. *The Power of Intuition*. New York, NY: Currency Books.

Klein, G. A. 2006. The strengths and limitations of teams for detecting problems. *Cognition Technology and Work*, 8, 227–236.

Klein, G. A. 2007. Flexecution part 2: Understanding and supporting flexible execution. *IEEE Intelligent Systems*, 6, 108–112.

Klein G. A. 2009. *Streetlights and Shadows: Searching for the Keys to Adaptive Decision Making*. Cambridge, MA: MIT Press.

Klein, G.A., Calderwood, R. and Clinton-Cirocco, A. 1986. Rapid decision making on the fire ground. *In Proceedings of the 30th Annual Human Factors Society Meeting*. Dayton, OH: Human Factors Society, 576–580.

Klein, G. A., Moon, B., and Hoffman, R. R. 2006. Making sense of sensemaking 2: A Macrocognitive Model. *IEEE Intelligent Systems*, 21(5), 88–92.

Klein, G. A., Orasanu, J., Calderwood, R., and Zsambok, C. E.. (Eds.). 1993. *Decision Making in Action: Models and Methods*. Norwood, NJ: Ablex Publishing.

Klein, G. A., Philips, J. K., Rall, E. L., and Peluso, D. A. 2007. A data/frame theory of sensemaking. In R. R. Hoffman (Ed.), *Expertise Out of Context.Proceedings of the 6th International Conference on Naturalistic Decision Making*. Mahwah: Erlbaum, 113–155.

Klein, G. A. and Pierce, L. G. 2001. Adaptive teams. *Proceedings of the 6th ICCRTS Collaboration in the Information Age Track 4: C2decision Making and Cognitive Analysis*. Washington D.C.: CCRP Press. http://www.dodccrp.org/6thICCRTS

Klein, G. A., Ross, K., Moon, B., Klein, D. E., Hoffman, R., and Hollnagel, E. 2003. Macrocognition. *IEEE Intelligent Systems*, 18, 81–85.

Klein, G. A., Wiggins, S., and Dominguez, C. O. 2010. Team sensemaking. *Theoretical Issues in Ergonomics Science*, 11, 304–320.

Klir, G. J. 1969. *An Approach to General Systems Theory*. New York, NY: Nostrand.

Kontogiannis, T. 1996. Stress and operator decision making in coping with emergencies. *International Journal of Human-Computer Studies*, 45, 75–104.

Kontogiannis, T. 1999a. Training effective human performance in the managing of stressful emergencies. *Cognition Technology and Work*, 1, 7–24.

Kontogiannis, T. 1999b. User strategies in recovering from errors in man machine systems. *Safety Science*, 32, 49–68.

Kontogiannis, T. 2010a. A contemporary view of organizational safety: Variability and interactions of organizational processes. *Cognition Technology and Work*, 12, 231–249.

Kontogiannis, T. 2010b. Adapting plans in progress in distributed supervisory work: Aspects of complexity, coupling, and control. *Cognition Technology and Work*, 12, 103–118.

Kontogiannis, T. 2011. A systems perspective of managing error recovery and tactical replanning of operating teams in safety critical domains. *Journal of Safety Research*, 42, 73–85.

Kontogiannis, T. and Malakis, S., 2009. A proactive approach to human error detection and identification in aviation and air traffic control. *Safety Science*, 47, 693–706.

Kontogiannis, T. and Malakis, S. 2012a. A systemic analysis of patterns of breakdown in organizational accidents: A case from helicopter emergency medical service (HEMS) operations. *Reliability Engineering and System Safety*, 99, 193–208.

Kontogiannis, T. and Malakis, S. 2012b. Recursive modeling of loss of control in human and organizational processes: A systemic model for accident analysis. *Accident Analysis & Prevention*, 48, 303–316.

Kontogiannis, T. and Malakis, S. 2013a. Remaining safe by working at the edge of compliance and adaptation: Reflective practices in aviation and air traffic control. *Theoretical Issues in Ergonomics Science*, 14(6), 565–591.

Kontogiannis, T. and Malakis, S. 2013b. Strategies in coping with complexity: Development of a behavioural marker system for air traffic controllers. *Safety Science*, 57, 27–34.

Koopman, P. and Hoffman, R.R. 2002. Work-arounds, make-work, and kludges. *IEEE Intelligent Systems–Human Centered Computing*, 18(6), 70–75.

Kopardekar, P. and Magyarits, S. 2003. Measurement and prediction of dynamic density. *Paper Presented at the 5th USA/Europe Air Traffic Management Research and Development Seminar.* Budapest, Hungary.

Koros, A., Della Roco, P. S., Panjwani, G., Ingurgio, V., and D'Arcy, J. F. 2006. *Complexity in Aerodrome Traffic Control Towers: A Field Study. Part 2. Controller Strategies and Information Requirements.* Report No. DOT/FAA/TC-06/22. Springfield, VA: National Technical Information Service (NTIS).

Kranz, G. 2001. *Failure is not an Option: Mission Control from Mercury to Apollo 13 and Beyond.* New York, NY: Berkley Books.

Lancaster, J. A. and Casali, J. G. 2008. Investigating pilot performance using mixed-modality simulated data link. *Human Factors*, 50(2), 183–193.

Lanir, Z. 1983. *Fundamental Surprise: The National Intelligence Crisis.* Tel Aviv: Center for Strategic Studies.

LaPorte, T. R. 1988. The United States air traffic system: Increasing reliability in the midst of rapid growth. In R. Mayntz and T. Hughes (Eds.), *The Development of Large scale Technical Systems.* Boulder, CO: Westview Press, 215–244.

LaPorte, T. R. and Consolini, P. M. 1991. Working in practice but not in theory: Theoretical challenges of high-reliability organizations. *Journal of Public Administration Research and Theory*, 1, 19–47.

Laudeman, I., Shelden, S., Branstrom, R., and Brasil, C. 1998. *Dynamic Density: An Air Traffic Management Metric.* NASA-TM-1988–11226. Moffet Field, CA: NASA Ames Research Center.

Lave, J. and Wenger, E. 1991. *Situated Learning: Legitimate Peripheral Participation.* New York, NY: Cambridge University Press.

Lenny, M. G., Ashby, K., and Fitzharris, M., 2008. Analysis of general aviation crashes in Australia using the Human Factors Analysis and Classification System. *International Journal of Aviation Psychology*, 18, 340–352.

Leplat, J. 1987. Occupational accident research and systems approach. In J. Rasmussen, K. Duncan, and J. Leplat (Eds.), *New Technology and Human Error.* New York, NY: Wiley, 181–191.

Leveson, N. 2004. A new accident model for engineering safer systems. *Safety Science*, 42, 237–270.

Leveson, N. 2011. *White Paper on the Use of Safety Cases in Certification and Regulation*. http://sunnyday.mit.edu/SafetyCases.pdf

Leveson. N., Dulac, N., Marais, K., and Carroll, J. 2009. Moving beyond normal accidents and high reliability organizations (HRO): A systems approach to safety in complex systems. *Organization Studies*, 30, 227–249.

Leveson, N. G. 2012. *Engineering a Safer World: Systems Thinking Applied to Safety*. Cambridge, MA: The MIT Press.

Lewis, M. 2000. Exploring paradox: Toward a more comprehensive guide. *Academy of Management Review*, 25(4), 760–776.

Li, W. C. and Harris, D. 2006. Pilot error and its relationship with higher organizational levels: HFACS analysis of 523 accidents. *Aviation, Space, and Environmental Medicine*, 77, 1056–1061.

Li, W.-C., Harris, D., and Yu, C. S. 2008. Routes to failure: Analysis of 41 civil aviation accidents from the Republic of China using the human factors analysis and classification system. *Accident Analysis and Prevention*, 40(2), 426–434.

Licu, T., Cioran, F., Hayward, B., and Lowe, A. 2007. Systemic occurrence analysis methodology (SOAM)—"Reason" based organizational methodology for analyzing accidents. *Reliability Engineering and System Safety*, 92, 1162–1169.

Lipshitz, R. 1997. On-line coping with uncertainty: Beyond the reduce, quantify and plug heuristics. In R. Flin, E. Salas, M. Strub, and L. Martin (Eds.), *Decision Making Under Stress: Emerging Themes and Applications*. Aldershot: Ashgate, 149–160.

Loft, S., Sanderson, P., Neal, A., and Mooij, M. 2007. Modeling and predicting mental workload in en-route air traffic control: Critical review and broader implications. *Human Factors*, 49, 376–399.

Loukopoulos, L. D., Dismukes, R. K., and Barshi, I. 2001. Cockpit interruptions and distractions: A line observation study. *Proceedings of the 11th International Symposium on Aviation Psychology*. Columbus, OH: Ohio State University Press.

Malakis, S. 2009. *Decision Making in Air Traffic Control and Team Performance in Emergency Scenarios*. PhD Thesis. Chania: Technical University of Crete.

Malakis, S. and Kontogiannis, T. 2012. Refresher training for air traffic controllers: Is it adequate to meet the challenges of emergencies and abnormal situations? *International Journal of Aviation Psychology*, 22(1), 59–77.

Malakis, S. and Kontogiannis, T. 2013. A sensemaking perspective on framing the mental picture of air traffic controllers. *Applied Ergonomics*, 44, 327–339.

Malakis, S. and Kontogiannis, T. 2014. Exploring team sensemaking in air traffic control (ATC): Insights from a field study in low visibility operations. *Cognition Technology and Work*, 16, 211–227.

Malakis, S., Kontogiannis, T., and Kirwan, B. 2010a. Managing emergencies and abnormal situations in air traffic control (Part I): Taskwork strategies. *Applied Ergonomics*, 41, 620–627.

Malakis, S., Kontogiannis, T., and Kirwan, B. 2010b. Managing emergencies and abnormal situations in air traffic control (Part II): Teamwork strategies. *Applied Ergonomics*, 41, 628–635.

Malakis, S., Kontogiannis, T., and Psaros, P. 2014. Monitoring and evaluating failure-sensitive strategies in air traffic control simulator training. *Proceedings of the 7th International Conference on PErvasive Technologies Related to Assistive Environments*. Rhodes, Greece: ACM.

Marais, K. and Saleh, J. H. 2008. Conceptualizing and communicating organizational risk dynamics in the thoroughness-efficiency space. *Reliability Engineering and System Safety*, 93, 1710–1719.

Marais, K., Saleh, J. H., and Leveson, N. G. 2006. Archetypes for organizational safety. *Safety Science*, 44, 565–582.

Marca, D. and McGowan, C. 1987. *Structured Analysis and Design Technique*. New York, NY: McGraw-Hill.

Masalonis, A. J., Callaham, M. B., and Wanke, C. R. 2003. *Dynamic Density and Complexity Metrics for Real Time Traffic Flow Management*. McLean, VA: MITRE.

Masson, M. and Morier, Y. 2011. *Methodology to Assess Future Risks. European Aviation Safety Plan (EASp). Action EME 1.1 of the European Aviation Safety Plan (EASp)*, EASA, and the FAST - Presented to ECAST 4–12. Koln, Germany: EASA.

McDonald, N. 2006. Organizational resilience and industrial risk. In E. Hollnagel, D. D. Woods, and N. Leveson (Eds.), *Resilience Engineering: Concepts and Precepts*, Aldershot: Ashgate, 205–221.

McDonald, N., Morrison, R., Leva, M. C., Atkinson, B., Mattei, F., and Cahill, J. 2011. Operational modeling and data integration for management and design. In C. Cacciabue, M. Hjalmdahl, A. Luedtke, and C. Riccioli (Eds.), *Human Modelling in Assisted Transportation: Models, Tools and Risk Methods*, Milan: Springer, 55–63.

McDonald, N., Ulfvengren, P., and Ydalus, M. 2012. A methodology for managing system change- An airline's development of a SMS and a performance management process. *11th Probabilistic Safety Assessment and Management Conference (PSAM11) and the Annual European Safety and Reliability Conference (ESREL 2012)*. Helsinki.

Meadows, D. 1999. *Leverage Points: Places to Intervene in a System*. Hartland, VT: Sustainability Institute.

Means, B., Salas, E., Crandall, B., and Jacobs, T. O. 1993. Training decision makers for the real world. In G. A. Klein, J. Orasanu, R. Calderwood, C. E. Zsambok (Eds.), *Decision Making in Action: Models and Methods*, Norwood, NJ: Ablex Publishing, 306–326.

Mearns, K., Kirwan, B., Reader, T. W., Jackson, L., Kennedy, R., and Gordon, R. 2013. Development of a methodology for understanding and enhancing safety culture in Air Traffic Management. *Safety Science*, 53, 123–133.

Mesarovic, M. D. and Takahara, Y. 1975. *General Systems Theory: Mathematical Foundations*. New York, NY: Academic Press.

Metzger, U. and Parasuraman, R. 2001. The role of the air traffic controller in future air traffic management: An empirical study of active control versus passive monitoring. *Human Factors*, 43, 519–528.

Metzger, U. and Parasuraman, R. 2005. Automation in future air traffic management: Effects of decision aid reliability on controller performance and mental workload. *Human Factors*, 47, 35–49.

Mogford, R. H. 1997. Mental models and situation awareness in air traffic control. *International Journal of Aviation Psychology*, 7, 331–341.

Mogford, R. H., Guttman, J., Morrow, S. L., and Kopardekar, P. 1995. *The Complexity Construct in Air Traffic Control: A Review and Synthesis of the Literature*. No. DOT/FAA/CT-TN95/22. Atlantic City, NJ: William Hughes Technical Center, Federal Aviation Administration.

Morel, G., Amalberti, R., and Chauvin, C. 2008. Articulating the differences between safety and resilience: The decision-making process of professional sea-fishing skippers. *Human Factors*, 50(1), 1–16.

Morrow, D. G., Lee, A. T., and Rodvold, M. 1993. Analysis of problems in routine controller-pilot communication. *International Journal of Aviation Psychology*, 3, 285–302.

Nathanael, D. and Marmaras, N. 2008. On the development of work practices: A constructivist model. *Theoretical Issues in Ergonomics Science*, 9, 359–382.

NATS, 2008. *Low Altitude Windshear.* Aeronautical Information Circular 84/2008. Hounslow Middlesex: NATS.

Neal, A., Flach, J., Mooij, M., Lehmann, S., Stankovic, S., and Hasenbosch, S. 2011. Envisaging the future air traffic management system. *International Journal of Aviation Psychology*, 21, 16–34.

Niessen, C., Eyferth, K., and Bierwagen, T. 1997. Modeling cognitive processes of experienced air traffic controller. In S. Bagnara, E. Hollnagel, M. Mariani, and L. Norros (Eds.), *Sixth European Conference on Cognitive Science Approaches to Process Control, Time and Space in Process Control, 23–26* September 1997, Baveno, Italy.

Nikolic, M. J. and Sarter, N. B. 2007. Flight deck disturbances management: A simulators study of diagnosis and recovery from breakdowns in pilot-automation coordination. *Human Factors*, 49, 553–563.

Norman, D. A. 1990. The "problem" of automation: Inappropriate feedback and interaction, not "over-automation." *Philosophical Transactions of the Royal Society of London B*, 327, 585–593.

Norman, D. A. 1993. *Things that Make us Smart: Defending Human Attributes in the Age of the Machine*. Reading, MA: Addison-Wesley.

NTSB. 1995. *Flight into Terrain during Missed Approach, USAir Flight 1016, DC-9-31, N954VJ, Charlotte/ Douglas International Airport, Charlotte, North Carolina, July 2, 1994*, Report No. PB95-910403, NTSB/AAR, DCA94MA065, Washington DC.

Oprins, E., Burggraaff, E., and Van Weerdenburg, H. 2006. Design of a competence-based assessment system for air traffic control training. *International Journal of Aviation Psychology*, 16(3), 297–320.

Orasanu, J.M., 1993. Decision Making in the Cockpit. In: E.L. Wiener, R.G. Kanki, and R.L. Helmreichm (Eds), *Cockpit Resource Management*. San Diego, CA: Academic Press, 137–172.

Papazoglou, I. A., Bellamy, L. J., Hale, A. R., Aneziris, O. N., Ale, B. J. M., Post, J. G. and Oh, J. I. H. 2003. I-Risk: Development of an integrated technical and management risk methodology for chemical installations. *Journal of Loss Prevention in the Process Industries*, 16, 575–591.

Paraskevas, A. 2006. Crisis management or crisis response system? A complexity science approach to organizational crises. *Management Decision*, 44, 892–907.

Paté-Cornell, M. E. and Murphy, D. M., 1996. Human and management factors in probabilistic risk analysis: The SAM approach and observations from recent applications. *Reliability Engineering and System Safety*, 53, 115–126.

Patey, R., Flin, R., Fletcher, G., Maran, N., and Glavin, R. 2005. Developing a taxonomy of anaesthetists' non-technical skills (ANTS). In K. Hendrics (Ed.), *Advances in Patient Safety: From Research to Implementation*. Rockville: Agency for Healthcare Research & Quality.

Patterson, E. S., Watts-Perotti, J., and Woods, D. D. 1999. Voice loops as coordination aids in space shuttle mission control. *Computer Supported Cooperative Work*, 8, 353–371.

Patterson, E. S., Woods, D. D., Cook, R. I., and Render, M. L. 2007. Collaborative cross-checking to enhance resilience. *Cognition, Technology & Work*, 9, 155–162.

Pawlak, W. S., Brinton, C. R., Crouch, K., and Lancaster, K. M. 1996. A framework for the evaluation of air traffic control complexity. *Presentation at the AIAA Guidance, Navigation and Control Conference*. San Diego, CA.

Perrow, C. 1984. *Normal Accidents: Living with High-Risk Technologies*. Princeton, NJ: Princeton University Press.

Pettersen, K. A., and Aase, K. 2008. Explaining safe work practices in aviation line, maintenance. *Safety Science*, 46, 510–519.

Prevot, T., Homola, J. R., Martin, L. K., Mercer, J. S., and Cabrall, C. D. 2012. Toward automated air traffic control—Investigating a fundamental paradigm shift in human-systems interaction. *International Journal of Human-Computer Interaction*, 28, 77–98.

Rantanen, E. M. and Nunes, A. 2005. Hierarchical conflict detection in air traffic control. *International Journal of Aviation Psychology*, 15, 339–362.

Rasmussen, J. 1983. Skill, rules, knowledge: Signals, signs, and symbols and other distinctions in human performance models. *IEEE Transactions on Systems, Man, & Cybernetics*, 13(3), 257–267.

Rasmussen, J. 1986. *Information Processing and Human-machine Interaction: An Approach to Cognitive Engineering*. New York: North-Holland.

Rasmussen, J. 1993. Deciding and doing: decision making in natural context. In: G., Klein, J., Orasanu, R., Calderwood, and C. Zsambok (Eds), *Decision Making in Action: Models and Methods*. Norwood, NJ: Ablex Publishing Corporation.

Rasmussen, J. 1994. Risk management, adaptation and design for safety. In B. Brehmer and N. E. Sahlin (Eds.), *Future Risks and Risk Management.* Dordrecht: Kluwer Academic, 1–36.

Rasmussen, J. 1997. Risk management in a dynamic society: A modeling problem. *Safety Science*, 27, 183–213.

Rasmussen, J. and Svendung, I. 2000. *Proactive Risk Management in a Dynamic Society.* Karlstad: Swedish Rescue Service Agency.

Reason, J. 1997. *Managing the Risks of Organizational Accidents.* Aldershot: Ashgate.

Reason, J. 2001. Score Your Safety Culture. *Flight Safety Australia*, (January-February, 2001).

Reason, J. 2004. Beyond the organizational accident: The need for "error wisdom" on the frontline. *Quality & Safety in Health Care*, 13, Supplement II, 28–33.

Reason, J. 2008. *The Human Contribution: Unsafe Acts, Accidents and Heroic Recoveries.* Aldershot: Ashgate.

Reason, J. T. 1990.*Human Error.* Cambridge: Cambridge University Press.

Reason, J. T., Hollnagel, E., and Pariès, J. 2006. *Revisiting the "Swiss Cheese" model of accidents.* EEC Note No. 13/06. Brussels: Eurocontrol.

Redding, R. E., Ryder, J. M., Seamster, T. L., Purcell, J. A., and Cannon, J. R. 1991. *Cognitive Task Analysis of En-route Air Traffic Control: Model Extension and Validation.* FAA Report, ERIC Document Reproduction Service No. ED 340 848. McLean, VA.

Reiman, T. 2010. Understanding maintenance work in safety critical organizations–managing performance variability. *Theoretical Issues in Ergonomics Science*, 11, 1–28.

Rentsch, J. R. and Woehr, D. J. 2004. Quantifying congruence in cognition: Social relations modeling and team member schema similarity. In E. Salas and S. M. Fiore (Eds.), *Team Cognition: Understanding the Factors that Drive Process and Performance.* Washington, DC: American Psychological Association.

Reynolds, T. G., Histon, J. M., Davison, H. J., and Hansman, *R. J. 2002. Structure, intent and conformance monitoring in ATC. Proceedings of the Air Traffic Management (ATM) Workshop on ATM System Architectures and CNS Technologies*, 22–26 September 2002, Capri, Italy.

Rignér, J., Ulfvengren, P., and Kay, A. 2009. Measuring safety performance–Strategic risk data (airline safety and human factors issues) in the EASS, *21st Annual European Aviation Safety Seminar*, 16–18 March 2009, Cyprus.

Rizzo, A., Ferrante, D., and Bagnara, S. 1994. Handling human error. In J. M. Hoc, P. C. Cacciabue, and E. Hollnagel (Eds.), *Expertise and Technology: Cognition & Human Computer Interaction.* Hillsdale, NJ: Lawrence Erlbaum Associates, 195–212.

Roberts, K. H. 1993. *New Challenges to Understanding Organizations.* New York, NY: Macmillan.

Robson, D. 2008. *Human Being Pilot: Human Factors for Aviation Professionals.* Australia: Aviation Theory Centre.

Rochlin, G.I. 1997. *Trapped in the Net: The Unanticipated Consequences of Computerization*. Princeton, NJ: Princeton University Press.

Rochlin, G., LaPorte, T. R., and Roberts, K. H. 1987. The self-designing high reliability organization: Aircraft carrier flight operations at sea. *Naval War College Review, (Autumn 1987)*, 76–90.

Roth, E. M., Malin, J. T., and Schreckenghost, D. L. 1997. Paradigms for Intelligent Interface Design. In M. Helander, T. Landauer, and P. Prabhu (Eds.), *Handbook of Human-Computer Interaction*. Second Edition. Amsterdam: North-Holland, 1177–1201.

Salas, E., Bowers, C. A., and Rhodenizer, L. 1998. It is not how much you have but how you use it: Toward a rational use of simulation to support aviation training. *International Journal of Aviation Psychology*, 8, 197–208.

Salas, E., Cooke, N. J., and Rosen, M. A. 2008. On teams, teamwork, and team performance: Discoveries and developments. *Human Factors*, 50, 540–547.

Salas, E., Driskell, J. E., and Hughes, S. 1996. Introduction: The study of stress and human performance. In J. E. Driskell and E. Salas (Eds.), *Stress and Human Performance*. Mahwah, NJ: Lawrence Erlbaum Associates, 1–45.

Salas, E., Sims, D. E., and Burke, C. S. 2005. Is there a "big five" in teamwork?. *Small Group Research*, 36, 555–599.

Santos-Reyes, J. and Beard, A. N. 2006. A systemic analysis of the Paddington railway accident. *Proceedings of the Institution of Mechanical Engineers, Part F: Journal of Rail and Rapid Transit*, 220(2), 21–151.

Santos-Reyes, J. and Beard, A. N. 2008. A systemic approach to managing safety. *Journal of Loss Prevention in the Process Industries*, 21, 15–28.

Santos-Reyes, J. and Beard, A. N. 2009. A SSMS model with application to the oil and gas industry. *Journal of Loss Prevention in the Process Industries*, 22, 958–970.

Sarter, N. B. and Alexander, H. M. 2000. Error types and related error detection mechanisms in the aviation domain: An analysis of aviation safety reporting system incident reports. *International Journal of Aviation Psychology*, 10, 189–206.

Sarter, N. B. and Woods, D. D. 1995. How in the world did we get into that mode? Mode errors and awareness in supervisory control. *Human Factors*, 39, 5–19.

Sarter, N. B. and Woods, D. D. 2000. Team play with a powerful and independent agent: A full-mission simulation study. *Human Factors*, 42, 390–402.

Sarter, N. B., Woods, D. D., and Billings, C. 1997. Automation surprises. In G. Salvendy (Ed.), *Handbook of Human Factors & Ergonomics*, Second Edition. New York, NY: Wiley.

Scholn, D. 1983. *The Reflective Practitioner: How Practitioners Think in Action*. New York, NY: Basic Books.

Schulman, P. R. 1993. The negotiated order of organizational reliability. *Administration and Society*, 25, 353–372.

Schwaninger, M. 2006. System dynamics and the evolution of the systems movement. *Systems Research and Behavioral Science*, 23(5), 583–594.

Seamster, T. L., Redding, R. E., Cannon, J. R. and Purcell, J. A. 1993. Cognitive task analysis of expertise in air traffic control. *International Journal of Aviation Psychology*, 3, 257–283.

Sellen, A. J. 1994. Detection of everyday errors. *Applied Psychology: An International Review*, 43, 475–498.

Senge, P. 2006. *The Fifth Discipline: The Art and Practice of the Learning Organization*. London, UK: Random House.

Serfaty, D. and Entin E. E. 1996. Team adaptation and coordination training. In R. Flin, E. Salas, M. Strub, and L. Martin (Eds.), *Decision Making under Stress: Emerging Themes and Applications*, Aldershot: Ashgate, 170–184.

Serfaty, D., Entin, E., and Volpe, C. 1993. Adaptation to stress in team decision making and coordination. *Proceedings of the Human Factors Society 37th Annual Meeting*. Santa Monica, CA: Human Factors and Ergonomics Society.

Sharples, S., Stedmon, A., Cox, G., Nicholls, A., Shuttleworth, T., and Wilson, J. 2007. Flightdeck and air traffic control collaboration evaluation (FACE): Evaluating aviation communication in the laboratory and field. *Applied Ergonomics*, 38, 399–407.

Shepherd, A. 2001. *Hierarchical Task Analysis*. London: Taylor & Francis.

Shorrock, S. T. and Isaac, A. 2010. Mental imagery in air traffic control. *International Journal of Aviation Psychology*, 20(4), 309–324.

Shorrock, S. T. and Kirwan, B. 2002. Development and application of a human error identification tool for air traffic control. *Applied Ergonomics*, 33, 319–336.

Shorrock, S. T., Leonhardt, J., Licu, T., and Peters, C. 2014. *Systems Thinking for Safety: Ten Principles. A White Paper*. Bretigny: Eurocontrol.

Simon, H. A. 1957. *Models of Man: Social and Rational. Mathematical Essays on Rational Human Behavior in a Social Setting*. New York: Wiley.

Skyttner, L. 2005. *General Systems Theory: Problems, Perspectives, Practice*. Second Edition. London: World Scientific Publishing.

Smith, P. J., McCoy, E., and Layton, C. 1997. Brittleness in the design of cooperative problem-solving systems: The effects on user performance. *IEEE Transactions on Systems, Man and Cybernetics*, 27, 360–371.

Smith, P. J, Spencer, A. L., and Billings, C. E. 2007. Strategies for designing distributed systems: Case studies in the design of an air traffic management system. *Cognition, Technology & Work*, 9, 39–49.

Snook, S. A. 2000. *Friendly Fire: The Accidental Shoot-down of US Black Hawks over Northern Iraq*. New York, NY: Doubleday.

Sperandio, J. C. 1978. The regulation of working methods as a function of workload among air traffic controllers. *Ergonmics*, 21, 195–202.

Stacey, R., Griffin, D., and Shaw, P. 2000. *Complexity and Management: Fad or Radical Challenge to Systems thinking?* London: Routledge.

Stanton, N., Salmon, P. M., Walker, G. H., Baber, C., and Jenkins D. P. 2005. *Human Factors Methods: A Practical Guide for Engineering and Design*. Aldershot: Ashgate.

Stein, E. S., Della Rocco, P. S., and Sollenberger, R. L. 2006. *Dynamic Resectorization in Air Traffic Control: A Human Factors Perspective*. Report DOT/FAA/TC - TN06/19. Washington, DC: Human Factors Research and Engineering Division, Federal Aviation Administration, 20591.

Sterman, J. D. 2000. *Business Dynamics: Systems Thinking and Modeling for a Complex World*. Boston, MA: Irwin McGraw-Hill.

Stolzer, A., Halford, D., and Goglia, J. 2008. *Safety Management System in Aviation*. Aldershot: Ashagte.

Stout, R. J., Cannon-Bowers, J.A., Salas, E., and Milanovich, D.M. 1999. Planning, shared mental models and coordinated performance: An empirical link is established. *Human Factors*, 41, 61–71.

Stringfellow, M. 2010. *Accident Analysis and Hazard Analysis for Human and Organizational Factors*. PhD Dissertation. Boston, MA: Department of Aeronautics and Astronautics, Massachusetts Institute of Technology.

Stroeve, S.H., Sharpanskykh, A., and Kirwan, B. 2011. *Agent-based Organisational modellling for analysis of safety culture at an air navigation service provider.* Report no. NLR-TP-2011-030, NLR The Netherlands.

Strybel, T. Z. and Vu, K. P. 2013. Measuring the impact of NextGen operating concepts for separation assurance on pilot situation awareness and workload. *International Journal of Aviation Psychology*, 23(1), 1–26.

Swezey, R. W. and Salas, E. 1992. Guidelines for use in team-training development. In R. W. Swezey and E. Salas (Eds.), *Teams: Their Training and Performance*. Norwood, NJ: Ablex, 219–245.

Thomas, M. J. W. 2004. Predictors of threat and error management: Identification of core non-technical skills and implications for training systems design. *International Journal of Aviation Psychology*, 14, 207–231.

Thomas, M. J. W. and Petrilli, R. 2004. *Error Management Training: An investigation of expert flight crews' error management strategies during normal operations and flight crew training.* ATSB Aviation Safety Research Grant Scheme Project 2004/0050, Centre for Applied Behavioral Science, University of South Australia.

Thomke, S. 2001. Enlightened experimentation. The new imperative for innovation. *Harvard Business Review*, 79(2), 66–75.

Toft, B. and Reynolds, S. 1994. *Learning from Disasters*. Oxford: Butterworth-Heinemann.

TSB. 2003. *Aviation Investigation Report: In-Flight Fire Leading to Collision with Water Swissair Transport Limited McDonnell Douglas MD-11 HB-IWF Peggy's Cove, Nova Scotia 5 nm SW 2 September 1998. Number A98H003.* Quebec: Transportation Safety Board of Canada.

Tucker, A. and Edmondson, A. 2003. Why hospitals don't learn from failures: Organizational and psychological dynamics that inhibit system change. *Californian Management Review*, 45(2), 55–72.

Turner, B. 1978. *Man-Made Disasters: The Failure of Foresight.* London: Butterworth-Heinmann.

Turner, B. A. and Pidgeon, N. F. 1997. *Man-Made Disasters.* Second Edition. London: Butterworth-Heinemann.

Ulfvengren, P. 2010. *ICAO SMS and the HILAS Approach.* HILAS project 516181, Aeronautics and Space Research Area 3, Improving Aircraft Safety and Security. Brussels: European Commission.

Ulfvengren, P., Leva, M. C., McDonald, N., and Ydalus, M. 2013. Airline framework for predictive safety performance management. In S. J. Landrey (Ed.), *Advances in Human Aspects of Aviation.* Boca Raton, FL: CRC Press, 511–527.

van Avermaete, J. and Kruijsen, E. (Eds.). 1998. *NOTECHS. The Evaluation of Non- Technical Skills of Multi-Pilot Aircrew in Relation to the JAR-FCL Requirements.* Final Report NLR-CR-98443. Amsterdam: National Aerospace Laboratory (NLR).

Vicente, K. 1999. *Cognitive Work Analysis.* London: Lawrence Erlbaum Associates.

von Bertalanffy, L. 1950. An outline of general system theory. *British Journal for the Philosophy of Science*, 1(2), 134–165.

Vu, K. P. and Strybel, T. Z. 2011. Factors influencing the decisions and actions of flight crews and air traffic controllers in three plausible NextGen environments. In Y. Yi (Ed.), *Cultural Factors in Systems Design Decision Making and Action.* Boca Raton, FL: CRC Press, 281–301.

Wahlstrom, B. and Rollenhagen, C. 2012. Safety management–A multi-level control problem. *Safety Science*, 69, 3–17.

Walmsley, R., Anderson, T., Brendish, C., McDermid, J., Rolfe, M., Sultana, J., Swan, M., and Toms, M. 2015. *NATS System Failure 12 December 2014 – Final Report Independent Enquiry.* Final Report dated 13 May 2015.

Walters, A. 2002. *Crew Resource Management is No Accident.* Wallingford, CT: Aries.

Weick, K. E. 1995. *Sensemaking in Organizations.*Thousand Oaks, CA: Sage.

Weick, K. E. 2007. *Making Sense of the Organizations.* Malden, MA: Blackwell Publishing.

Weick, K. and Sutcliffe, K. 2001. *Managing the Unexpected.* San Francisco: Jossey Bass.

Weick, K. E., Sutcliffe, K. M., and Obstfeld, D. 1999. Organizing for high reliability: Processes of collective mindfulness. *Research in Organizational Behavior*, 21, 81–123.

Weir, D. 2004. Sequences of failure in complex socio-technical systems: Some implications of decision and control. *Kybernetes*, 33, 522–537.

Weitzman, D. O. 1993. Identifying information processing requirements for air traffic control problem solving: Designing for diversity. *Proceedings of the Human Factors and Ergonomics Society 37th Annual Meeting.* Santa Monica, CA: Human Factors and Ergonomics Society, 103–107.

Whitfield, D. and Jackson, A. 1982. The air traffic controller's picture as an example of mental model. In G. Johannsen, and J. E. Rijnsdorp (Eds.), *Proceedings of the IFAC Conference on Analysis, Design and, Evaluation of Man-Machine Systems*. London: Pergamon Press, 45–52.

Wickens, C. 1992. *Engineering Psychology and Human Performance*. Second Edition. New York, NY: Harper-Collins.

Wickens, C. D. and Colcombe, A. 2007. Dual-task performance consequences of imperfect alerting associated with a cockpit display of traffic information. *Human Factors*, 49, 839–850.

Wickens, C. D., Mavor, A. S., and McGee, J. P. 1997. *Flight to the Future: Human Factors in Air Traffic Control*. Washington, DC: National Academy Press.

Wiegmann, D. A. and Shappell, S. A. 2003. *A Human Error Approach to Aviation Accident Analysis. The Human Factors Analysis and Classification System*. Burlington, VT: Ashgate Publishing.

Wildavsky, A. 1988. *Searching for Safety*. London: Transaction Publishers.

Willems, B. and Koros, A. 2007. *Advanced Concept of the National Airspace System: Human Factors Considerations for Air Traffic Control*. DOT/FAA/TC-TN-07/21. US Department of Transport. Atlantic City International Airport, NJ: FAA.

Wing, D. 2008. Performance basis for airborne separation. *Proceedings of the 26th International Congress of the Aeronautical Sciences*, 14–19 September 2008, Anchorage, AK.

Wing, D., Prevot, T., Lewis, T., Martin, L., Johnson, S., Cabrall, C., Commo, S., Homola, J., Chandra, M. S., Mercer, J. and Morey, S. 2013. Pilot and controller evaluations of separation function allocation in air traffic management. *10th USA/Europe Air Traffic Management Research and Development Seminar (ATM 2013), 10–13 June 2013*, Chicago, IL.

Wioland, L. and Amalberti, R. 1996. When errors serve safety: Towards a model of ecological safety. *Proceedings of the First Conference on Cognitive Systems Engineering in Process Control*, November 1996, Japan: Kyoto University.

Woods, D. D. 1988. Coping with complexity: The psychology of human behavior in complex systems. In L. P. Goodstein, H. B. Andersen, and S. E. Olsen (Eds.), *Tasks, Errors and Mental Models*. New York, NY: Taylor and Francis.

Woods, D. D. 1994. Cognitive demands and activities in dynamic fault management: Abduction and disturbance management. In: Standon, N. (Ed.), *Human Factors of Alarm Design*. London: Taylor & Francis.

Woods, D. D. 1995. Towards a theoretical base for representation design in the computer medium: Ecological perception and aiding human cognition. In J. Flach, P. Hanock, J. Caird, and K. Vicente (Eds.), *Global Perspectives on the Ecology of Human Machine Systems*. Hillsdale: Lawrence Erlbaum Associates, 157–188.

Woods, D. D. 2006. Essential characteristics of resilience. In E. Hollnagel, D. D. Woods, and N. G. Leveson (Eds.), *Resilience Engineering: Concepts and Precepts*. Aldershot: Ashgate.

Woods, D. D. and Branlat, M. 2010. Holnagell's test: Being in control of highly interdependent multi-layered networked systems.*Cognition, Technology & Work*, 12, 95–101.

Woods, D. D. and Cook, R. 2006. Incidents - markers of resilience or brittleness. In E. Hollnagel, D.D. Woods, and N. G. Leveson (Eds.), *Resilience Engineering: Concepts and Precepts*. Aldershot: Ashgate, 69–76.

Woods, D. D. and Cook, R. I. 2002. Nine steps to move forward from error. *Cognition, Technology & Work*, 4, 137–144.

Woods, D. D., Dekker, D., Cook, R., Johannesen, L., and Sarter, N. 2010. *Behind Human Error.* Second Edition. Farnham: Ashgate Publishing.

Woods, D. D. and Hollnagel, E. 2006. *Joint Cognitive Systems: Patterns in Cognitive Systems Engineering*. Boca Raton, FL: CRC Press.

Woods, D. D., Patterson, E. S., and Cook, R. I. 2007. Behind human error: Taming complexity to improve patient safety. In P. Carayon (Ed.), *Handbook of Human Factors and Ergonomics in Health Care and Patient Safety*. Hillsdale, NJ: Lawrence Erlbaum Associates, 459–476.

Woods, D. D., Patterson, E. S., and Roth, E. M. 2002. Can we ever escape from data overload? A cognitive systems diagnosis. *Cognition, Technology & Work*, 4, 22–36.

Wood, S. D. and Kieras, D. E. 2002. Modeling human error for experimentation, training, and error-tolerant design. *Proceedings of the Interservice/Industry Training, Simulation, and Education Conference*. Orlando, FL, 1075–1085.

Woods, D. D. and Roth, E. M. 1988. Cognitive engineering: Human problem solving with tools. *Human Factors*, 30, 415–430.

Woods, D. D. and Sarter, N. 2000. Learning from automation surprises and going sour accidents. In N. Sarter and R. Amalberti (Eds.), *Cognitive Engineering in the Aviation Domain*. Hillsdale, NJ: Lawrence Erlbaum, 327–354.

Xiao, Y., Milgram, P., and Doyle, D. J. 1997. Capturing and modeling planning expertise in anesthesiology: Results of a field study. In C. E. Zsambok, and G. Klein (Eds.), *Naturalistic Decision Making*. Mahwah, NJ: Lawrence Erlbaum, 197–205.

Xing, J. M. and Manning, C. 2005. *Complexity and Automation Displays of Air Traffic Control: Literature Review and Analysis*. Report DOT/FAA/AM-05/4. FAA Civil Aeromedical Institute, Oklahoma City.

Zhang, J. and Norman, D. A. 1994. Representations of distributed cognitive tasks. *Cognitive Science*, 18, 87–122.

Zsambok, C. E. and Klein, G. A. (Eds.). 1997. *Naturalistic Decision Making*. Mahwah, NJ: Lawrence Erlbaum.

Index

F

G

H

M

N

O

P

W